ACS SYMPOSIUM SERIES **458**

Biotechnology of Amylodextrin Oligosaccharides

Robert B. Friedman, EDITOR
American Maize-Products Company

Developed from a symposium sponsored
by the Divisions of Carbohydrate Chemistry
and of Agricultural and Food Chemistry
at the 198th National Meeting
of the American Chemical Society,
Miami Beach, Florida,
September 10–15, 1989

American Chemical Society, Washington, DC 1991

Library of Congress Cataloging-in-Publication Data

Biotechnology of amylodextrin oligosaccharides /   Robert Friedman, editor.

p.   cm.—(ACS symposium series; 458)

"Developed from a symposium sponsored by the Divisions of Carbohydrate Chemistry and of Agricultural and Food Chemistry at the 198th National Meeting of the American Chemical Society, Miami Beach, Florida, September 10–15, 1989"

Includes bibliographical references and indexes.

ISBN 0–8412–1993–1

1. Amylodextrins—Biotechnology—Congresses.
2. Oligosaccharides— Biotechnology—Congresses.

I. Friedman, Robert, 1938–   . II. American Chemical Society. Division of Carbohydrate Chemistry. III. American Chemical Society. Division of Agricultural and Food Chemistry. IV. American Chemical Society. Meeting (198th: 1989: Miami Beach, Fla.) V. Series.

[DNLM: 1. Amylose—chemical synthesis—congresses. 2. Biotechnology—congresses. 3. Dextrins—chemical synthesis—congresses. 4. Oligosaccharides—chemical synthesis—congresses. QU 83 B616 1989]

TP248.65.A46B56   1991
661'.8—dc20
DNLM/DLC   91–13893
for Library of Congress   CIP

The paper used in this publication meets the minimum requirements of American National Standard for Information Sciences—Permanence of Paper for Printed Library Materials, ANSI Z39.48–1984. ∞

TP248
.65
A46 B56
1991
CHEM

ACS Symposium Series

M. Joan Comstock, *Series Editor*

1991 ACS Books Advisory Board

Foreword

THE ACS SYMPOSIUM SERIES was founded in 1974 to provide a medium for publishing symposia quickly in book form. The format of the Series parallels that of the continuing ADVANCES IN CHEMISTRY SERIES except that, in order to save time, the papers are not typeset, but are reproduced as they are submitted by the authors in camera-ready form. Papers are reviewed under the supervision of the editors with the assistance of the Advisory Board and are selected to maintain the integrity of the symposia. Both reviews and reports of research are acceptable, because symposia may embrace both types of presentation. However, verbatim reproductions of previously published papers are not accepted.

Contents

Preface

BIOTECHNOLOGY is one of the oldest technologies, though, at the same time, it is also one of the newest. Since the age of Noah, biotechnology has been employed as a means of producing products that are both desirable and specific in function. In this contemporary age of biochemistry and molecular biology, however, biotechnology has assumed a totally different dimension of applicability regarding specificity and usefulness. As the demand increases for enhanced biocompatibility in products and for abundantly available raw materials from naturally renewable resources, more and more people have been turning to biotechnology to produce products and raw materials for the next decades.

With their general availability in the vegetable kingdom, carbohydrates have proved to be a valuable substrate for biotechnology. Indeed, the production of corn sweeteners is a triumph of modern commercial biotechnology. Carbohydrates are also functionally important in the animal kingdom, but the vegetable-derived polysaccharides are much more accessible and available in higher quantities. In particular, starch-derived polysaccharides and oligosaccharides have proved to be exceptionally valuable in a broad spectrum of industrial and food applications.

Because the main repeating monomer of these biopolymers is glucose, the diversity of functionality of these substances must result from structural differences. It was recognized fairly early on that changing the structure of starch polysaccharides can dramatically affect the behavior of those materials. For this reason, chemically modified starches and depolymerized starches have served competently over the years.

Contemporary society, however, has stressed the importance of the perception of healthfulness. To meet the demands of this perception, a reduction in chemical modification will probably be required. To replace those necessary functionalities, biotechnology will be called on to perform the required structural changes. To achieve this end, however, a more fundamental understanding of starch structures as well as function–structure relationships of these compounds will be required. Furthermore, enzyme systems must be understood along with the basic biochemical and genetic mechanisms that control their production and behavior. The purpose of this book, as well as the symposium from which it was derived, is to focus attention on the many facets of the biotechnology of amylodextrin polysaccharides. It is hoped that not only will the potential usefulness of this group of substances be illuminated, but also the broad extent of fundamental scientific information will be revealed.

This book is designed to be useful to a broad array of researchers who might find oligosaccharide biopolymers of interest. As such, it must cover biochemistry and enzymology as well as those unique structural characteristics that require novel analytical tools. New aspects of usefulness must also be addressed. Consequently, the book has three basic parts:

- The first section deals with the basic biochemical aspects of biotechnology of amylodextrin oligosaccharides. It includes an introduction to genetic engineering as well as enzyme structure and enzymology.

- The second section focuses on applications of specific new analytical tools that are essential to characterize adequately these new types of materials. These oligosaccharides are characterized as polymeric materials.

- The third area addresses specific fields of usefulness for these polysaccharides.

Several years ago, it would have been difficult to find more than a few contributions on the subject. Soon it will be necessary to develop entire books on specific areas of oligosaccharide biotechnology. For example, it is expected that genetic engineering will have a significant effect on the direction of development of these substances.

I acknowledge the financial assistance of the following organizations, which made the symposium possible: Allied-Signal Corporation; American Maize-Products Company; Aqualon Company; Corn Products Company; DCA Food Industries, Inc.; Janssen Pharmaceutica; Miles, Inc.; National Starch and Chemical Corporation; Novo Laboratories; Pfizer, Inc.; Pioneer Hi-Bred International; U.O.P.; and Wyatt Technology Corporation. In addition, I would like to acknowledge the assistance and support of the officers of American Maize-Products Company, in particular William Ziegler III, Patric J. McLaughlin, and Frances R. Katz, for their support and encouragement in the development of this book. The assistance of Sherree Jackson and Gloria Kras in the preparation of the text is also acknowledged. I would like to thank Alfred French for the graphic idea incorporated in the cover design. Finally, I would like to thank Ellen and the boys for their inspiration and vast amounts of encouragement.

ROBERT B. FRIEDMAN
American Maize-Products Company
Hammond, IN 46320–1094

September 9, 1990

BIOTECHNOLOGY AND BIOCHEMISTRY

Chapter 1

Helical and Cyclic Structures in Starch Chemistry

J. Szejtli

Cyclodextrin Research and Development Laboratory, Cyclolab, 1026 Budapest, Endrődi S. 38/40, Hungary

Two opposing hypotheses attempted to describe the confor- mation of the α-1,4-linked glucopyranoside polymers in ne- utral aqueous solutions: the random coil and the segmented helix structure hypotheses. The former one was based mainly on hydrodynamic studies of amylose solutions, the latter on many very different observations, but mainly on the formation and properties of amylose-helix complexes. The formation of cyclodextrins, catalysed by cyclizing enzy- mes, delivers further proof for the helical structure, and simultaneously is the source of a new technology: the molecular encapsulation of different compounds by cyc- lodextrin complexation. The significance of the cyclodext- rins and their derivatives in commercial applications will be discussed.

When two D-glucopyranose units are linked by α-1,4 glucosidic linkage, the formed disaccharide is called maltose, while a β-1,4 glucosidic linkage between them results in cellobiose. Inspecting a molecular model of such a diglucoside, it becomes evident, that the rotation around the C_1-O and O-C_4, bond is hindered, i.e. the values ϕ and ψ rotational angles (the so-called bond conformation, Fig.1.) are restricted to a relatively narrow domain, because of collision of the hydrogens and hydroxyls of the interconnected glucopyranoside units. Maltose can exists only in the cis-configuration (Fig.2.b.) and cellobiose only in the trans-configuration (Fig.2.c.). That configuration, in which the plane of one glucopyranose ring lies at right angle to the other, is sterically impossible (Fig.2.a.). (1,2). This cis-configuration (=maltose) is, however, not a planar structure, because the bond angle of the bridge-oxygen is between 113 and 119°, depending on the steric strain of the whole structure. (3). Constructing a long-chain from these disaccharides, the repetition of these fundamental structural elements results in the case of cis-type configuration (maltose) a helical turn, or in case

0097–6156/91/0458–0002$06.00/0
© 1991 American Chemical Society

of the trans-type configuration a zig-zag conformation i.e. cellulose (Fig. 3).

The amylose helix

Amylose and cellulose are the most thoroughly studied polysaccharides, nevertheless, some properties of amylose are until today not understood. For some time two conflicting hypotheses existed concerning the molecular conformation of amylose in aqueous solution. According to one hypothesis the dissolved molecule forms a coil, (Fig.4.a.) in which helical segments are connected by disordered segments. Adherents to the other hypothesis claim that the molecule forms a completely random coil, without any helically ordered segments. (Fig.4.b.). Neither the completely stretched uncoiled structure, nor the rigid, rodlike totally helical form (Fig.4.c.) has ever been seriously considered as a possibility, and their existence can be easily disproved by viscosimetric studies. The nature of the coiled form, i.e. whether it is completely disordered or contains ordered, helical segments has been vividly disputed in several dozens of papers. (4).

According to the actual accepted hypothesis the amylose molecule in aqueous solution behaves as a flexible random coil, (Fig.4.a.) which, however, consists of extended helical segments (Fig.5.a.) connected by deformed non-helical segments. Upon contact with an appropriate complex forming agent - like iodine, or fatty acids - the helical segments become "tighter". (Fig.5.b.)

Some thirty years ago this inclusion complex formation seemec to be some sort of scientific curiosity only. This type of association of different molecules is never of strictly stoichiometric composition. The ratio of the components in the isolated complex depends on many factors, moreover the amylose is a macromolecular colloid, by which is more difficult to perform any work than with well defined crystalline substances. The study of cyclodextrin inclusion complexes, and quite recently the isolation and study of pure maltooligomers up to DP= 30 (5), clarified some scientific problems about the amylose and cyclodextrin inclusion complexes, but, first of all, revealed an anormous field for industrial utilization of cyclodextrins and their inclusion complexes. (6)

The cyclic dextrins

The cyclodextrins are produced, by an enzymic reaction, from starch, more exactly from amylose and from the α-1,4 linked segments of amylopectin. (Fig.6.) The applied enzyme is a biocatalyst: it does nothing else, but to help to reach the thermodynamically most stable conformation of the polyglucan: i.e. the helical segments are converted to closed rings. The easy formation of cyclodextrins with high yield is an indirect proof for the existence of the helical structure of amylose in the aqueous solution.

Three different cyclic dextrins can be produced: (Fig.7.) the 6 membered αCD, the 7 membered βCD and the 8 membered γCD. All of these are industrially produced crystalline products, with a purity of over 99 %.

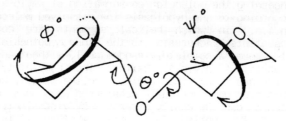

Fig.1.   Bond conformation of a disaccharide: rotation around the
         C_1-O and O-C' bonds.

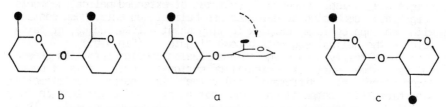

    b   a   c

Fig.2.   Only the "cis" and the "trans" conformations can exist.

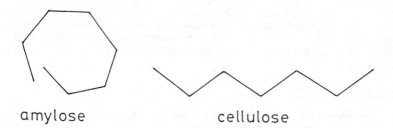

   amylose   cellulose

Fig.3.   Repeating the "cis" (=maltose) conformation results in a
         helical structure while the "trans" (=cellobiose) leads
         to a zig-zag chain.

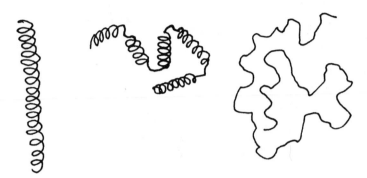

Fig.4.   The rigid, rod-like helix, the "segmented" flexible coil-
         like helical structure, and the random coil.

Fig.5.   The extended helix is contracted to a tight-helix upon
         inclusion complex formation - e.g. with iodine.

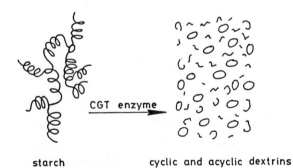

starch   cyclic and acyclic dextrins

Fig.6.   Degradation of starch to a mixture of cyclic and acyclic
         dextrins by cyclodextrin glycosyl-transferase enzyme.

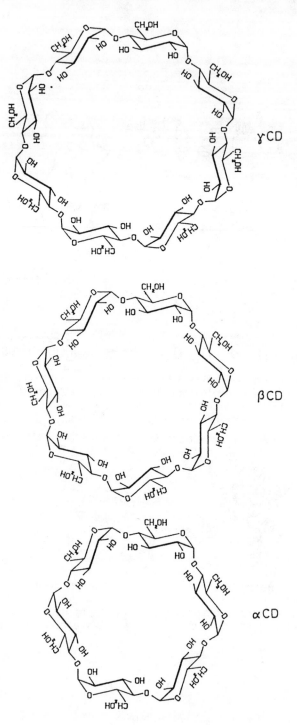

Fig.7. Structure of the cyclodextrins.

The inclusion complex formation

In this cyclic structure an interesting, and practically very important assymetry of the glucopyranose units is revealed: the internal cavity has an apolar character, because it is "lined" by hydrogen atoms, and ether-oxygen atoms. (Fig.8.) One rim of the molecule consists of the primary hydroxyl groups, the other rim of the secondary hydroxyl groups. The dimensions of these "empty cylinders" are considerable. (Fig.9.) Dissolving cyclodextrin in water, the cavity-included water molecules will contact an apolar cavity-surface, which is energetically unfavorable. Adding any substance to an aqueous cyclodextrin solution which has also an apolar - water repellent - character, and which can tightly fit into the cavity, an inclusion complex is formed(Fig.10).The cyclodextrin "host" can accomodate molecules of very different "guest" compounds. The association constants are very different, (Fig.11.) but generally such inclusion complexes can be isolated in microcrystalline form. The composition of the isolated complexes again depends on various factors, and only occassionally are of strictly stoichiometric ratios, but in most cases a nearly 1:1 (CD:guest), or 2:1, or 3:2 etc. composition is found. (Fig.12.) In such molecularly encapsulated form the physical-chemical properties of the "guest" substances are strongly modified.

Industrial aspects of molecular encapsulation by CDs

These effects can be utilized for many industrial purposes. For example benzaldehyde is a liquid compound, which very rapidly oxidized to benzoic acid by atmospheric oxygen. The βCD-benzaldehyde complex can be stored even in pure oxygen atmosphere, without significant oxidation. Volatile liquids become stable, also. This is the base of the stabilization of flavours and fragrances by CD-complexation. Upon contact with water, these complexes immediately begin to dissociate, i.e. the entrapped substances will be released rapidly. Such products are marketed already in several countries.

Unstable, poorly soluble drugs can be stabilised, or their bioavailability can be improved by CD-complexation. Several drugs are marketed in CD-complexed form already. With appropriate cyclodextrins, injectable aqueous solutions can be prepared from insoluble drugs.

Light sensitive substances can be protected against decomposition by CD-complexation. For example, light-sensitive pyrethroids are insecticides which can be stabilized for an extended period of time. In biotechnological processes the CDs can act as solubilizers (e.g. in microbiological steroid conversion) or as toxicity reducing agents (e.g. in waste-water detoxication).

The total number of CD-papers and patents - which is already over 5000 - shows an explosionlike increase. Actually yearly some 650 new papers and patents are published. (7). More than 800 patents are dedicated to some sort of industrial utilization of cyclodextrins. The CD production began just 10 years ago, practically on laboratory scale, but now it increases rapidly. The approximately 1000 ton/year production is expected to grow up to a several ten thousand ton/year CD-market within several years.

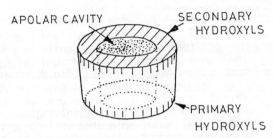

Fig.8. Functional schematic representation of a cyclodextrin "cylinder".

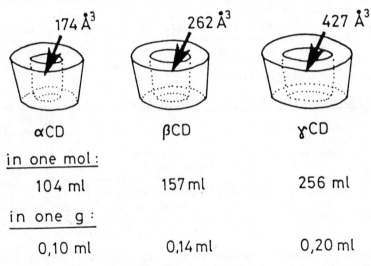

Fig.9. The cavity volumes in the cyclodextrin "capsules"

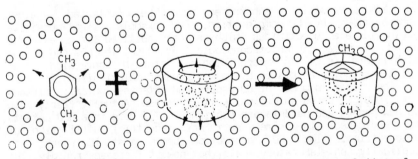

Fig.10. The "host-guest" interaction: molecular encapsulation of p-xylene by β-cyclodextrin.

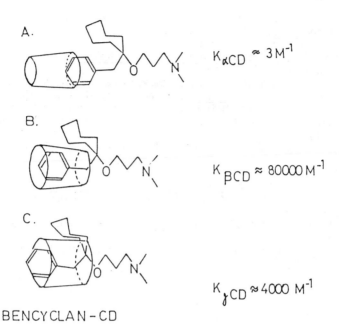

$K_{\alpha CD} \approx 3\,M^{-1}$

$K_{\beta CD} \approx 80000\,M^{-1}$

$K_{\gamma CD} \approx 4000\,M^{-1}$

BENCYCLAN - CD

Fig.11. The assumed structures, the determined association constants for the bencyclan (a cerebral vasodilator) cyclodextrin complexes.

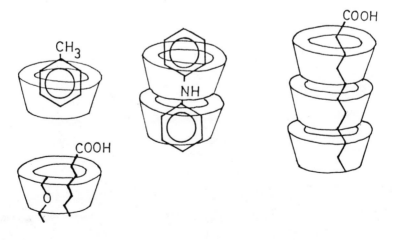

Fig.12. Various stoichiometries of cyclodextrin complexes.

Literature Cited

1. Eliel, E.L., Allinger N.L., Angyal S.J., Morrison G.A. Conformational Analysis. Interscience Publishers, New York, London, Sydney 1965. (Third Printing 1967).

2. Stoddart J.F. Stereochemistry of Carbohydrates. Wiley-Intersciences, New York. 1971. p.249.

3. Rao V.S.R., Sundararajan P.R., Ramakrishnan C., Ramachandran G.N. Conformation of Biopolymers. (Papers read at an International Symposium held at the University of Madras. 1967. jan. 18-21). ed.: Ramachandran G.N. Academic Press, London, New York 1967. p. 721-33.

4. Szejtli J., Richter M., Augustat S. Biopolymers, 1967, 5, 5-16, 17-26.
(1967)

5. Vetter D. Workshop on Supramolecular Organic Chemistry and Photochemistry, Universität des Saarlandes, Saarbrücken, 27 aug.-1. sept. 1989.

6. Szejtli J. Cyclodextrin Technology, Kluwer Academic Publisher, Dordrecht, 1988.

7. Cyclodextrin News, published by FDS Publications, Trowbridge U.K. Vol. 3. 1989.

RECEIVED October 19, 1990

Chapter 2

In Vitro Gene Manipulation

An Introduction

Michael Bagdasarian

Michigan Biotechnology Institute, Lansing, MI 48909
and Department of Microbiology, Michigan State University,
Lansing, MI 48824

Isolation, transfer and *in vitro* manipulation of genes is the strategy of choice not only in research of poorly characterized genetic systems and proteins but also in engineering of new organisms and new biological catalysts for biotechnology, medicine, environmental protection or agriculture. The basic strategies of engineering genes are outlined.

Isolation of genes, their manipulation *in vitro* and expression in different organisms by gene cloning has rapidly become one of the most powerful and versatile research strategies available to modern biology. This set of methods, also referred to as *genetic engineering*, is of particular value for poorly characterized genetic systems, such as microorganisms isolated from natural environments or cells of higher eukaryotic organisms, because these methods are universal. They are not limited by the relatedness of the organisms under investigation to any well characterized laboratory strains and they do not depend upon the availability of detailed information on the structure and organization of their genomes.

In the past two decades of an explosive development of genetic engineering techniques many excellent articles and laboratory manuals on *in vitro* gene manipulation have appeared. Some of the best of them, however, are multiarticle or multivolume editions useful in a molecular cloning laboratory, but not as an introduction into the field. This article will attempt to present a short overview of the strategies used in isolation and manipulation of genes, concentrating particularly on newer developments. The reader is referred to other publications for more detailed descriptions and experimental protocols *(1 - 6)*.

Gene cloning procedures involve four essential steps: (1) generation of DNA fragments that are suitable for cloning; (2) chemical linkage of these fragments to DNA molecules of self-replicating genetic elements called vectors; (3) introduction of thus generated recombinant molecules into suitable host cells and (4) detection

0097–6156/91/0458–0011$06.00/0

of clones containing the new hybrid DNA species. The basic steps used in a cloning procedure are represented schematically in Fig. 1.

Generation of DNA Fragments for Cloning

In the majority of cases this is achieved by treatment of the DNA sample with restriction endonucleases (7 - 10). Endonucleases that create single-strand protruding ends, often called cohesive ends, are preferred because single strand portions of cohesive termini will form complementary duplex structures with the termini of the same type and thus facilitate ligation to the vector molecules (see subsequent paragraphs). However, DNA fragments having blunt ends can also be used for cloning (10).

A large selection of restriction endonucleases with different specificities is available now from many commercial sources and new enzymes are continuously being introduced into the market. It is therefore always possible to find a suitable enzyme or a combination of enzymes that enables one to excise a fragment of DNA carrying a particular gene. In many cases, however, it is not known in advance which endonuclease-sensitive sites might exist within the coding sequence of a gene to be isolated. In this case it is desirable to obtain DNA fragments by a partial cleavage with an enzyme that cuts very frequently. This procedure generates a mixture of fragments that resemble those obtained by completely random fragmentation methods such as   mechanical shearing. The enzymes used most frequently in this procedure are *Sau*3A or *Mbo*I. The staggered ends produced by this enzyme are complementary to the protruding ends generated by *Bam*HI digestion. As shown below the fragments thus obtained may be inserted into the *Bam*HI site present in many cloning vectors (see also Fig. 1).

Cohesive ends generated by:

```
Sau3A or MboI                        BamHI

5'.NNN        GATC.NNN.3'    5'.NNN.G        GATCC.NNN.3'
3'.NNN.CTAG       .NNN.5'    3'.NNN.CCTAG        G.NNN.5'
```

This strategy is particularly useful in those cases where a gene library is prepared such as to carry representatives of every gene of a particular genome.

Cloning Vectors

Vectors are well characterized molecules of DNA to which a fragment of DNA, carrying the genetic marker under investigation, may be linked, and which will allow the introduction into, and stable maintenance of the resulting recombinant DNA molecule in, a suitable host. Vectors are constructed from plasmid or bacteriophage replicons by addition of DNA sequences that provide the vector with specific properties required for its use. The basic functional parts of a vector are: *(i)* a replicon, consisting of an origin of replication and genes that allow the replication of this DNA molecule as an extrachromosomal element; *(ii)* selection marker,

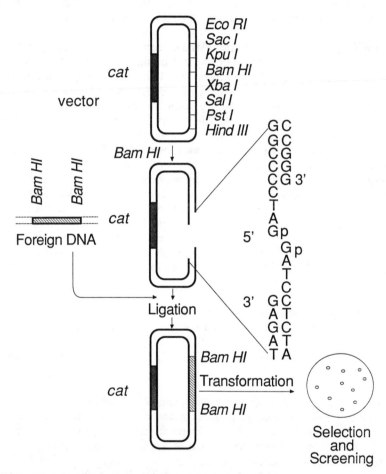

Figure 1. Schematic representation of basic steps involved in a gene cloning procedure. *cat*, chloramphenicol transacetylase.

usually a resistance to toxic substance such as antibiotic or heavy metal ion; *(iii)* cloning sites, one or several restriction endonuclease sensitive sites that are unique in the vector molecule. Many types of vectors have additional features that facilitate screening for inserted fragments, controlled expression of cloned genes or selection for specific functions, such as for example promoters. Figure 2 presents a physical and genetic map of a vector that, in addition to the basic elements contains also a built-in system for controlled expression of inserted genes. Examples of the most frequently used vectors and their basic properties are listed in Table I.

General Purpose Cloning Vectors. Plasmids derived from pBR322 [for reviews see *(11, 12)*] or from pUC series *(13, 14)* are still the most frequently used vectors for general-purpose cloning. In the case of pBR322 and similar vectors the fragments of foreign DNA are inserted into sites within the Ap^R or Tc^R gene and the clones containing the recombinant plasmids may be detected by screening for the insertional inactivation of either of these genes. The vectors of the pUC series are even more convenient than pBR vectors since they carry unique cloning sites for as many as 13 different restriction endonucleases. These sites are located within a gene encoding a fragment of ß-galactosidase. Insertions in any of the unique cloning sites result in the inactivation of the enzyme and the appropriate colonies may be screened on plates containing 5-bromo-4-chloro-3-indoxyl-ß-D-galactoside (X-Gal). These vectors are present in bacterial cells in multiple copies of 40-70 copies per cell. This results in the synthesis of high amounts of the gene products which may be a desirable situation. In many cases, however, high concentrations of foreign gene products are toxic to the host cell. These problems may be solved by the use of low copy number vectors, such as pACYC177 and pACYC184 *(15)* or the derivatives of RSF1010 vectors described below.

Bacteriophage Cloning Vectors. Bacteriophage replicons have been widely used for construction of cloning vectors that have specific advantages over plasmid-based vectors. The most versatile and highly developed vectors are derived from bacteriophage λ [for reviews see *(16, 17)*] and bacteriophage M13 [for reviews see *(14, 18, 19)*].

    Replacement Vectors from Bacteriophage λ By extensive manipulation *in vivo* and *in vitro* bacteriophage genomes have been created that contain a set of cloning sites on each side of the non-essential part of its genome. This portion, called stuffer, is excised with the appropriate restriction enzyme and replaced by a fragment of foreign DNA. The ligated mixture is packaged *in vitro* into preformed λ heads and the packaged DNA is introduced into *Escherichia coli* by infection. Since only those molecules that have a correct length (between the two *cos* sites) can be successfuly packaged this procedure selects for the recombinant DNA molecules and against those containing only the religated vector. The strategy used for cloning in bacteriophage λ vectors are shown in Fig. 3. A variety of λ vectors have been developed to serve different purposes such as different unique cloning sites, different sizes of fragments that may be introduced as insert, expression of the

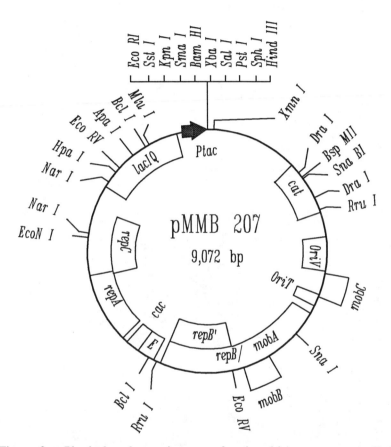

Figure 2.   Physical and genetic map of a broad-host-range, controlled-expression plasmid vector. *oriV*, origin of vegetative replication; *oriT*, origin of transfer replication; *rep*, replication genes; *mob*, genes essential for function of *oriT*; *cac* and E, genes regulating *rep* expression; *Ptac*, hybrid *trp-lac* promoter; *lacI*^{Q}, gene overproducing the Lac repressor; *cat*, chloramphenicol transacetylase.

Table I. Properties of the Most Frequently Used Cloning Vectors

Vector	Size (kb)	Markers (detection)	Cloning sites	Insertional change	Reference
General Type Cloning Vectors					
pBR322	4.4	Ap	ScaI, PvuI, PstI, AseI	$Ap^R \to Ap^S$	(12, 26)
		Tc	EcoRI, ClaI, AvaI, HindIII, EcoRV, BamHI, SphI, EcoNI, NruI, BspMI	None	
pBR325	5.7	Ap	PstI, AvaI	$Tc^R \to Tc^S$ $Ap^R \to Ap^S$	(12, 27)
		Tc	HindIII, BamHI, SalI	None $Tc^R \to Tc^S$	
		Cm	EcoRI	$Cm^R \to Cm^S$	
pUC18	2.7	Ap, $lacZ\alpha$	HindIII, SphI, PstI, SalI, AccI, HincII, XbaI, BamHI, SmaI, KpnI, SacI, EcoRI	$lacZ\alpha^+ \to lacZ\alpha^-$	(14, 19)
pUC19	2.7	Ap, $lacZ\alpha$	Same as pUC18, reverse orientation	$lacZ\alpha^+ \to lacZ\alpha^-$	(14, 19)
pACYC177	3.5	Ap	PstI	$Ap^R \to Ap^S$	(28)
		Km	BamHI, HindIII, SmaI, XhoI	$Km^R \to Km^S$	
pACYC184	3.9	Cm	EcoRI	$Cm^R \to Cm^S$	(28)
		Tc	BamHI, SalI, HindIII	$Tc^R \to Tc^S$	

Bacteriophage-derived Vectors

EMBL 3[a]	44 (14)	Plaque	SalI, BamHI, EcoRI	$spi^+ \rightarrow spi^-$	(16, 21)
λDASH	44 (14)	Plaque	XbaI, SalI, EcoRI, BamHI, HindIII, SacI, XhoI	$spi^+ \rightarrow spi^-$	(21)
λgt11	43.7	Plaque	EcoRI	None	(20)
M13mp18		Plaque	EcoRI, SacI, KpnI, SmaI, BamHI, XbaI, SalI, PstI, SphI, HindIII	$lacZ\alpha^+ \rightarrow lacZ\alpha^-$	(17-19)
M13mp19		Plaque	Same as M13mp18 but reverse orientation	$lacZ\alpha^+ \rightarrow lacZ\alpha^-$	(17-19)

Cosmid Vectors

pHC79	6.4	Ap Tc	PstI, EcoRV, ClaI, BamHI, SalI	$Ap^R \rightarrow Ap^S$ $Tc^R \rightarrow Tc^S$	(28)
pHSG274	3.4	Km	EcoRI, HindIII, BstEII	None	(29)
pMMB33	13.8	Km, mob^+	BamHI, HindIII, SalI BamHI, EcoRI, SacI	None	(30)

Continued on next page

Table I. Continued

Vector	Size (kb)	Markers (detection)	Cloning sites	Insertional change	Reference
pLAFR5	21.5	Tc, mob^+	EcoRI, SmaI, BamHI, SalI, PstI, HindIII	None	(31)
pcos2EMBL	6.1	Tc, Km	BamHI, SalI XhoI,	$Tc^R \rightarrow Tc^S$ None	(36)
Promoter Cloning Vectors					
pGA39	5.1	Ap	HindIII, PstI, SmaI BamHI, BglII, HindIII, MboI, PstI	$Tc^S \rightarrow Tc^R$	(32)
pGA46	4.8	Ap			(32)
pKT240	12.9	Km, Ap	EcoRI, HpaI	$Tc^S \rightarrow Tc^R$ $Sm^S \rightarrow Sm^R$	(35)
Broad-host-range Vectors					
pMMB66EH	8.8	Ap, P_{tac} $lacI^Q$, mob^+	EcoRI, SmaI, BamHI, SalI, PstI, HindIII	None	(34)
pMMB66HE	8.8	Ap, P_{tac}	HindIII, PstI, SalI,		

Plasmid	Size (kb)	Markers	Restriction sites	Inactivation	Reference
pMMB207	9.0	Cm, P_{tac} $lacI^Q$, mob^+	EcoRI, SacI, KpnI, SmaI, BamHI, XbaI, SalI, PstI, SphI, HindIII	None	Morales and Bagdasarian unpublished
pRK2501	11.1	Km	HindIII, XhoI	$Km^R \to Km^S$	
		Tc	SalI	$Tc^R \to Tc^S$	(37)
pRK310	20.4	Tc	EcoRI, BglII,	None	
			BamHI, HindIII, PstI	$lacZ\alpha^+ \to lacZ\alpha^-$	(38)
			BglII	None	

[a]The size of the "stuffer" fragment is indicated in brackets

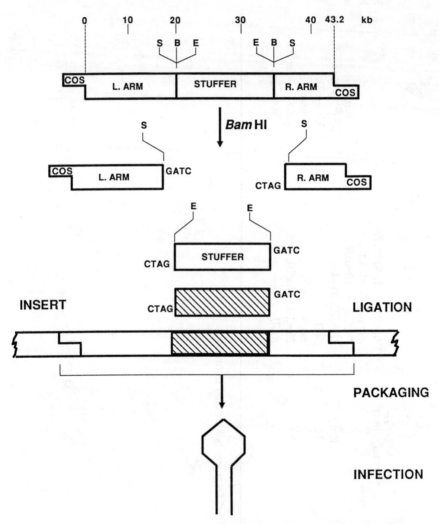

Figure 3.   Schematic representation of steps involved in cloning with λ bacteriophage vectors. cos, cohesive ends of λ DNA; B, *Bam*HI; E, *Eco*RI; S, *Sal*I; kb, kilobase.

cloned gene or easy conversion of the DNA to single strand form for sequence determination. Thus, λgt10 and λgt11 are convenient for screening of large numbers of recombinant phages by hybridization to a nucleotide gene probe or by immunodetection, respectively *(20)*, λDASH and λFIX contain a T3 or T7 promoter on either side of the insert and provides the possibility to synthesize large amounts of respective mRNAs or gene products and λZAP may be used for quick conversion of the cloned DNA into an M13 clone used for sequencing [for a review see *(21)*].

Vectors from Filamentous Phages. Their principal advantages is in generation of single-strand DNA for nucleotide sequence determination *(22)* and for site-directed mutagenesis of cloned genes *(23)*. Although bacteriophage-derived vectors are were developed mainly for special purposes, they still may be successfully used for general type cloning [for a review on filamentous phage vectors see *(25)*].

Special Purpose Cloning Vectors. Several types of vectors have been constructed to exhibit special properties not present in other cloning vehicles. These properties facilitate the use of these vectors in special cases. However, they can often be used for general purpose cloning.

Cosmids. These are plasmid vectors containing one or two *cos* sites of bacteriophage λ. They have the advantage of a small size, the replication mode of plasmids and the ability to be packaged *in vitro* into bacteriophage λ heads. Their main use is in generation of gene libraries since they allow cloning of large fragments of DNA, up to 45 kb, thus reducing the number of recombinant clones representing the entire library. Due to the λ packaging mechanism recombinant cosmids smaller than 40 kb do not get successfully packaged. This selects for only those recombinant molecules that contain a large insert of foreign DNA [for a review see *(33)*].

Promoter and Terminator Cloning Vectors. The principle used in construction of promoter-probe vectors is to place unique restriction sites upstream of a gene with a phenotype that is easily selected, such as e. g. antibiotic resistance, or scored, such as ß-galactosidase or galactokinase, from which the natural promoter has been deleted. By applying a similar principle, vectors for cloning transcription terminators are generated by placing cloning sites between a promoter and the coding sequence of a gene. Overproduction of galactokinase is often detrimental to the host cell rendering the *galK* promoter-cloning plasmids *(39)* difficult to use. On the other hand ß-galactosidase may be overproduced to high concentrations and is easily determined in a quantitative assay. Many different vectors for cloning "punctuation signals", including translational initiation sites, have been developed to satisfy different purposes. For a review see *(40)*.

Controlled-expression Vectors. The ability to control expression of cloned genes is particularly important in the following circumstances:

1. If the product of the gene is not tolerated well by the host organism and the gene needs to be maintained under repressed conditions until its induction is desired.
2. If a particular gene product is needed in large quantities for isolation and purification.
3. If effects of particular gene products on cellular metabolism are studied.

Several different regulatory systems have been used to achieve regulated expression of the cloned genes. Among the most widely used are: *trp* promoter and *trp*R regulatory gene *(41)*; *lac* promoter and the *lac*I or *lac*I^Q gene, such as for example the promoters present in the M13mp series or the pUC series of plasmids *(14, 17-19)*; the hybrid *trp-lac* (*tac*) promoter *(42, 43)*, the vector pMMB207, presented in Fig. 2 is an example of a *tac* promoter based broad-host range vector carrying its own regulatory gene *(34, 44)*. Convenient expression of many cloned genes was achieved with a system combining the promoters of bacteriophage λ and their thermosensitive repressor protein provided by the cI_{857} gene *(45, 46)*. Very high expression of genes may be obtained by the use of the bacteriophage T7 promoter and the gene for T7 RNA polymerase cloned into a different expression system (such as λP_L or *tac*) and introduced into the same host. The T7 promoter gives virtually no expression in the absence of the T7 polymerase and allows therefore to clone genes that are highly toxic to the host cell or obtain an exclusive labelling of a particular gene product *(47)*.

Broad-host-range Cloning Vectors. Microorganisms other than *E. coli* exhibit a wealth of metabolic properties of great scientific interest and of considerable environmental, medical and commercial importance. As their extraordinary ability to carry out recycling of carbon and nitrogen, degradation of toxic waste, enhancement of animal and plant productivity and catalysis in various biotechnological processes is recognized, genetic manipulation in these organisms is becoming of utmost importance.

Many of the vector systems discussed in the previous paragraphs are specific for *E. coli* as host bacterium. The most serious limitations of the *E. coli* cloning system are:

1. Poor recognition by *E. coli* of transcriptional and translational signals from many other species. As a consequence, heterologous genes are often expressed poorly in this host.
2. Impossibility to study in *E. coli* the function of genes as part of metabolic or regulatory pathway that is absent in this host (for example hydrocarbon degradation, pathogenicity, secretion, nitrogen fixation etc.).
3. Toxicity of heterologous genes in *E. coli* (such as genes specifying secretion proteins, membrane components).
4. The inability of the narrow-host-range *E. coli* vectors to replicate in other bacterial species.

Plasmids of the incompatibility groups C, N, P, Q and W exhibit a host range much wider than those commonly used as replicons for *E. coli*-specific vectors. Their properties have been reviewed in several recent articles that provide lists of vectors and hosts *(34, 48, 49)*. Vectors based on IncP and IncQ replicons are among the most highly developed broad-host-range vectors available to date. For some of them, such as pMMB66 and pMMB207 series the entire nucleotide sequence has been determined *(34)*.

Suicide Vectors and Genetic Engineering *In Vivo*.   For delivery of transposable elements or to select for clones where the recombination of genes occurred between the incoming replicon and a resident replicon, such as for example the chromosome, it is desirable to have vectors that may be transferred to a donor cell, but cannot be maintained in it. A classical example of such "suicide" vectors are λ bacteriophages with nonsense mutations in the replication genes that are also defective in their lysogenization and recombination functions. If non-permissive host cells are infected with such bacteriophage that carries a transposon and selection for the transposon markers is applied, only those cells in which a transposition has occurred will survive and will give rise to a colony *(50)*. A similar strategy was followed to construct a transposon-delivery system for Gram-negative bacteria other than *E. coli*. Mobilizable derivatives of pBR322, carrying different transposons, were constructed and transferred to bacteria in which this plasmid can not replicate, such as *Pseudomonas* or *Rhizobium*. High efficiency of transposon insertion could be achieved by this method *(51)*. An extension of this technique was the incorporation of an antibiotic resistance marker and a functional pBR322 origin into the Omegon transposable element. After transposition, a restriction digest of the total DNA from the host cells may be subjected to religation and transformation into *E. coli*. The sequences flanking the transposon insertion can thus be easily cloned *(52)*.

Introduction of Recombinant DNA Into New Hosts

Purified DNA is introduced into bacterial or yeast cells by transformation. The procedures to make microbial cells competent to take up purified DNA usually involve treatment with Ca^{++} or Mg^{++} ions in the cold *(53, 54)*. Although the molecular mechanism of transformation is not well understood the efficiency of transfer in *E. coli* and a few other Gram-negative and Gram-positive species is good enough to be used in gene cloning technology. On the other hand many species are transformed poorly if at all and the restriction systems make it virtually impossible to introduce into them recombinant DNA isolated from other species. Cloning vectors that contain transfer origins and may be transferred by conjugative mobilization directly from one cell to another are of great advantage in such cases. Conjugative transfer is more efficient than transformation by several orders of magnitude and some molecules of the recombinant plasmids are modified by the host modification system before they get digested by restriction enzymes. The vectors of the pMMB series, exemplified in Fig. 2 have this advantage which has been used to transfer genes, inserted into these vectors, to *Pseudomonas* and other Gram-negative species *(30, 52)*.

A new technique of DNA transfer into living cells has recently emerged and is being rapidly developed. It consists of placing the cells in a high voltage electric field for a short period of time. This method, known as electroporation, seems to be applicable to a wider range of species than is calcium-induced competence and in many cases gives higher efficiencies of DNA transfer [for a review see (55)].

Engineering and Synthesis of Genes *In Vitro*. Protein Engineering

The development of chemical synthesis of short oligonucleotides with defined sequence made it possible to synthesize *in vitro* genes with desired substitutions of any nucleotide. This technique, known as site-directed (or oligonucleotide-directed) mutagenesis, is used to generate proteins with desired substitution of amino acids by subcloning the modified genes and producing new proteins specified by these genes. Since the changes of amino acids may be combined and predicted and since it is anticipated that we will be able, in the near future, to improve substantially our ability to predict the properties of mutant proteins obtained by this method, this strategy has often been called protein engineering. The actual synthesis of the mutant gene is an easy process. The gene to be mutated is inserted into a vector that produces single-stranded DNA, e. g. one of the M13 bacteriophage vectors, the oligonucleotide with one or several mismatched bases is annealed to the single-strand template and the complementary strand is completed by the action of DNA polymerase in the presence of deoxyribonucleotides. After transformation into a permissive host plaques of the bacteriophage are selected and checked for the presence of the mutated gene. It is obvious that the main problem is this screening. Several methods have been developed to distinguish bacteriophages containing the copy of the mutant strand. One of the most effective is the method that synthesizes the daughter strand with the use of modified nucleotides making it resistant to certain restriction nucleases. The parental strand that remains sensitive can thus be eliminated and the progeny of the bacteriophage, after transformation, contains predominantly the mutated gene (24, 56).

Synthesis of Genes by Polymerase Chain Reaction. Another strategy that takes advantage of specific oligonucleotide primers is the *in vitro* amplification of specific DNA fragments from very small amounts of an impure sample. Since the first demonstration of the technique by Mullis and Faloona (57) it has become one of the most popular procedures in molecular biology and virtually has transformed our approach to isolation of genes [for review see (58-60).

Polymerase chain reaction (PCR) consists of repeated cycles of *in vitro* DNA synthesis from a template by using two primer oligonucleotides that hybridize to opposite strands of the fragment to be amplified. The use of thermostable Taq DNA polymerase enables to run the required steps of denaturation, annealing and DNA chain elongation in an automatic thermocycler, now commercially available from several sources, and thus greatly reduce the time and effort. Recent modifications of the technique allows amplification of DNA fragments with the use of only one specific primer. The other primer is selected to be complementary to sequences in the vector such as for example one of the universal sequencing primers available commercially *(61)*.

Concluding Remarks

The development of new techniques in molecular biology proceeds at an astonishing pace. It is impossible to give a comprehensive review of the subject in a short article. I have, therefore, outlined the main strategies that have been already tested and are widely adopted by the scientific community, giving references to review articles, books and experimental manuals whenever possible to enable the reader to look through a wide range of methods and references before decideing on the experimental strategy most useful for his particular purpose.

Acknowledgments. Work in the author's laboratory was supported in part by grants from U. S. Department of Agriculture (#89-01053 to MBI) and from Research Excellence Fund of the State of Michigan.

Literature cited

1. Ausubel, F.M.; Brent, R.; Kingston, R.E.; Moore, D.D.; Seidman, J.G.; Smith, J.A.; K. Struhl. *Current Protocols in Molecular Biology;* John Wiley: New York, NY, 1987.
2. Davis, L.G.; Dibner, M.D.; Battley, J.F. *Basic Methods in Molecular biology;* Elsevier: New York, NY, 1986.
3. *DNA Cloning, a Practical Approach;* Glover, D.M., Ed.; IRL Press: Oxford, 1985.
4. Walker, J.M.; *Methods in Molecular Biology. New Nucleic Acid Techniques;* Humana Press: Clifton, NJ, 1988; Vol. 4.
5. Zyskind, J.W.; S. I. Bernstein, I.S.; *Recombinant DNA Laboratory Manual.* Academic Press: New York. 1989.
6. Sambrook, J.; Fritsch, E.F.; Maniatis, T.; *Molecular Cloning. A Laboratory Manual;* Cold Spring Harbor Laboratory Press: Cold Spring Harbor, NY. 1989.
7. *Genetic Engineering. Principles and Methods;* Setlow, J.K.; Hollaender, A., Eds.; Plenum Press: New York, NY, 1980; Vol. 2.
8. Timmis, K.N. In: *Genetics as a Tool in Microbiology*; Glover, S.W.; Hopwood, D.A., Ed.; Microbiol. Symp. 31; Cambridge Univ. Press: Cambridge, England, 1981; pp. 49-109.
9. *Recombinant DNA*; Wu, R., Ed.; Methods in Enzymology Vol. 68; Academic Press: New York, NY, 1979.
10. *Advanced Molecular Genetics*; Pühler, A.; Timmis, K.N., Eds.; Springer-Verlag: Berlin, Germany, 1984.

11. Balbás, P.; Soberón, X.; Merino, E.; Zurita, M.; Lomeli, H.; Valle, F.; Flores, N.; Bolivar, F. *Gene* **1986**, *50*, 3-40.
12. Balbás, P.; Soberón, X.; Bolivar, F.; Rodriguez, R.L. In *Vectors*; Rodriguez, R.L.; Denhardt, D.T., Eds.; Butterworths: Boston, MA, 1988, pp 12 3-42.
13. Viera, J.; Messing, J. *Gene*, **1982**, *19*, 259-268.
14. Yanisch-Perron, C.; Viera, J.; Messing, J. *Gene*, **1985**, *33*, 103-119.
15. Rose, E.E. *Nucl. Acid Res.* **1988**, *16*, 355-356.
16. Kaiser, K.; Murray, N.E. In *DNA Cloning. A practical Approach*; Glover, D., Ed.; IRL Press: Oxford, England, 1985, Vol. 2, pp 1-47.
17. Smith, G.P. In *Vectors*; Rodriguez, R.L.; Denhardt, D.T., Eds.; Butterworths: Boston, MA, 1988, pp 61-83.
18. Messing, J. *Methods in Enzymology*, **1983**, *101*, 20-77.
19. Viera, J.; Messing, J. *Methods in Enzymology*, **1987**, *153*, 3-11.
20. Huynh, T.V.; Young, R.A.; Davis, R.W. In *DNA Cloning. A practical Approach*; Glover, D., Ed.; IRL Press: Oxford, England, 1985, Vol. 2, pp 49-78.
21. Sorge, J.A. In *Vectors*; Rodriguez, R.L.; Denhardt, D.T., Eds.; Butterworths: Boston, MA, 1988, pp 43-60.
22. Sanger, F.; Nicklen, S.; Coulson, A.R. *Proc. Natl. Acad. Sci. USA* **1977**, *74*, 5463-5467.
23. Smith, M. *Ann. Rev. Genet.* **1985**, *19*, 423-462.
24. Sayers, J.R.; Schmidt, W.; Eckstein, F. *Nucl. Acid Res.* **1988**, *16*, 791-802.
25. Smith, G.P. In *Vectors*; Rodriguez, R.L.; Denhardt, D.T., Eds.; Butterworths: Boston, MA, 1988, pp 61-83.
26. Bolivar, F.; Rodriguez, R.L.; Greene, P.J.; Betlach, M.C.: Heyneker, H.L.; Boyer, H.W.; Crosa, J.H.; Falkow, S. *Gene*, **1977**, *2*, 95-113.
27. Bolivar, F. *Gene*, **1978**, *4*, 121-136.
28. Hohn, B.; Collins, J. *Gene*, **1980**, *11*, 291-298.
29. Brady, G.; Jantzen, H.M.; Bernard, H.U.; Brown, R.; Schütz, G.; Hashimoto-Gotoh, T. *Gene*, **1984**, 223-232.
30. Frey, J.; Bagdasarian, M.; Feiss, D.; Franklin, F.C.H.; Deshusses, J. *Gene*, **1983**, *24*, 299-308.
31. Keen, N.T.; Tamaki, S.; Kobayachi, D.; Trollinger, D. *Gene*, **1988**, *70*, 191-197.
32. Chang, A.C.Y.; Cohen, S.N. *J. Bacteriol*, **1978**, *134*, 1141-1156.
33. Hohn, B; Koukolikova-Nicola; Lindenmaier, W; Collins, J. In *Vectors*; Rodriguez, R.L.; Denhardt, D.T., Eds.; Butterworths: Boston, MA, 1988, pp 113-127.
34. Morales, V.; Bagdasarian, M.M.; Bagdasarian, M. In *Pseudomonas. Biotransformations, Pathogenesis and Evolving Biotechnology*; Silver, S.; Chakrabarty, A.M.; Iglewski, B.; Kaplan, S., Eds.; ASM, Washington, DC, 1990, pp 229-241.
35. Bagdasarian, M.M.; Aman, E.; Lurz, R.; Rückert, B.; Bagdasarian, M. *Gene*, **1983**, *26*, 273-282.
36. Poustka, A.; Rackwitz, H.R.; Frischauf, A.M.; Hohn, B.; Lehrach, H. *Proc. Natl. Acad. Sci. USA*, **1984**, *81*, 4129-4133.

37. Kahn, M.; Kolter, R.; Thomas, C.; Figurski, D.; Meyer, R.; Remaut, E.; Helinski, D.R. *Methods in Enzymology*, **1979**, *68*, 268-280.
38. Ditta, G.; Schmidhauser, T.; Yacobson, E.; Lu, P.; Liang, X.-W.; Finlay, D.R.; Guiney, D.; Helinski, D.R. *Plasmid*, **1985**, *13*, 149-153.
39. Rosenberg, M.; Chepelinski, A.B.; McKenney, K. *Science*, **1983**, *222*, 734-739.
40. Nierman, W.C. In *Vectors*; Rodriguez, R.L.; Denhardt, D.T., Eds.; Butterworths: Boston, MA, 1988, pp 153-177.
41. Tacon, W.C.A.; Bonas, W.A.; Jenkins, B.; Emtage, J.S. *Gene*, **1983**, *23*, 255-265.
42. de Boer, H.A.; Comstock, L.J.; Vasser, M. *Proc. Natl. Acad. Sci. USA*, **1983**, *80*, 21-25.
43. Amann, E.; Brosius, J.; Ptashne, M. *Gene*, **1983**, *25*, 167-178.
44. Fürste, J.P.; Pansegrau, W.; Frank, R.; Blöcker, H.; Scholz, P.; Bagdasarian, M.; Lanka, E. *Gene* **1986**, *48*, 119-131.
45. Remaut, E.; Stanssens, P.; Fiers, W. *Gene*, **1981**, *15*, 81-93.
46. Remaut, E.; Stanssens, P.; Fiers. W. *Nucl. Acid Res.* **1983**, *11*, 4677-4688.
47. Tabor, S.; Richardson, C.C. *Proc. Natl. Acad. Sci. USA*, **1985**, *82*, 1074-1078.
48. Shapiro, J.A. In *DNA Insertion Elements, Plasmids and Episomes*. Shapiro, J.A.; Adhya, S.L., Eds., Cold Spring Harbor Laboratory: Cold Spring Harbor, NY, 1977, pp. 601-670.
49. Schmidhauser, T.J.; Ditta, G.; Helinski, D.R. In *Vectors*; Rodriguez, R.L.; Denhardt, D.T., Eds.; Butterworths: Boston, MA, 1988, pp 287-332.
50. Kleckner, N. *Ann. Rev. Genet.*, **1981**, *15*, 341-404.
51. Simon, R.; Priefer, U.; Pühler, A. *Bio/Technology*, **1983**, *1*, 784-791.
52. Fellay, R.; Krisch, H.M.; Prentki, P.; Frey, J. *Gene*, **1989**, *76*, 215-226.
53. Cohen, S.N.; Chang, A.C.Y.; Hsu, L. *Proc. Natl. Acad. Sci. USA*, **1972**, *69*, 2110-2114.
54. Hanahan, D. *J. Mol. Biol.* **1983**, *166*, 557-580.
55. Chassy, B.M.; Mercenier, A.; Flickinger, J. *Trends Biotechnol.* **1988**, *6*, 303-309.
56. Vandeyar, M.A.; Weiner, M.A.; Hutton, C.J.; Batt, C.A. *Gene*, **1988**, *65*, 129-133.
57. Mullis, K.B.; Faloona, F.A. 1987. Specific synthesis of DNA *in vitro* via a polymerase-catalized chain reaction. Methods Enzym. 155:335-350.
58. Erlich, H. A. *PCR technology. Principles and applications for DNA amplification.* Stockton Press: New York, NY, 1989.
59. Innis, M.A.; Gelfand,D.H; Sninsky,J.J.; White, T.J. *PRC protocols. A guide to methods and aplications.* Academic Press: San Diego, CA, 1989.
60. *Polymerase Chain Reaction;* Erlich, H.A.; Gibbs, R.; Kazazian, H.H., Eds., Current Communications in Molecular Biology; Cold Spring Harbor Laboratories: Cold Spring Harbor, NY, 1989.
61. Shyamala, V.; Ames, G.F.-L. *Gene*, **1989**, *84*, 1-8.

RECEIVED October 3, 1990

Chapter 3

Structure–Function Relationships in Amylases

B. Svensson, M. R. Sierks, H. Jespersen, and M. Søgaard

Carlsberg Laboratory, Department of Chemistry, Gamle Carlsberg Vej 10, DK–2500 Copenhagen Valby, Denmark

Chemical modification has been used in conjunction with site-directed mutagenesis to identify and assign specific functional roles to amino acid residues involved in substrate binding and catalysis in glucoamylase from *Aspergillus niger* and in barley α-amylase. Kinetic properties of selected mutant enzymes thus indicated that the Trp120 of glucoamylase stabilized the substrate transition state, while Glu179 and Asp176 behaved as the general acid and base catalyst, respectively. In barley α-amylase Glu205 was proposed as the general acid catalyst while Trp276 and Trp277 were influenced by binding of β-cyclodextrin, and hence starch granules, at a distance from the catalytic site. Structure comparison guided by functional residues suggested that most starch-degrading enzymes share key active site features. A few also contain homologous raw starch binding domains.

Amylases degrade starch to a diversity of oligodextrins according to their individual substrate specificities and action patterns (1,2). Classification of activity can be made by the following distinctions: i) endo- versus exo-mode of attack, ii) inversion versus retention of the product anomeric configuration, iii) poly-, oligo-, or disaccharide/disaccharide-analogue substrate preference, and iv) α-(1→4), α-(1→6) or dual bond-type specificity.

In spite of their long history as well as widespread and abundant occurrence, only a few amylases or related enzymes have been thoroughly investigated with regard to structure and mechanism. Since the development of cloning and sequencing techniques, more than 70 primary structures representing enzymes from 12 different classes (Table I) have been reported, the first sequences, of murine pancreatic and salivary α-amylases, being described in 1980 (3). These enzymes are multidomain single polypeptide-chain proteins with one exception, glucoamylase from *Saccharomyces cerevisiae (var. diastaticus)*, which consists of a large and a small subunit (4). They vary in size from approx. 400 (5) to 1100 (6) amino acid residues. Complete protein sequences including post-translational modifications have been determined for porcine pancreatic α-amylase (7,8), Taka-amylase A (9), glucoamylase 1 and 2 from *Aspergillus niger* (10, 11), and β-amylase from soybean (12) and sweet potato (13).

Three-dimensional structures have been reported for Taka-amylase A (TAA) (14),

Table I. Complete Sequence of Starch-Hydrolases and
Related Enzymes

Enzyme Class	Number of Available Sequences	
α-Amylase	ca. 40	(a)
β-Amylase	6	(b)
Glucoamylase	6	(c)
α-Glucosidase	2	(d)
Isomaltase	1	(e)
Maltogenic α-amylase	1	(f)
Maltotetraohydrolase	2	(g)
Isoamylase	1	(h)
Pullulanase	1	(i)
Cyclodextrin glucanotransferase	10	(j)
Amylomaltase	1	(k)
Branching Enzyme	1	(l)

First report:
a. Hagenbüchle, O. et al. Cell 1980, 21, 179-87. b. Kreis, M. et al. Eur.J.Biochem. 1987, 169, 517-25. c. Svensson, B. et al. Carlsberg Res.Commun. 1983, 48, 529-44. d. Hong, S.H.; Marmur, J. Gene 1986, 41, 75-84. e. Hunziker, W. et al. Cell 1986, 16, 227-34. f. Diderichsen, B.; Christiansen, L. FEMS Microb.Lett. 1988, 56, 53-60. g. Fujita, M. et al. J.Bacteriol. 1989, 171, 1333-39. h. Amemura, A. et al. J.Biol.Chem. 1988, 263, 9271-75. i. Katsuragi, N. et al. J.Bacteriol. 1987, 169, 2301-6. j. Takano, T. et al. J.Bacteriol. 1986, 166, 1118-22. k. Lacks, S.A. et al. Cell 1982, 31, 327-36. l. Baecker, P.A. et al. J.Biol.Chem. 1986, 261, 8738-43.

porcine pancreatic α-amylase (15), and *Bacillus circulans* (16) and *Bacillus stearothermophilus* (17) cyclodextrin glucanotransferases (CGTases). A low resolution structure for soybean β-amylase has been described (18). Molecular details on enzyme-substrate interactions have been proposed based on a difference Fourier map between native and maltotriose-analogue-soaked pancreatic α-amylase (15) and a model fitting of a seven-glucosyl-residues segment of amylose to the active site cleft of TAA (14), respectively. The α-amylases, like the lysozymes, are suggested to contain two catalytic carboxyl groups and an extended substrate binding cleft composed of an array of subsites (19). Although the physical structure of the α-amylase-ligand complexes are not as well described as in lysozyme, the apparent number of subsites and site of cleavage have been defined for several amylases using enzyme kinetic techniques (20).

Despite little, if any, homology existing between many of the starch-degrading enzymes listed in Table I, they are still likely to constitute a protein superfamily. At least one of three kinds of related structural characteristics has been recognized among them: i) predominant α/β-barrel catalytic domain folding (14-17,21,22; Jespersen, MacGregor, Sierks and Svensson, unpublished data), ii) short regional sequence similarities, perhaps reflecting functionally important folding modules (23-25), and iii) homologous terminal domains thought to bind starch granules (11,26-28). Therefore, insight acquired from structure/function relationships for one member of the family may likely be applied to certain others. The analysis of data from such related enzymes could lead to a tertiary structure template based on which substrate specificities and action patterns can be tailored.

Elucidating Roles of Active Site Residues

The first essential residue to be identified in an enzyme from the amylase superfamily was the acarbose-protected Trp120 in glucoamylase from *A. niger* (23). Interestingly, it pointed out a short regional sequence similarity with TAA where the corresponding Trp83 was proposed to participate in substrate binding at a distance from the site of catalysis (14). This suggests that an exo- and an endo-α-glucanase showing no overall sequence homology can still have functional structural elements in common. In fact, the segment Tyr75-Trp83 creates most of the surface of the substrate binding area to one side of the catalytic residues in TAA (not shown). Chemical modification experiments were performed to identify the two catalytic carboxyl groups in glucoamylase that had been inferred from earlier kinetic studies (29). Acarbose (Figure 1) efficiently protected the active site against inactivation, enabling differential labelling with the water-soluble carbodiimide, ethyl-3-(4-azonia-4,4-dimethylpentyl)carbodiimide (EAC) (Figure 2). Sequence analysis of the [^{3}H]EAC-substituted glucoamylase indicated that Asp176, Glu179, and Glu180 comprise the two essential acidic groups (30,31). The individual roles in the glucoamylase mechanism of these three carboxylic acid residues in the critical acidic cluster were in turn elucidated by the analysis of kinetic properties of mutant enzymes containing the corresponding amides (32) (Table II). The drop in k_{cat} for Glu179$\rightarrow$Gln glucoamylase and increase in K_m for Glu180$\rightarrow$

Table II. Kinetic Parameters and Change in Activation Energy
for Glucoamylase Mutants

Substrate	Enzyme				
	Asp176 $\rightarrow$Asn	Glu179 $\rightarrow$Gln	Glu180 $\rightarrow$Gln	Trp120 $\rightarrow$Tyr	Wild-type
Maltose					
k_{cat} (s^{-1})	0.73	-	1.5	0.12	14
K_m (mM)	6.2	-	41	0.63	1.7
k_{cat}/K_m (s^{-1}mM^{-1})	0.12	-	0.037	0.19	8.3
$\Delta\Delta G^a$ (kJ/mol)	11.5	-	14.6	10.2	-
Maltoheptaose					
k_{cat} (s^{-1})	8.2	0.047	31	0.44	84
K_m (mM)	0.63	0.15	9.4	0.059	0.22
k_{cat}/K_m (s^{-1}mM^{-1})	13	0.32	3	7.5	390
$\Delta\Delta G^a$ (kJ/mol)	9.1	-	12.8	10.6	-
Isomaltose					
k_{cat} (s^{-1})	0.052	-	0.18	0.0059	0.49
K_m (mM)	135	-	95	3.2	36
k_{cat}/K_m (s^{-1}mM^{-1})	0.00039	-	0.0019	0.0018	0.014
$\Delta\Delta G^a$ (kJ/mol)	9.6	-	5.2	5.3	-

a. $\Delta\Delta G = -RT\ln [(k_{cat}/K_m)_{mut}/(k_{cat}/K_m)_{wt}]$ (33). (Compiled with permission from Refs. 32, 34. Copyright 1989, Oxford University Press).

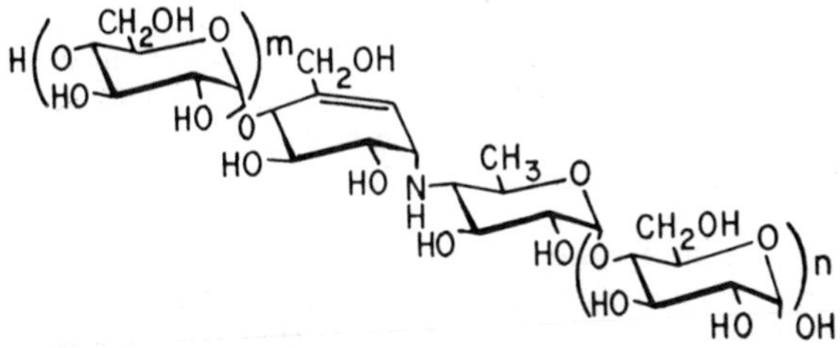

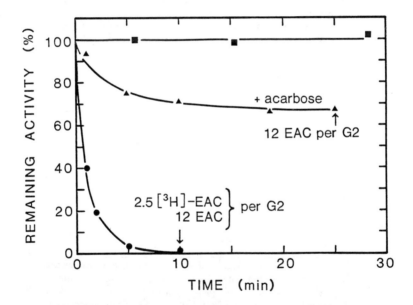

Figure 1. Pseudooligosaccharide inhibitors of various amylases and glucosidases. Acarbose (BAYg5421): m=0,n=2. Aplanin is a maltooligodextrin mixture (average DP=12) containing an acarbose core.

Figure 2. *A. niger* glucoamylase G2 (4.5 mg·ml^{-1} in 0.05 M Mes buffer pH 6.0) was treated with 0.10 M EAC in the presence of 0.6 mM acarbose (▲). The inhibitor was removed as described (30) and the derivative inactivated by 0.02 M [^{3}H]EAC (●). The control G2 enzyme incubated without EAC added retained full activity (■).

Gln glucoamylase suggested these residues were involved in catalysis and substrate binding, respectively. The pK_a value of the general acid decreased in both the Glu180→Gln and Asp176→Asn enzymes compared to the wild-type (Table III). Glu179 was concluded to be the general acid catalytic group responsible for initial protonation of the oxygen in the glucosidic bond to be cleaved. Due to the effects on k_{cat}/K_m and k_{cat} values by Asp176→Asn, Asp176 was concluded to act as the catalytic base (Table II; ref. 32). The relative increases in activation energy, $\Delta\Delta G$, for hydrolysis of α-(1→4)- and α-(1→6)-linked substrates caused by mutation of Glu180 (Table II) indicate that this residue stabilizes the transition state complex at subsite 2 (32). A schematic diagram of these interactions is given in Figure 3. Trp120 was similarly found to stabilize the transition state complex in subsite 2 from a more distant subsite (34).

The glucoamylase Glu180 involved in substrate binding can be aligned to acidic amino acid residues in related enzymes (Figure 4; ref. 25). In TAA and porcine pancreatic α-amylase it matches aspartic acid residues suggested by crystallographic analysis to be binding and catalytic, respectively (14,15). Mutagenic studies in *B. subtilis* α-amylase suggested the equivalent residue to be important for binding (35). The aligned Asp505 in isomaltase, however, was suggested as the catalytic base by affinity labelling with conduritol-B-epoxide (24,36). Evidence obtained from mutation of barley α-amylase isozyme 1 Glu205 (37) and *B. subtilis* α-amylase Glu208 (35), which both align with TAA Glu230 (25), identifies these groups as likely candidates for the general acid catalyst. Also an α-glucosidase (38) was affinity labelled by conduritol-B-epoxide, but the modified residue remains to be identified. Affinity labelling of soybean β-amylase (39) using 2,3-epoxypropyl-α-D-glucoside, identified Glu186 as the catalytic base. This catalytic base has still not enabled alignment of the β-amylase sequence with those from the other enzyme classes.

Table III. pK_a Values for Catalytic Groups
in Glucoamylases

	Enzyme		
	Asp176 →Asn	Glu180 →Gln	Wild- type
$pK_{1,E}$	n.d.	2.2	2.7
$pK_{2,E}$	5.3	4.9	5.9
$pK_{1,ES}$	n.d.	1.7	2.4
$pK_{2,ES}$	5.3	4.9	6.0

The pH dependence of k_{cat}/K_m and k_{cat} for maltose hydrolysis was determined in the pH range 2.4-7.1 at 45°C and used to estimate pK_E (free enzyme), and pK_{ES} (substrate-enzyme complex), respectively (20). Derived from Ref. 32. n.d. = not determined.

Tailoring Substrate Specificity

In an attempt to correlate the dual bond-type specificity of glucoamylase with certain structural elements, comparison of the active site sequences shown in Figure 4 led to the postulation that Asn507 of isomaltase and Asn182 of glucoamylase were associated with the ability to effect an exo-attack on α-(1→6) glucosidic bonds. The Asn182→Ala mutant was constructed to make glucoamylase conform more closely to the α-(1→4) hydrolases. Comparison of the relative substrate specificity of this and the Trp178→Arg mutant, made

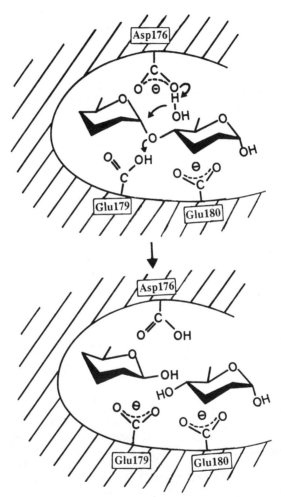

Figure 3. Tentative schematic representation of the roles of the three groups in the active site acidic cluster from *A. niger* glucoamylase.

on the basis of the same rationale, with that of other mutant and wild-type glucoamylases indicated that only Asn182→Ala had increased selectivity for maltose without losing activity (Tables II and IV). This has potential technical applications since the high glucose syrups produced with glucoamylase suffer from reduced glucose yields due to accumulation of isomaltose generated in the reverse reaction (40).

Table IV. Effect of Mutation on Glucoamylase
Substrate Specificity

| Enzyme | k_{cat}/K_m (s^{-1}mM^{-1}) | | Maltose |
	Maltose	Isomaltose	Isomaltose
Asn182→Ala	8.4	0.0062	1370
Trp178→Arg	1.22	0.0010	1230
Tyr116→Ala	2.9	0.0048	620
Wild-type	8.3	0.014	620
Leu177→His	0.49	0.00088	550
Asp176→Asn	0.12	0.00039	304
Trp120→Tyr	0.19	0.0019	100
Glu180→Gln	0.037	0.0019	19.1

The activities were measured at pH 4.5 and 50°C.

The dual specificity of glucoamylase was addressed also by a different approach using chemical modification in combination with ligand protection. A special binding mode was suggested earlier for α-(1→6)- versus α-(1→4)-linked substrates in glucoamylase from *Rhizopus sp.* (41). We confirmed this result for the *A. niger* enzyme and identified, essentially as before (23), a binding region by sequence analysis after N-bromosuccinimide oxidation of tryptophans in the presence of isomaltose. Trp170 has been demonstrated to be specifically protected by isomaltose against N-bromosuccinimide (B. Svensson, unpublished data). Acarbose and isomaltose added together protected one more group than either one individually (Figure 5). A possible explanation is that two binding areas are found in connection with the same catalytic site, rather than one flexible binding site accomodating the different substrates. It cannot be excluded, however, that a separate binding site exists for α-(1→6)-linked compounds outside of the catalytic region.

Fungal glucoamylases show 25-35% overall sequence homology, with five short highly conserved stretches (42). Interestingly, Trp120 belongs to one of these and Trp170, Asp176, Glu179 and Glu180 to another. Distant, but clear sequence similarity exists between three of these five regions and certain other starch-degrading enzymes. However, as discussed below, the majority of these enzymes presumably contains an α/β-barrel domain, which cannot be predicted using currently available algorithms for glucoamylase or for a few other of these enzymes. The conformation of the active site elements from glucoamylase thus still remains to be unveiled.

```
Taka-amylase A           197 - N Y S I D G L R I D T V K H
Pancreatic α-amylase     188 - D I G V A G F R L D A S K H
Glucoamylase             172 - Q T G Y D - L W E E V N G S
Isomaltase               496 - E V N Y D G L W I D M N E V
Maltase                  205 - D H G V D G F R I D T A G L
Pullulanase              666 - D Y K I D G F R F D L M G Y
Isoamylase               264 - T M G V D G F R F D L A S V
Maltogenic α-amylase     216 - Q L V A H G L R I D A V K H
Maltotetraohydrolase     184 - Q Y G A G G F R F D F V R G
CGTase                   220 - G M G V D G I R F D A V K H
Amylomaltase             286 - F K I Y D I V R I D H F R G
Branching enzyme         395 - R F G I D A L R V D A V A S
```

Figure 4. Alignment of short segments from starch-hydrolases and related enzymes as guided by Glu180 in glucoamylase, Asp505 of isomaltase (24), and the putative catalytic Asp197 of porcine pancreatic α-amylase (15) and substrate binding Asp206 of Taka-amylase A (14). The sequences for the remaining enzymes are taken from the Refs. in Table I.

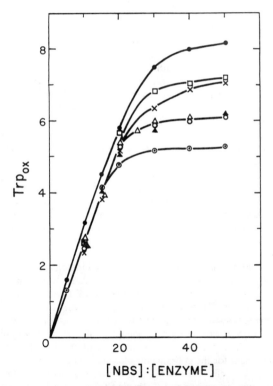

Figure 5. N-Bromosuccinimide (NBS) treatment of glucoamylase Gl (12 μM in 50 mM sodium acetate pH 4.3) in the absence (●) and in the presence of 50 mM isomaltose (O), 50 mM isomaltotriose (×), 56 mM maltose (□), 0.15 mM acarbose (△), 50 mM isomaltose plus 24 mM gluconolactone (O), 50 mM isomaltose plus 56 mM maltose (▲) and isomaltose plus 0.15 mM acarbose (⊙).

Architecture of the Catalytic Domain

Crystallographic studies show that TAA (14), porcine pancreatic α-amylase (15), *B. circulans* CGTase (16) and *B. stearothermophilus* CGTase (17) are multidomain proteins with an α/β-barrel catalytic domain. In this folding motif, 8 β-strands and 8 α-helices alternate along the polypeptide chain to yield an inner cylinder of 8 parallel β-strands surrounded by the 8 helices (Figure 6). The α/β-barrel is one of the most common enzyme 3D folds, seen first in triose phosphate isomerase (43) and since then in almost 20 enzymes of diverse specificity and mechanism. The active site is always located at the C-terminal ends of the barrel β-strands. The loops linking the β-strands to the subsequent helical segments create the substrate binding site.

A computational procedure adapted to α-amylases of known 3D structure has been used to predict barrel elements in the related enzymes (21). In this way, secondary structures were defined throughout the polypeptide chains of a maltase (22), two types of exo-α-amylases and debranching enzymes (Jespersen, MacGregor, Sierks and Svensson, unpublished results), as well as more α-amylases (21) and CGTases (22). In general, except for β-strands 4 and 6, which both are intimately involved in enzyme activity, very little sequence similarity was found (14,15,21,25). The variation in loop structure is assumed to account for substrate specificities such as strict bond-type specificity (α-amylases and the debranching enzymes), dual bond-type specificities (thermophilic α-amylase/pullulanase (44,45)), product pattern differences (cereal (46) versus mammalian α-amylases (47)) and the substrate fine structure preferences that distinguish pullulanase and isoamylase (48). A superficial pattern emerges from comparison of α/β-barrel loop lengths (Table V) as deduced from the results of the secondary structure predictions.

Several common features can be drawn from the α/β-barrel starch-degrading enzymes. They all contain an L_3 loop (Table V), defined by crystallography as a separate small domain with several β-strand elements (14-16), which participates both in substrate binding and provides two of the Ca^{2+}-ligands in the two α-amylases (14,15). The putative general acid catalytic group (discussed above) is conserved at the end of the fifth strand, and the base catalyst is suggested to be the conserved Asp of the L_7 (14,15,25,35,37). A third conserved carboxylic acid residue in the L_4 is likely to participate in binding of a glucosyl residue at the catalytic site (25,32,35). Several additional side-chains engaged in protein-substrate interactions have been described for TAA (14) and pancreatic α-amylase (15). Certain loops are assigned a specific subsite location in TAA as listed in Table V. Most of the loops (l_{1-7}) extending from the C-terminus of helices seem to connect through a short turn to the following strand. Assuming prediction accuracy of ± 2 residues (21) a few clearly longer ones include l_4 in pancreas α-amylase and yeast maltase, l_5 in pullulanase, l_6 in barley α-amylase and both of the debranching enzymes. Some of these six cases may qualify as omega loops (49).

In general, the loops that participate in substrate binding at the C-terminus of the β-strands are long. L_1, however, is short in α-amylases with a relatively short active site cleft, comprising only 5 subsites as in pancreatic α-amylase (47) and saccharifying bacterial α-amylase (20). L_2 is longest in the two debranching enzymes. The length of L_3 varies a lot without a clear correlation to either specificity or action pattern. L_4 might be an omega loop. L_5 is longer in barley α-amylase, pullulanase and maltotetraohydrolase. The polypeptide chain segment folding to yield the part of the barrel comprising strands 5 through 8 is in fact longer for the debranching enzymes compared to the others. L_6 consists of three elements: L_6', a helix (H) and L_6''. L_6 is longest in the two debranching enzymes. L_7 is particularly long in isoamylase and *Klebsiella* CGTase, as is L_8 in pancreatic α-amylase and the debranching enzymes, while barley α-amylase and yeast maltase are predicted to have very short L_8.

Assuming the glucosyl residues connected by the α-(1→4) and α-(1→6) bonds to be

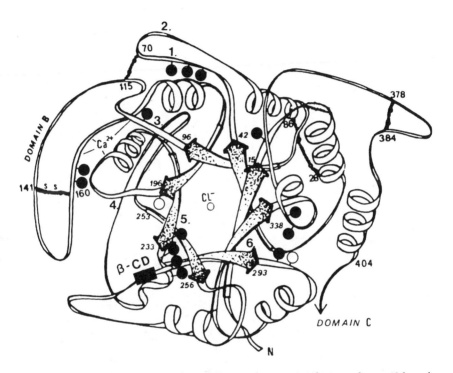

Figure 6. Schematic representation of the porcine pancreatic α-amylase α/β-barrel catalytic domain (adapted with permission from Buisson, Duée, Payan, Haser, Food Hydrocolloids 1987, 1, 399-406. Copyright 1987, Oxford University Press). Catalytic (○), binding (15) and additional binding residues in Taka-amylase A (14) are indicated (●), as are 6 conserved regions (1.-6.) (25) and a position for a β-cyclodextrin binding site (β-CD) unique to cereal α-amylases (50).

Table V. Amino Acid Residues in Loops of α/β-Barrel Domains of Starch-Degrading Enzymes

Subsite Assignment in TAA[a]:

```
                    2           4
Subsite        1    4    4      5
Assignment     2    6    5      6              4    2
in TAA[a]  1   3    7    6      7              5    3
```

Enzyme	L_1	l_1	L_2	l_2	L_3	l_3	L_4	l_4	L_5	l_5	L_6'	H	L_6''	l_6	L_7	l_7	L_8
TAA	28	4	34	1	63	6	8	3	5	5	0	9	8	8	14	2	27
PPA	5	7	35	2	75	3	7	15	9	6	3	10	12	6	25	4	53
aBl	16	4	28	1	66	10	8	4	22	6	2	7	2	15	16	2	9
aBam	10	4	37	1	108	6	8	9	5	8	0	8	8	4	17	3	20
aBsu	5	4	37	1	54	6	16	3	5	5	1	8	6	7	20	2	30
aShy	5	5	29	1	62	7	9	4	6	5	1	9	5	7	16	3	24
CBci	35	10	39	2	72	3	14	4	12	5	3	8	6	10	18	3	25
CKpn	34	8	37	5	78	3	13	8	14	5	2	8	8	10	39	3	34
CAlB	34	10	39	2	72	3	12	4	11	5	2	9	6	10	18	3	25
CBst	35	10	38	2	72	3	12	5	11	5	5	5	9	10	18	3	25
MSce	16	4	28	2	83	5	14	37	8	6	6	7	4	12	17	8	14
IaPa	21	4	47	3	56	5	9	9	8	5	17	6	17	24	53	4	38
PuKa	28	5	103	4	48	5	11	3	30	21	10	18	2	27	27	3	46
MαBs	36	5	36	5	76	6	10	4	11	5	3	10	8	9	16	3	26
MTPs	15	6	35	5	62	4	8	8	18	6	0	7	3	11	23	4	10

a) Substrate cleavage is between subsites 4 and 5 in Taka-amylase A, Ref. 14. L_{1-8} signify peptide chain loops in the α/β-barrel as indicated by number of amino acid residues in segments from the C-termini of β-strands 1-8 to the N-termini of the following helices 1-8 and for l_{1-7} from C-termini of helices 1-7 to the N-termini of the following β-strands 2-8, H represents the number of residues in the extra helix.

The lengths were defined by the 3D structures of Taka-amylase A (TAA) (14) and porcine pancreatic α-amylase (PPA) (15), and from the predicted structures of barley α-amylase 1 (aBl), *Bacillus amyloliquefaciens* (aBam), *Bacillus subtilis* (aBsu) and *Streptomyces hygroscopicus* (aShy) α-amylases, *Bacillus circulans* (CBci), *Klebsiella pneumonia* (CKpn), *Bacillus sp.* strain 1011 (CAlB) and *Bacillus stearothermophilus* (CBst) cyclodextrin glucanotransferases, *Saccharomyces cerevisiae* maltase (MSce) (38,39), *Pseudomonas amyloderamosa* isoamylase (IaPa), *Klebsiella aerogenes* pullulanase (PuKa), *Bacillus stearothermophilus* maltogenic α-amylase (MαBs) and *Pseudomonas stutzeri* maltotetraose-forming amylase (MTPs) (Jespersen, MacGregor, Sierks and Svensson, unpublished data).

cleaved are accomodated in the respective enzymes at equivalent subsites, all enzymes listed in Table V use subsites equivalent to 4 and 5 in TAA (14). By substrate and product structure analysis both debranching enzymes are suggested to lack subsites analogous to 6 and 7 in TAA and pullulanase to lack subsite 1. The larger L_5 and L_7 for the debranching enzymes may be responsible for binding the main chain of the substrate across the C-terminal end of the β-barrel (Jespersen, MacGregor, Sierks and Svensson, unpublished results).

Although α/β-barrel enzymes bind substrate at the C-terminus of the barrel (14,15), chemical modification of barley α-amylase isozyme 2 led to identification of Trp276-Trp277 as a β-cyclodextrin binding site (50) which is equivalent to Met287-Pro288 located to l_6 in the 3D structure of pancreatic α-amylase (15). Since β-cyclodextrin does not affect the catalytic site, but competitively inhibits binding of cereal α-amylases onto granular starches (51,52), evidence is provided for the first time for a specific ligand binding at the N-terminal end of an α/β-barrel. A linear pseudomaltooligodextrin (Aplanin) of DP approx. 12 (Figure 1) also protects this β-cyclodextrin binding site in addition to an essential tryptophan, Trp206 in isozyme 2 near the proposed catalytic Glu204 (25,50). The relatively long binding site for linear substrates characteristic for cereal α-amylases (46) is speculated to be in part due to Trp276-Trp277, aromatic groups being as a rule important for binding of sugars (53).

Domain Level Organisation

For at least two important amylase classes, β-amylases and glucoamylases, we have been unable to predict the catalytic domain folding to be of the α/β-barrel type. Glucoamylase, however, shows local similarity to some other enzymes near functional residues (23,31), whereas the affinity labelled part of β-amylase (39) as yet seemed not to resemble any other member of the amylase superfamily. Glucoamylases and β-amylases, like α-amylases and CGTases (14-18,26) contain more than one domain. The C-terminal domain of glucoamylase has been ascribed raw starch binding capacity, and 21-45% sequence identity exists between it and putative domains recognized in a number of enzymes of varying specificity including a bacterial β-amylase (26) (Figure 7). Sequence comparison of the different enzymes of the superfamily assisted by hydrophobic cluster analysis yields a picture of a highly diverse domain level organization (Figure 8). The functional role so far has been determined only for two of these domains, while a third may be a linker segment similar to the highly glycosylated region in glucoamylases. Others, in the cases for example of CGTases and debranching enzymes, might participate in substrate and acceptor binding, in transferase reactions or in main chain binding, respectively. The crystallographic analysis of the three-dimensional domain organisation (17), the interesting fine structure of the active site region, and tertiary structure details of the C-terminal domain, perhaps involved in binding onto granular starch, are currently underway for a few enzymes.

Prospects

The recognition of the amylases and related enzymes as constituting a protein superfamily enables a deeper understanding of the structure/function relationships in the individual enzymes. Together they offer an attractive system for engineering, whether it be tinkering of α/β-barrel loops to modulate substrate specificity, changing the electrostatic properties of the catalytic site environment, tailoring of pH activity profiles (32), or grafting of domains to provide altered and/or supplementary functional features such as facile raw starch adsorption (26).

The technical importance of the majority of the discussed enzymes surely motivates "surgery" of the outlined nature aiming at alteration of the functional properties. The well explored protein engineering approach of enzyme hybrid formation had earlier largely addressed the thermal stability. We have here summarized evidence that justifies further attempts to engineer enzymatic properties applying this as well as more rationally based design procedures.

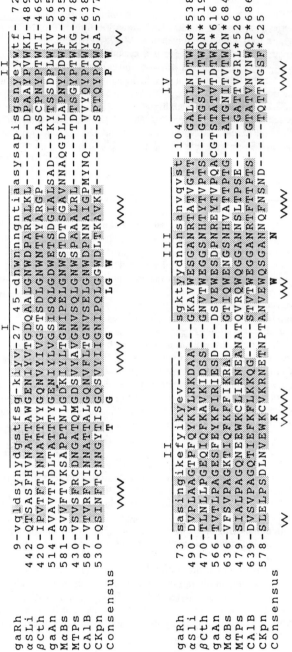

Figure 7. Comparison of C-terminal, putative raw starch binding sequences of α-amylase from *Streptomyces limosus* (αSli), β-amylase from *Clostridium thermosulfurogenes* (βCth), glucoamylase from *Aspergillus niger* (gaAn), maltogenic α-amylase from *Bacillus stearothermophilus* (MαBs), maltotetraose-forming amylase from *Pseudomonas stutzeri* (MTPs), CGTases from alkalophilic *Bacillus* sp. strain 1011 (CALB) and *Klebsiella pneumonia* (CKpn). The N-terminal starch-binding region from *Rhizopus oryzae* glucoamylase (gaRh) is shown in lower case. Residues from dominant exchange groups are shaded and predicted β-strand is indicated by ᴧ. (Reproduced with permission from Ref. 26. Copyright 1989, Portland Scientific Press).

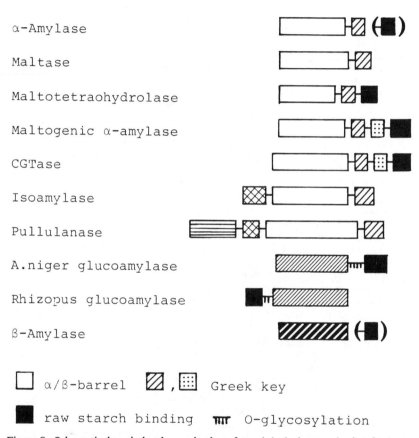

Figure 8. Schematic domain level organization of starch-hydrolases and related enzymes (see text for further details).

Legend of Symbols

k_{cat} (s^{-1}) = turn over number (apparent first order rate constant of the enzyme).
K_m (mM) = apparent dissociation constant of enzyme-substrate complex.
k_{cat}/K_m ($s^{-1}mM^{-1}$) = specificity constant (apparent second order rate constant).
$\Delta\Delta G$ (kJ/mol) = change in activation energy.
L_n (n=1,...8) = the peptide chain (loop) between β-strand n C-terminus and α-helix n N-terminus in the α/β-barrel.
l_n (n=1,...7) = the peptide chain (loop) between α-helix n C-terminus and β-strand ($n+1$) N-terminus in the α/β-barrel.

Literature cited

1. Handbook of amylases and related enzymes. Ed. The Amylase Research Society of Japan. Pergamon Press, 1988.
2. Robyt, J.F. In Starch: Chemistry and Technology; Whistler, R.L.; Bemiller, J.N.; Paschall, E.F., Eds. 2nd Edition, Academic Press, 1984, pp 87-123.
3. Hagenbüchle, O.; Bovey, R.; Young, R.A. Cell 1980, 21, 179-87.
4. Yamashita, I.; Suzuki, K.; Fukui, S. J. Bacteriol. 1985, 161, 567-73.
5. Khursheed, B.; Rogers, J.C. J. Biol. Chem. 1988, 263, 18953-60.
6. Katsuragi, N.; Takizawa, N.; Murooka, Y. J. Bacteriol. 1987, 2301-6.
7. Kluh, I. FEBS Lett. 1981, 136, 231-4.
8. Pasero, L.; Mazzei-Pierron, Y.; Abadie, B.; Chicheportiche, Y.; Marchis-Mouren, G. Biochim. Biophys. Acta 1986, 869, 147-57.
9. Toda, H.; Kondo, K.; Narita, N. Proc. Jap. Acad. 1982, 58B, 208-12.
10. Svensson, B.; Larsen, K.; Svendsen, I.; Boel, E. Carlsberg Res. Commun. 1983, 48, 529-44.
11. Svensson, B.; Larsen, K.; Gunnarsson, A. Eur. J. Biochem. 1986, 154, 497-502.
12. Mikami, B.; Morita, Y.; Fukazawa, C. Seikagaku (Tokyo) 1988, 60, 211-16.
13. Toda, H. Denpun Kagaku 1989, 36, 87-101.
14. Matsuura, Y.; Kusunoki, M.; Harada, W.; Kakudo, M. J. Biochem. 1984, 95, 697-702.
15. Buisson, G.; Duée, E.; Haser, R.; Payan, F. EMBO J. 1987, 6, 3908-16.
16. Hofmann, B.E.; Bender, H.; Schultz, G.E. J. Mol. Biol. 1989, 209, 793-800.
17. Kubota, M.; Matsuura, Y.; Sakai, S.; Katsube, Y. "Protein Engineering '89", 2nd International Conference, Kobe, August 1989, abstract SII-P12.
18. Aibara, S.; Yamashita, H.; Morita, Y. Agric. Biol. Chem. 1984, 48, 1575-9.
19. Johnson, L.N.; Cheetham, J.; McLaughlin, P.J.; Acharya, K.R.; Barford, D.; Phillips, D.C. Curr. Top. Microbiol. Immunol. 1988, 139, 81-134.
20. Hiromi, K.; Ohnishi, M.; Tanaka, A. Mol. Cell. Biochem. 1983, 51, 79-95.
21. MacGregor, E.A. J. Prot. Chem. 1988, 7, 399-415,
22. MacGregor, E.A.; Svensson, B. Biochem. J. 1989, 259, 145-52.
23. Clarke, A.J.; Svensson, B. Carlsberg Res. Commun. 1984, 49, 559-66.
24. Hunziker, W.; Spiess, M.; Semenza, G.; Lodish, H.F. Cell 1986, 46, 227-34.
25. Svensson, B. FEBS Lett. 1988, 230, 72-6.
26. Svensson, B.; Jespersen, H.; Sierks, M.R.; MacGregor, E.A. Biochem. J. 1989, 264, 309-11.
27. Svensson, B.; Pedersen, T.G.; Svendsen, I.; Sakai, T.; Ottesen, M. Carlsberg Res. Commun. 1982, 47, 55-69.
28. Svensson, B.; Clarke, A.J.; Svendsen, I. Carlsberg Res. Commun. 1986, 51, 61-73.
29. Hiromi, K.; Takahashi, K.; Hamauzu, Z.; Ono, S. J. Biochem. 1966, 59, 469-75.
30. Svensson, B.; Møller, H.; Clarke, A.J. Carlsberg Res. Commun. 1988, 53, 331-42.

31. Svensson, B.; Clarke, A.J.; Svendsen, I.; Møller, H. Eur. J. Biochem. 1990, 188, 29-38.
32. Sierks, M.R.; Ford, C.; Reilly, P.J., Svensson, B. Protein Eng. 1990, 3, 193-198.
33. Wilkinson, A.J.; Fersht, A.R.; Blow, D.M.; Winter, G. Biochemistry 1983, 22, 3581-6.
34. Sierks, M.R.; Ford, C.; Reilly, P.J.; Svensson, B. Protein Eng. 1989, 2, 621-5.
35. Takase, K.; Matsumoto, T.; Mizuno, H.; Yamane, K. "Protein Engineering '89", 2nd International Conference, Kobe, August 1989, abstr. SIII-PO8.
36. Quaroni, A.; Semenza, G. J. Biol. Chem. 1976, 3250-3.
37. Søgaard, M. M. Sc. Thesis, Copenhagen University, May 1989.
38. Yang, S.; Ge, S.; Zeng, Y.; Zhang, S. Biochim. Biophys. Acta 1985, 828, 236-40.
39. Nitta, Y.; Isoda, Y.; Toda, H.; Sakiyama, F. J. Biochem. 1989, 105, 573-6.
40. Nikolov, Z.L.; Maegher, M.M.; Reilly, P.J. Biotech. Bioeng. 1989, 34, 694-704.
41. Ohnishi, M.; Wada, S.; Yamada, T.; Tanaka, A.; Hiromi, K. J. Jpn. Soc. Starch Sci. 1983, 30, 57-61.
42. Itoh, T.; Ohtsuki, I.; Yamashita, I.; Fukui, S. J. Bacteriol. 1987, 169, 4171-6.
43. Banner, D.W.; Bloomer, A.C.; Petsko, G.R.; Phillips, D.C.; Pogson, C.I.; Wilson, I.A.; Corran, P.H.; Furth, A.J.; Milman, J.D.; Offord, R.E.; Priddle, J.D.; Waley, S.G. Nature 1975, 255, 609-14.
44. Plant, A.R.; Clemens, R.M.; Morgan, H.W.; Daniel, R.M. Biochem. J. 1987, 246, 537-41.
45. Melasniemi, H. Biochem. J. 1988, 250, 813-18.
46. MacGregor, E.A.; MacGregor, A.W. Carbohydr. Res. 1985, 142, 233-36.
47. Seigner, C.; Prodanov, E.; Marchis-Mouren, G. Biochim. Biophys. Acta 1987, 913, 200-9.
48. Kainuma, K.; Kobayashi, S.; Harada, T. Carbohydr. Res. 1978, 61, 345-57.
49. Fetrow, J.S.; Zehfus, M.R.; Rose, G.D. Bio/Technology 1988, 6, 167-171.
50. Gibson, R.M.; Svensson, B. Carlsberg Res. Commun. 1987, 52, 373-9.
51. Weselake, R.G.; Hill, R.D. Cereal Chem. 1984, 60, 98-101.
52. Kuzovlev, V.A.; Fursov, O.V.; Darkanbaev, T.B. Prikl. Biokh. Mikrobiol. 1988, 24, 636-41.
53. Quiocho, F.A. Ann. Rev. Biochem. 1986, 55, 287-315.

RECEIVED September 9, 1990

Chapter 4

Substrate-Based Investigations of the Active Site of CGTase

Enzymatic Syntheses of Regioselectively Modified Cyclodextrins

Sylvain Cottaz and Hugues Driguez

Centre de Recherches sur les Macromolécules Végétales, Centre National de la Recherche Scientifique, B.P. 53 X, 38041 Grenoble cedex, France

New insights have been obtained on the specificity of the catalytic site of CGTase of *Bacillus macerans* by enzymatic conversion of modified maltosyl fluorides. The catalytic capability of this enzyme led to the first regioselective synthesis of modified cyclodextrin in good yield. Furthermore, under coupling conditions, new acceptors led to reaction products suggesting at least four subsites for the catalytic site of this enzyme.

Cyclodextrin glycosyltransferase (EC 2.4.1.19) is an enzyme which catalyzes the reversible transfer reaction of glucosyl units between maltodextrins :

$$G_n + G_m \xrightleftharpoons{\text{disproportionation}} G_{n-y} + G_{m+y}$$

$$G_n \xrightleftharpoons[\text{coupling}]{\text{cyclization}} CD_x + G_{(n-x)}$$

0097–6156/91/0458–0044$06.00/0

The cyclisation reaction of linear malto-oligosaccharides into α,β and γ-cyclodextrins (CD) and the coupling reaction which leads to modified linear dextrins on the reducing end are special disproportionation reactions.

The binding of an oligosaccharide to a protein involves a number of hydrogen bonds, hydrophobic interactions and Van der Waals contacts between the sugar and amino acid residues. This area of contact defines the global binding site. It is convenient to express the binding of an oligosaccharide as the sum of the interactions between each monosaccharide unit and the corresponding binding site. These individual binding sites or subsites prove very useful for studying structure-activity relationships or for the localization of the catalytic site (usually situated between two subsites).

In order to obtain information about the number of subsites in the catalytic site of CGTase from *Klebsiella pneumoniae*, Bender (1) studied the cyclization and disproportionation reactions with various maltooligo-saccharides from DP 2-8 and a maltodextrin DP 19. Since maltooctaose proved to be the smallest oligosaccharide for direct cyclization and that DP 19 is not a better substrate, it was suggested that the active site of this enzyme consists of 8 subsites.

To define the specificity of this active site only the coupling reaction was studied by using modified CDS (2,3) or modified acceptors (4-6). These approaches have shown that a bulky or charged groups on C-6 of a glucosyl unit prevent this modified glucosyl residue from fitting into subsites S and R of the donor part of the active site (Figure 1). These studies also demonstrated that subsites T and U may accomodate various modified glucosyl acceptors ; for instance, alkyl and aryl α- or β-D-glucosides, methyl α-D-xyloside, methyl 6-O-methyl-α-D glucoside, panose, sucrose and maltobiuronic acid are acceptors.

The aim of this work is to present a new approach based on the autocondensation reaction of modified maltosyl fluorides for the determination of the specificity of the active site, and on the coupling reaction on 1,6-anhydro-maltooligosaccharides to obtain information on the number of subsites.

MATERIALS and EXPERIMENTAL

Enzymes

CGTase was a gift of Amano Co Ltd. (Japan) and subtilisin was obtained from Sigma.

Chemicals

The details of the preparation and characterization of maltosyl fluorides **1-10** will be reported elsewhere. The corresponding acetylated derivatives gave elemental analysis and spectral data (^{13}C nmr and MS) in accord with the proposed structure. Catalytic deacetylation with sodium methoxide in methanol was performed just before incubations. The purity of the deacetylated compounds was checked by H.P.L.C. (column : NH_2 μ-Bondapack 10 μm ; eluent : acetonitrile-water 7:3).

    6'-*O*-Acetyl-α-maltosyl fluoride **10** was obtained from α-maltosylfluoride **11** (7) by transacetylation of vinyl acetate in pyridine in the presence of subtilisin.

    1,6-Anhydro-maltose **13** and 1,6-anhydro-maltotriose **14** were synthesized as already described (8,9).

Enzymatic Incubation with CGTase

Coupling Reactions for the Specificity of Acceptor Part of the Active Site. - Various potential substrates (5 mg) and CDS (5 mg) were incubated with CGTase (635 U/ml ; 20 μl) in phosphate buffer (0.1 M, pH 7, 1 ml) for 1 h at 45°C. The mixture was then analyzed by H.P.L.C. as already described.

Specificity of the Donor Part of CGTase. - Fluorides **1-11** (50 mM) in phosphate buffer were incubated with CGTase (20 $\mu l/ml$) for 12 h at 45°C, and then the mixture was analyzed by H.P.L.C.

RESULTS and DISCUSSION

The results of experiments wherein α-CD and CGTase were incubated with all the fluorides **1-11** are shown on the right part of table 1. In each case, H.P.L.C. analysis of the mixture showed that all the compounds are acceptors. These results confirm the poor specificity of the T and U subsites.

To appreciate the specificity of R and S subsites, the fluorides **1-11** were incubated with CGTase but in the absence of α-CD. All the modifications made at the 6 position of the reducing unit prevent binding in subsite S (left part of table 1). The only modifications allowed are the acetylation and the methylation of primary hydroxyl group of the non reducing unit ; the 4-thiomaltosyl fluoride **9** is also recognized. By preparative H.P.L.C. and characterization by mass spectrometry, using the FAB technique, it has been shown that alternating oligosaccharides are formed, with the starting modified disaccharide as the repeating unit. In two cases, for **8** and **10**, cyclic compounds **15** and **16**

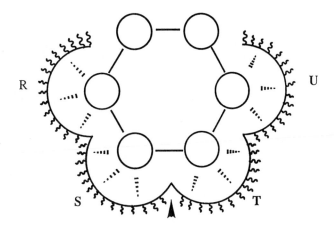

Figure 1. Postulated enzyme-substrate complex in subsites around the catalytic site (indicated by arrow).

Table I. Structure of the modified maltosyl fluorides 1-11 and their enzymatic behaviour under coupling and cyclization conditions

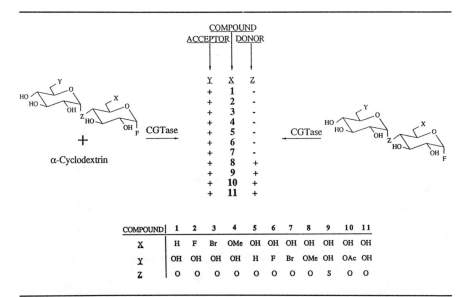

	COMPOUND	
ACCEPTOR	DONOR	
X	X	Z
+	1	-
+	2	-
+	3	-
+	4	-
+	5	-
+	6	-
+	7	-
+	8	+
+	9	+
+	10	+
+	11	+

COMPOUND	1	2	3	4	5	6	7	8	9	10	11
X	H	F	Br	OMe	OH	OH	OH	OH	OH	OH	OH
Y	OH	OH	OH	OH	H	F	Br	OMe	OH	OAc	OH
Z	O	O	O	O	O	O	O	O	S	O	O

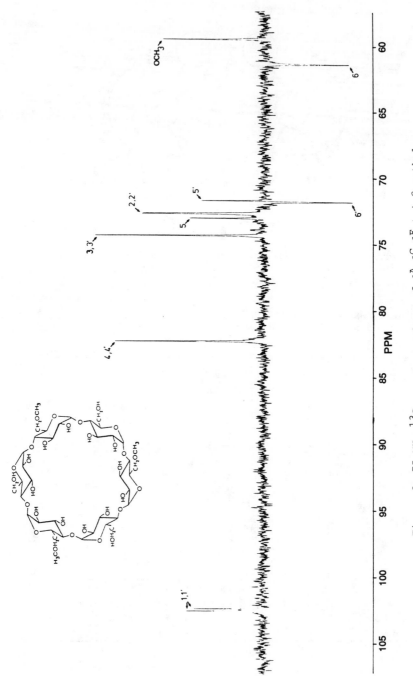

Figure 2. 75 MHz ^{13}C-n.m.r. spectrum of 6^A,6^C,6^E-tri-O-methyl cyclomaltohexaose in D$_2$O using DEPT sequence. Primed numbers refer to the carbons of the methylated substrates.

were also isolated. These compounds are $6^A,6^C,6^E$-tri-O-substituted α-CDS. ^{13}C-N.m.r. spectra of these compounds show a simple pattern compatible with a regular disaccharide repeating unit and also match the C_3 molecular symmetry (fig. 2). $6^A,6^C,6^E$-Tri-O-methyl cyclomaltohexaose was obtained in an overall yield of ca. 12% from the commercially available maltose (10).

The incubation of modified maltosyl fluorides confirmed some features already reported in the literature, but also bring some new information about the specificity of the active site of CGTase. The amino acids of subsites T and U do not establish essential bondings with primary hydroxyls of maltosyl residues. On the donor part, the subsite S does not accept any modification of these positions, but R is less specific. The same specificity was also found in *Taka*-amylase (11) and was confirmed when linear dextrins from the condensation of 6'-O-methyl-maltosyl fluoride were incubated with *Taka*-amylase to afford 6'-O-methyl-maltose.

To try to determine the number of subsites of the acceptor part of the active site we incubated 1,6-anhydro-maltose **13** and maltotriose **14** with CGTase in the presence of α-CD. Only compound **14** gave coupling products at a reasonable rate. It is suggested that two glucosyl residues are recognized on a coupling reaction so only two subsites constitute the acceptor part of the active site.

Reactions are in progress for the determination of the number of subsites on the donor part of the active site.

Literature Cited

1.   Bender, H. "Proc. 4th Intern. Symposium on Cyclodextrins". In Advances in Inclusion Science ; Huber, O., Szejtli, J., Eds ; Kluver Academic Publishers: Dordrecht, 1988 ; p 19.
2.   Kobayashi, S. ; Lee Ashraf, W.R. ; Braun, P. ; French, D. Starch/Stärke 1988, 40, 112-16.
3.   Nagamine, Y. ; Sumikawa, M. ; Omichi, K. ; Ikenaka, T. J. Biochem. 1987, 102, 767-75.
4.   French, D. ; Levine, M.L. ; Norberg, E. ; Nordin, P. ; Pazur, J.H. ; Wild, G.M. J. Am. Chem. Soc. 1954, 76, 2387-90.
5.   Kitahata, S. ; Okada, S. ; Fukui, T. Agric. Biol. Chem. 1978, 42, 2369-74.
6.   Wheeler, M. ; Hanke, P. ; Weill, E. Arch. Biochem. Biophys. 1963, 102, 397-9.
7.   Hehre, E.J. ; Mizokami, K. ; Kitahata, S. Denpun Kagaku 1983, 30, 76.

8.   Fujimaki, I. ; Ichikawa, Y. ; Kuzuhara, H.
     Carbohydr. Res. 1982, 101, ·148-51.
9.   Sakairi, N. ; Hayashida, M. ; Kuzuhara, H.
     Carbohydr. Res. 1989, 185, 91-104.
10. Cottaz, S. ; Driguez, H., J. Chem. Soc., Chem.
     Commun. 1989, 16, 1088-9.
11. Arita, H. ; Isemura, M., Ikenaka, T., Matsushima,
     Y. Bull. Chem. Soc. Jpn. 1970, 43, 818-23.

RECEIVED October 3, 1990

Chapter 5

Enzymatic Synthesis and Use of Cyclic Dextrins and Linear Oligosaccharides of the Amylodextrin Type

John H. Pazur

The Pennsylvania State University, University Park, PA 16802

The enzymatic synthesis of cyclic dextrins occurs
during the action of macerans amylase on starch by
a cyclizing mechanism while the synthesis of linear
oligosaccharides occurs during the action of the
enzyme on the dextrins and appropriate co-
substrates by coupling and homologizing reactions.
The cyclic dextrins are composed of 6 to 12 glucose
units all joined by α-$\underline{D}$-(1,4) glucosidic bonds.
Many organic compounds form inclusion complexes
with the dextrins modifying properties and making
the dextrins valuable for research and industrial
uses. The linear oligosaccharides are composed of
maltooligosaccharide moieties at the non-reducing
ends and possibly novel carbohydrate units at the
reducing ends. The cyclic dextrins and the linear
oligosaccharides are structurally related to
amylodextrin and starch in which a majority of the
glucose units are linked by α-$\underline{D}$-(1,4) glucosidic
bonds and a small number are linked by α-$\underline{D}$-(1,6)
bonds. Several types of oligosaccharides have been
synthesized with the aid of macerans amylase and
have been used in other types of enzymological
studies.

Cyclic dextrins are synthesized from starch ([1],[2]) enzymatically by
the action of macerans amylase, an amylolytic enzyme elaborated by
strains of microorganisms particularly of the Bacillus group ([3]).
Macerans amylase catalyzes several types of glucosyl transfer
reactions leading to the synthesis of cyclic dextrins and linear
oligosaccharides by the cyclizing, coupling and homologizing actions
of the enzyme ([4]). In the initial action on starch, the amylase
forms a number of cyclic dextrins. Three of these have been
isolated in pure form and characterized structurally. The latter
dextrins are composed of 6, 7 and 8 glucose units ([5]) known
individually as α-, β-, and γ-dextrin and collectively as
Schardinger dextrins ([2]). The dextrins have also been named to

0097–6156/91/0458–0051$06.25/0

indicate structural relationships to amylose (5) and to
maltooligosaccharides (6). These names are: cyclohexaamylose
(cyclomaltohexaose), cycloheptaamylose (cyclomaltoheptaose) and
cyclooctaamylose (cyclomaltooctaose). All of the glucose units in
each dextrin are linked by α-D-(1,4) glucosidic bonds and as a
result, cyclic non-reducing molecules are formed. Diagrammatic
structures for the α- and β-dextrins are shown in Figure 1, Frame A
and B (7). Frame C of the Figure shows a structure constructed from
molecular models for the α-dextrin molecule (4). Because of steric
factors, the cyclic dextrin molecules assume truncated cone shapes
and as a result possess cavities in the center of the molecules
(8,9). In such a structure there is restricted rotation about the
glucose residues and the hydrogen atoms on carbons 3 and 5 of the
residues are directed inward to the central cavity (6,10). The
cavity space is therefore apolar and hydrophobic bonds can form
readily with other hydrophobic substances forming stable inclusion
complexes. The complexing agent is located in the cavity of the
cyclic dextrin.

        Macerans amylase produces other cyclic dextrins from starch in
low yields (11). These dextrins have been isolated and preliminary
structural data have been obtained. These data indicate that the
dextrins are composed of 9 to 12 glucose units some of which form a
cyclic ring and have oligosaccharide side chains linked by α-D-(1,6)
bonds to the residues of the core. However, the complete structures
of the new dextrins have not yet been determined.

        Macerans amylase has been used to synthesize linear
oligosaccharides by the coupling (12) and homologizing pathways
(13). In the coupling pathway, the ring of the cyclic dextrin is
opened by the enzyme at a single α-D-(1,4) glucosidic bond and the
resulting glucosyl segment is transferred to a co-substrate molecule
which may be glucose or an appropriate glucose derivative. A linear
oligosaccharide terminated in the co-substrate molecule is produced.
A variety of glucose containing compounds can function as co-
substrates (14). The enzymatically synthesized oligosaccharides are
similar to amylodextrin, which is a glucose polymer of about 20
units joined by α-D-(1,4) linkages, but the oligosaccharides are
generally of lower molecular weight than the dextrin (15).

        The coupled products once formed are converted to a series of
homologous oligosaccharides by the homologizing reactions in which a
redistribution of the glucose units occurs (13). In homologizing
reactions glucose segments consisting of one or more residues are
transferred from the non-reducing end of the initial coupled product
to a co-substrate molecule. The transfer occurs to position 4 of a
glucose unit of the co-substrate. The transfer process continues
until an equilibrium is reached and the synthesis of a homologous
series of oligosaccharides has occurred. Each oligosaccharide is
made up of the co-substrate molecule at one terminus and a
maltooligosaccharide moiety at the other terminus. The desired
oligosaccharide can be isolated by chromatographic methods and used
for structural and enzymatic studies.

        The applications of the cyclic dextrins are based largely on
the ability of the dextrins to form inclusion complexes with
appropriate organic substances or inorganic compounds (6). The
complexing agents are held in the central cavities of the dextrin

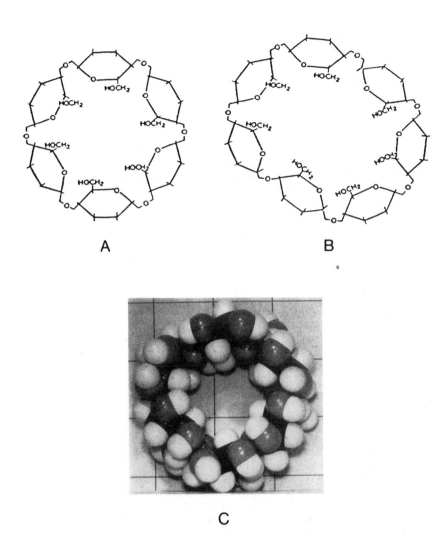

Figure 1. Diagrammatic structure for α-dextrin (A) and for β-dextrin (B) and for a molecular model for α-dextrin (C). (Reproduced with permission from Refs. 4 and 8. Copyright 1957 Academic Press and 1967 John Wiley and Sons.)

molecules by weak hydrophobic bonds. These bonds can be dissociated readily by heat and the complexing agent can be removed by boiling the solution. The dextrin is then recovered in pure form by crystallization from water or alcohol (16). Many of the properties of the dextrins and of the complexing agents or "guest molecules" in the adduct may be altered by the formation of the inclusion complex. These are enhancement of solubility of poorly soluble compounds, stabilization of labile compounds, modification of flavors, elimination of odors and modification of chemical reactivity in products such as food items (17), agrichemicals (18), pharmaceuticals, (19) and health and beauty aides (17). The dextrins and their inclusion complexes have been used in studies to elucidate enzyme-substrate reactions (20,21), drug receptor interactions (22), modification of rates of chemical reactions (23), and the mechanism of complex formation (6). Other potential applications of the cyclic dextrins are being developed at the present time (24).

Applications of the linear oligosaccharides derived from the dextrins include the development of a mapping procedure for elucidating the action mechanisms of amylases (25,26,27), the preparation of oligosaccharides with sucrose moieties (28) which can replace sucrose in foods, confectionaries and beverages used for retarding dental caries formation (29), the preparation of novel oligosaccharides for use in the measurement of furanose-pyranose interconversions in oligosaccharides (30) and most recently the preparation of adsorbents useful for analytical and purification purposes. Adsorbents with carbohydrate ligands useful for the separation of stereoisomers which are difficult to separate by other means (31,32) have been made. Also affinity adsorbents for the purification of enzymes, antibodies and abnormal proteins of diseased tissues (33) have been developed.

Macerans amylase is produced by strains of bacteria of Bacillus macerans (35), Bacillus circulans (3), Bacillus stearothermophilus (36), Bacillus mageterium (37), alkalophilic Bacillus sp. (38), and Klebsiella pneumoniae (39). The amylase from Bacillus macerans has been used most often as the enzyme source for synthesizing cyclic dextrins and linear oligosaccharides. This enzyme has been purified (40) and obtained in homogeneous form (41,42). The early studies on the preparation, characterization and reactions of the cyclic dextrins have been reviewed (4) and later studies on inclusion complex formation have been discussed in more recent articles (6,43). In the present article, new developments in enzymology of the cyclic dextrins, the preparation of novel linear oligosaccharides therefrom, and the industrial and research applications of the dextrins and oligosaccharides are discussed.

Materials and Methods

Preparation of Macerans Amylase. Macerans amylase was separated from cultures of Bacillus macerans in 1939 by Tilden and Hudson (35). The enzymatic hydrolysis of starch by this organism and the detection of the crystalline dextrins was first reported by Villiers (1) and characterization of the dextrins was achieved by Schardinger (2) later. Cultures of the macerans amylase may be obtained from

the American type collection (ACTC-8517). The organism grows
readily on a medium of 5 gm of potato slices, 1.5 g of calcium
carbonate and 0.2 g of ammonium sulfate in 50 ml of water (44).
Since the macerans amylase was produced in the cell culture
following primary growth, the culture was incubated for 5 days at
37°C. At the end of this time the cells were removed by
centrifugation and the clear filtrate was collected. The filtrate
was assayed for macerans amylase activity and units of enzyme were
calculated by the Tilden-Hudson method (40). Cultures grown under
the above conditions yielded supernatant solutions which contained
from 1 to 2 units of enzyme activity per ml which were generally
suitable for use in most types of experiments. If solutions
containing higher enzyme activities are required, the initial
supernatant can be concentrated by lyophilization.

Macerans amylase from Bacillus macerans has been purified to
molecular homogeneity by chromatography on several types of
adsorbents (41). The organism was grown in batch wise fashion in
several 500 ml shaker flasks. Seventy ml of medium were used in
each flask which contained corn steep liquor (1%), soluble starch
(1%), ammonium sulfate (0.5%) and calcium carbonate (0.5%). The
organism was grown with agitation on a Shaker at a temperature of
30°C for 70 hrs. At the end of this period the cultures were
centrifuged and the supernatant was collected. The supernatant
solutions were combined and treated with solid ammonium sulfate to
30% saturation. This solution was cooled to 5°C and chromatographed
on a column containing 150 g of corn starch granules and 60 g of
filter aid. The adsorbed enzyme was eluted with 1200 ml of 0.33 M
Na_2HPO_4 buffer. The fractions of the eluate containing protein were
collected and combined. Additional solid ammonium sulfate was added
and the resulting precipitate was collected by filtration, dissolved
in 80 ml of distilled water and dialyzed against running water. The
dialyzed solution was applied to a DEAE-Sephadex column equilibrated
with 0.001 M tris-HCl buffer (pH 7.3). The macerans amylase was
eluted with a linear gradient of sodium chloride from 0 to 0.5 M in
the tris buffer. Fractions were assayed for ability to synthesize
cyclic dextrins from a starch substrate; those fractions with enzyme
activity were combined and concentrated to 5 ml. The concentrated
solution was then subjected to chromatography on a Sephadex G-75
column with 0.01 M phosphate buffer of pH 7.0. The macerans amylase
was eluted with the same buffer. The fractions were assayed for
macerans amylase on a starch substrate and those fractions
containing enzyme activity were combined and concentrated by
lyophilization. The sample was checked for purity by
ultracentrifugation and electrophoresis. A single ultracentrifuge
peak indicated homogeneity in molecular size. An electrophoretic
pattern revealed that a single protein band was present in the
purified preparation and this band exhibited macerans amylase
activity. The increase in enzyme activity was 200 fold in the
overall purification process. The action of the purified enzyme on
starch and cyclic dextrins was identical to that of the enzyme in
the initial culture filtrates (41).

Preparation of Cyclic Dextrins with Selective Precipitants. Perhaps
the best preparative method for cyclic dextrins involves the use of

selective precipitants such as toluene, trichloroethylene or bromobenzene to precipitate the dextrins as crystalline complexes from
enzymolysates of starch (16). The complexes are isolated by
centrifugation and individual dextrins obtained by fractionation
with different precipitants. The maximum yield of a particular
cyclic dextrin depends on incubation conditions, the length of
incubation period and the type of isolation procedure. Different
precipitants are employed for isolating the individual dextrins. To
prepare α-dextrin, macerans amylase is added to a 5% starch paste
and the digest is incubated for a preliminary period of 24 hrs. The
insoluble material which may form in this period is removed by
centrifugation and discarded. The digest is then incubated for an
additional 48 to 72 hrs. Alpha dextrin can be obtained from this
digest by concentrating the solution to about 15% solids and
precipitating the dextrin as the trichloroethylene-dextrin complex.
The complex is dissolved in water at a final concentration of 2%
solids, boiled to remove the trichloroethylene and bromobenzene is
added to precipitate β and γ-dextrin. The precipitated material is
removed by filtration and the filtrate is concentrated to 40% of the
original volume. The α-dextrin is precipitated from the
concentrated solution with trichloroethylene. The complex is
isolated by filtration and air-dried. The dried complex is
dissolved in about 5 parts of hot water and 1.5 volumes of hot n-
propyl alcohol are added and crystalline α-dextrin is obtained from
the solution (Fig. 2).

The β-dextrin is isolated from an enzymolysate of a 5% starch
paste incubated for 5 to 7 days with macerans amylase. The
enzymolysis is carried out in the presence of a precipitant such as
toluene for precipitating the cyclic dextrin and an inhibitor such
as thymol for inhibiting bacterial contamination. At the end of the
enzymolysis period, the complex of β-dextrin and toluene is
collected by centrifugation in a Sharples centrifuge and boiled in
water to remove the toluene. The concentration of solids is
adjusted to 20%. The β-dextrin solution is heated to dissolve the
dextrin, filtered to remove insoluble material and allowed to cool
to room temperature. Large crystals of β-dextrin are formed and
these are collected on a filter and air dried.

The γ-dextrin is isolated from an enzymolysate of 5% potato or
corn starch paste and macerans amylase prepared in a manner similar
to that for β-dextrin. However the reaction is allowed to proceed
in the absence of precipitant although thymol is added to inhibit
extraneous bacterial growth. The final digest is concentrated to
about one-fourth the original volume and cyclic dextrins are
precipitated with trichloroethylene. The precipitate is dissolved
in hot water to about 25% solids concentration and allowed to stand
at room temperature for 24 hrs to remove the less soluble β-dextrin.
The precipitated dextrin is removed by filtration and the clear
filtrate is diluted to about 3% solids. The γ-dextrin is
precipitated with bromobenzene. The γ-dextrin is collected,
dissolved in boiling water and concentrated to about 20% solids.
This solution is treated with 1.5 volumes of n-propyl alcohol.
Crystalline γ-dextrin is formed in a few hours and is collected and
air dried.

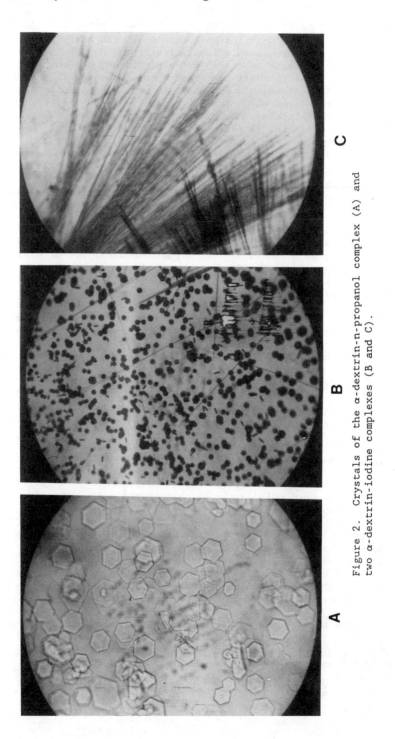

Figure 2. Crystals of the α-dextrin-n-propanol complex (A) and two α-dextrin-iodine complexes (B and C).

Preparation of Cyclic Dextrins by Acetylation. Other methods for isolating the cyclic dextrins have been used. An early procedure utilized the differences in solubilities of the dextrin acetates (45). An enzymolysate of starch and macerans amylase was prepared by methods similar to those described above. The cyclic dextrins were precipitated from this enzymolysate with a complexing agent and the dextrin complexes which formed were collected by filtration and air-dried. The mixture of dextrins was acetylated with acetic anhydride in pyridine. A preferential crystallization of the dextrin derivatives was used to separate the derivatives and the individual dextrin acetates were obtained. Deacetylation of the derivatives is required to obtain the crystalline dextrins.

Preparation of Cyclic Dextrins by Crystallization from Aqueous Solution. Two of the cyclic dextrins, α- and β-, have been isolated in large amounts from digests of starch by procedures which do not utilize complexing agents (38). Cylodextrin glucosyltransferase from an alkalophilic Bacillus is added to liquified starch and pH of the digest is adjusted to 9. The digest is maintained at this pH and at a temperature of 65°C for several days. After the enzymolysis has proceeded for a sufficient length of time, the mixture is decolorized, deionized and concentrated. Beta dextrin crystallizes from the concentrated solution and is recovered by filtration. The filtrate is then treated with an alpha amylase to remove reducing oligosaccharides. From the resulting solution α-dextrin is isolated by use of gel filtration and ion exchange chromatography. The fractions from the column containing carbohydrate components are used to crystallize α-dextrin. The procedures described in the previous sections or modifications thereof have been used to prepare cyclic dextrins on a laboratory scale (46) and to manufacture cyclic dextrins on an industrial scale (47).

Preparation of Cyclic Dextrins with More than Eight Glucose Residues. Four new cyclic dextrins which contain more than eight glucose residues have been isolated from starch-amylase digests using precipitation and solvent fractionation methods (48). These dextrins have been only partially characterized structurally. In the isolation procedure most of the α-, β-, and γ-dextrins in the enzymolysate were first removed by precipitation of these dextrins with tetrachloroethane or tetrachloroethylene. The filtrate was treated with crystalline sweet potato β-amylase to hydrolyze a small amount of contaminating linear reducing oligosaccharides. The solution of cyclic dextrins was then applied to a cellulose column maintained at an elevated temperature of 75°C. The cyclic dextrins were separated by using a gradient of the solvent water, ethanol and 1-butanol (49), the composition of which is recorded in Figure 3 in the results and discussion section. The fractions containing an individual carbohydrate component were pooled and evaporated to dryness in vacuo. Four cyclic dextrins composed of 9, 10, 11, and 12 glucose units were obtained and called λ, ε, ζ, and η, respectively. The yields of the dextrins expressed in percent of the starch used were λ, 0.4%, ε, 0.3%, ζ, 0.2%, and η, 0.2%.

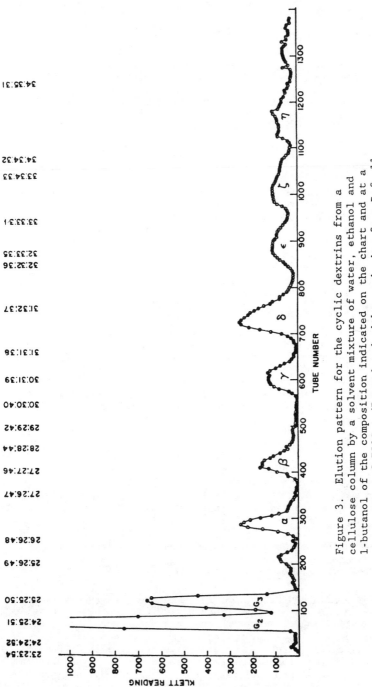

Figure 3. Elution pattern for the cyclic dextrins from a cellulose column by a solvent mixture of water, ethanol and 1-butanol of the composition indicated on the chart and at a temperature of 75°C. (Reproduced with permission from Ref. 11. Copyright 1961 Academic Press.)

Preparation of Linear Oligosaccharides from Cyclic Dextrins and Co-
substrates. Macerans amylase catalyzes coupling and homologizing
reactions and effects the synthesis of oligosaccharides with novel
sugar units at the reducing end from a cyclic dextrin and an
appropriate co-substrate. The amylase can act further on linear
oligosaccharides promoting homologizing or glucose redistribution
reactions on the initial coupled product. In this process glucose
units at the non-reducing ends of maltooligosaccharides synthesized
in the coupling reaction are transferred to acceptor molecules. A
series of homologous oligosaccharides is produced. In experiments
demonstrating such reactions, glucose and α-dextrin were incubated
with macerans amylase at room temperature. Aliquots of the digest
were analyzed at 0 and 24 hrs for reaction products by paper
chromatography in a solvent system of n-butyl alcohol-ethyl alcohol
water (4:1:1 by volume). A series of maltooligosaccharides of
glucose units joined by α-(1,4) glucosidic bonds was present in the
digest. These compounds were isolated and characterized by chemical
and enzymatic methods. Other series of oligosaccharides terminated
at the reducing end with ^{14}C labeled glucose, with sucrose, with
arabinose, or with p-aminophenyl glucoside were prepared by similar
methods to the above.

Identifying Amylases by the Mapping Procedure with Oligosaccharides-
1-^{14}C. Oligosaccharides labeled with ^{14}C were prepared by the
general procedure described in the previous section and were used to
develop an oligosaccharide mapping analytical procedure (25). The
procedure is useful for identifying and characterizing the action
mechanism of amylases from fungi, bacteria, cereal grains and
mammalian tissues. A mixture of labeled oligosaccharides was placed
at the edge of a paper square (12 inch x 12 inch of Whatman No. 1)
and separated by chromatography in a solvent system of n-butyl
alcohol, pyridine and water (6:4:3 by volume). Generally 3 or 4
ascents of the solvent were used. The area of the dried
chromatogram containing the oligosaccharides was then sprayed
uniformly with a solution of the enzyme under test (glucoamylase, α-
amylase or β-amylase). The sprayed chromatogram was maintained at
room temperature for approximately 15 min in which time the paper
strips had dried and the enzyme was inactivated. The chromatogram
was then rolled in a cylinder and developed in the second direction
by 3 or 4 ascents of the same solvent system. The finished
chromatogram was dried and a radioautogram of the chromatogram was
prepared. The chromatogram was then stained with the silver nitrate
reagent (50) to locate the reducing sugars which were produced by
amylase action on the oligosaccharides. Comparison of the results
on the radioautogram and the chromatogram is made and the
identification of amylase is achieved.

Coupling Sugars and Dental Caries Formation. Oligosaccharides
called coupling sugars were synthesized from sucrose and cyclic α-
dextrin utilizing the coupling and homologizing reactions of
macerans amylase (28). Coupling sugars have been used as a
replacement for sucrose in candies, jams and jellies. Studies have
been conducted on the effects of these products on the development
of dental plaques and cavities (29). In one study, a group of

experimental animals was divided into four sub-groups and each sub-
group was fed a different carbohydrate diet containing starch or
starch with sucrose or coupling sugar. The animals were examined
for number of cavities formed on each diet. Diets containing
coupling sugar reduced cavity formation in experimental animals
compared to control animals (51). On diets with sucrose, incidence
of cavities was 3.8 times greater than on control diets with no
sucrose, while the incidence of cavities on diets with coupling
sugar was only 1.3 times greater than the control. Other types of
experiments including inhibition of glucosyl transferase from oral
bacteria by coupling sugar have been performed (29,52). The results
of an inhibition experiment showed that inhibition of the
glucosyltransferase occurred with the coupling sugar but not with
sucrose. Further, dextran, known to be involved in plaque
formation, was synthesized in lower amounts from coupling sugar than
from sucrose.

Furanose-Pyranose Forms of a Trisaccharide of Glucose and Arabinose.
A trisaccharide of two glucose units and one arabinose unit was
prepared by the action of macerans amylase on α-dextrin and α-D-
glucopyranosyl-(1,3)-D-arabinose (30), the reaction yielded a
homologous series of oligosaccharides terminated in arabinose units.
The trisaccharide of the series was located on paper chromatograms
by the aniline oxalate spray reagent (53) and isolated in pure form
by a chromatographic procedure. On the basis of data from
methylation analysis and the specificity of the macerans amylase,
the structure of the oligosaccharide was determined to be α-D-
glucopyranosyl-(1,4)-α-D-glucopyranosyl-(1,3)-D-arabinose. The
trisaccharide has been used in experiments for determining the ratio
of furanose to pyranose ring forms.

Synthesis of Isomaltosyl-Sepharose and Maltosyl-Sepharose. The p-
aminophenyl-α-isomaltoside was synthesized from p-aminophenyl α-D-
glucoside and maltose utilizing the glucosyl transferase of
Aspergillus niger (54) while the p-aminophenyl α-maltoside was
prepared from p-aminophenyl α-D-glucoside and α-dextrin utilizing
macerans amylase (55). The aminophenyl glycosides were coupled to
cyanogen activated Sepharose (56). The Sepharose with the
carbohydrate ligands was packed into a column and equilibrated with
0.1 M phosphate buffered saline at pH 7.0. The affinity
chromatography of antiserum from a rabbit immunized with a
pneumococcal type 20 vaccine (33) or mouse ascites fluid containing
myeloma protein (34) was performed on the maltose or the isomaltose
adsorbents.
      The p-aminophenyl maltoside was attached to bovine serum
albumin by the carbodiimide reaction (57). This glycoconjugate has
been used for immunization of rabbits by subcutaneous multi-site
injection. Anti-maltose antibodies should be produced and can be
isolated by the affinity methods. The experiment is still in
progress.

Methylation, Gas-liquid Chromatography and Mass Spectrometry. The
methylation of the monosaccharides, oligosaccharides and reference
compounds was performed by the Hakomori method and analysis was done

by the Lindberg procedure. In the method used in this laboratory
(58) samples of 1 to 5 mg of the carbohydrates were thoroughly dried
and methylated in dimethyl sulfinyl carbanion with dry methyl
iodide. The methylated product was recovered by extraction with
chloroform. The material was subjected to a preliminary formolysis
with formic acid, then to hydrolysis in 0.25 M sulfuric acid,
reduction with sodium borohydride and finally acetylation with
acetic anhydride and pyridine.

The methylated and acetylated products were dissolved in
chloroform and subjected to GLC analysis in a Varian Aerograph
series 1400 chromatograph equipped with a stainless steel (6 ft x
1/8 inch) column containing 3% OV -225 on 80-100 mesh Supelcoport.
The GLC was performed at 190°C and the effluents were monitored with
a flame ionization detector. The individual derivatives from the
column of the gas chromatograph were subjected to fragmentation and
to analysis in a Dupont 21-490 mass spectrometer attached to the gas
chromatograph. Fragmentation was effected at source temperature of
250°C and an ionization potential of 70eV. The spectral data for
the reference arabinose and glucose derivatives were also obtained.

Results and Discussion

Cyclic dextrins possess unusual complexing properties due largely to
the hydrophobic nature of the central cavity of the cyclic
molecules. Inclusion complexes are readily formed with many organic
compounds and inorganic salts containing hydrophobic groups.
Diagrams showing the structures for the α- and β-dextrins are shown
in frames A and B of Figure 1. Frame C of this Figure depicts the
structure of α-dextrin constructed with molecular models showing the
orientation of the C-H bonds of the glucose units of the dextrin
molecule. The complexes which form are frequently crystalline and
highly insoluble. The complex forming property has been used to
devise methods for the isolation and purification of the dextrins.
Toluene, bromobenzene, cyclohexane, trichloroethylene, and
tetrachloroethane form insoluble complexes with the dextrins and
have been used in isolation procedures.

In Figure 2, photographs of crystals of some α-dextrin
complexes are shown. The hexagonal crystals of α-dextrin
crystalized from n-propyl alcohol are shown in Frame A. Frames B
and C of Figure 2 are photographs of two types of crystalline
iodine-dextrin complexes, hexagons and needles. The hexagons in
Frame B are brilliant blue in color while the clusters of needles in
Frame C are golden yellow. The time required for the initial
formation of the clusters of needles is used as the end point in an
assay method for measuring macerans amylase activity (40). Small
quantities of the higher molecular weight cyclic dextrins have been
isolated by a method utilizing high temperature cellulose
chromatography. An elution pattern for these dextrins from the
cellulose column is shown in Figure 3. Since these dextrins are
synthesized in low yields, extensive characterization of these
compounds has not been achieved.

Macerans amylase not only produces cyclic dextrins from starch
by the cyclizing reaction but also produces linear oligosaccharides
from the dextrins and appropriate co-substrates. Figure 4, top

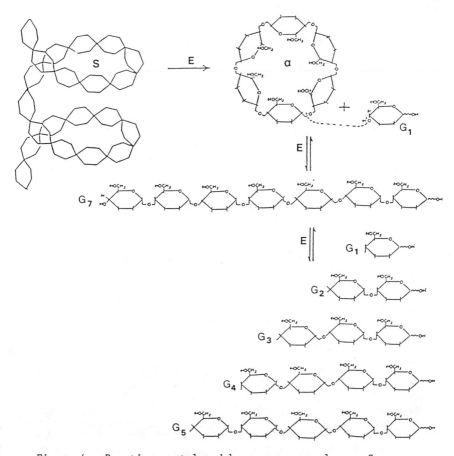

Figure 4. Reactions catalyzed by macerans amylase: S =
starch, E = enzyme, α = α-dextrin, G_1, G_2, G_3, etc. = glucose,
maltose, maltotriose, etc.

section, shows an equation for the formation of cyclic α-dextrin.
The helical parts of starch molecules are converted to the various
cyclic dextrins by the cyclizing action of the enzyme. The central
section of Figure 4 shows the equation for the coupling reaction.
In coupling reactions the ring of the dextrin is opened by the
enzyme and the oligosaccharide fragment is transferred to a co-
substrate molecule which may be glucose or glucose containing
oligosaccharides to yield linear oligosaccharides. The equation for
the synthesis of maltoheptaose is shown in the central section of
the Figure.

The lower section of Figure 4 shows the equations for the
occurrence of homologizing reactions of macerans amylase. In these
reactions segments of glucose units ranging in size from one glucose
to several glucose units are transferred to other co-substrates
until an equilibrium is attained and a series of homologous
oligosaccharides is synthesized. The equations for the synthesis of
the series of maltooligosaccharides from α-dextrin and glucose are
shown in Figure 4. Oligosaccharides with different functional
groups at the end of the maltooligosaccharide fragment have been
synthesized by proper selection of the co-substrate. Such
oligosaccharides have been useful for studies on the mechanism of
action of amylases, for the preparation of oligosaccharides for
affinity adsorbents and for the preparation of oligosaccharide
derivatives with potential technological uses.

The preparation of four types of oligosaccharide series is
described and these contain ^{14}C-glucose, glucosyl-(1,3)-arabinose,
sucrose, or p-aminophenyl glucoside at the reducing terminus of the
oligosaccharides. The new oligosaccharides have been detected by
chromatographic and radioautographic methods and isolated by
preparative chromatography. Photographs of the chromatographic
strips showing the products from ^{14}C glucose and from glucosyl-
(1,3)-arabinose with α-dextrin are reproduced in Figure 5. Lanes A
and B in the Figure show the products with ^{14}C glucose and α-dextrin
after incubation for 0 and 24 hrs. The members of this
oligosaccharide series have been identified as maltose, maltotriose,
maltotetraose, and maltopentaose. A radioautogram prepared from the
chromatogram showed that initially only glucose was radioactive but
at completion of the reaction all of the new oligosaccharides were
radioactive. Lane C and D of Figure 5 show the products which are
synthesized from glucosyl-(1,3)-arabinose and α-dextrin. These
products were detected with the aniline oxalate spray reagent which
reacts with oligosaccharides containing pentose units (53).
Analytical data from methylation, gas liquid chromatography and mass
spectra measurements confirmed that the arabinose of the
oligosaccharides was located at the reducing end of the
oligosaccharide. Sucrose as a co-substrate for the enzyme yielded
oligosaccharides which were terminated in a sucrose moiety while the
p-aminophenyl glucoside, yielded maltooligosaccharides terminated
with the p-aminophenyl glucosyl moiety.

Several applications have been developed using the new
oligosaccharides. Thus ^{14}C labeled maltooligosaccharides have been
used in a method for detecting and identifying various types of
amylases from biological sources. This method of analysis has been
called oligosaccharide mapping and also yields information on the

mechanism of action of the enzymes. Results of an oligosaccharide mapping experiment for glucoamylase are shown in Figure 6, Frame B and C, are photographs of the radioautogram and the chromatogram of the products of action of glucoamylase on the ^{14}C-oligosaccharides. Frame A of this Figure shows the separation of the radioactive oligosaccharide mixture which was not treated with the glucoamylase. The chromatogram pictured in Frame C has been stained with silver nitrate reagent for locating reducing products (50). Comparisons of the labeled products from the ^{14}C-oligosaccharides with the nature of reducing products yields information establishing that the pattern of action of the glucoamylase on the oligosaccharides is by a multi-chain mechanism.

On the basis of the above results it is likely that the hydrolysis of starch by glucoamylase occurs by a multi-chain mechanism in which single glucose units are removed from the non-reducing end of the starch chains. All of the starch chains are shortened progressively by one glucose unit. The action of the glucoamylase proceeds rapidly until an α-1,6-linkage is encountered. This type of linkage is slowly hydrolyzed by glucoamylase (59) and the slow phase of hydrolysis of these linkages is followed by the rapid removal of the internal α-(1,4) linked glucose units. Eventually the starch is converted quantitatively to glucose.

Coupling sugars are produced from sucrose and α-dextrin by the coupling reaction of macerans amylase. The coupling sugars are composed of oligosaccharides terminated at one end with sucrose moieties. Coupling sugars are not as sweet as sucrose, but can be used to replace sucrose in candies, jams, jellies, and other sweet products which when used in diets can cause a reduction of dental plaques and caries. Dextran can be synthesized from sucrose but not coupling sugar by the glucosyltransferase of oral bacteria (52) and participates in the development of dental plaques (29). Data obtained in feeding tests with diets containing starch, sucrose or coupling sugar show a significant reduction in the number of plaques and cavities in animals on a diet containing coupling sugar in place of sucrose (51). Additional data need to be obtained for the verification of these findings.

Oligosaccharides of glucose terminated at the reducing end with arabinose (30) have been prepared with the aid of macerans amylase. Such oligosaccharides have been used to measure furanose-pyranose transformations of reducing oligosaccharides. The structure of the oligosaccharide of glucose and arabinose showing the furanose and pyranose forms for the arabinose unit is shown in Figure 7. The ring forms of arabinose were determined by application of the methylation, gas liquid chromatography and mass spectrometry methods of analysis (58). The GLC chart for the methylation products from the trisaccharide is shown in Figure 8. The methylated derivatives were identified by retention times and the nature of the mass spectra fragments from the derivatives. From these data the ratio of furanose to pyranose forms for the arabinose unit of the oligosaccharide was calculated to be 3:2. Studies of this type are currently being extended to other oligosaccharides containing glucose, galactose, mannose or xylose as the reducing units. Differences in chemical reactivity and enzymatic susceptibility of oligosaccharides have been related to differences in the amounts of

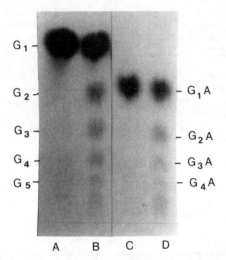

Figure 5. Photograph of a paper chromatogram of the products
of macerans amylase action on α-dextrin with glucose or with
glucosyl-(1,3)-arabinose at 0 time (A) and (C) and at 24 hr (B)
and (D): G_1, G_2, G_3, etc. = glucose, maltose, maltotriose,
etc., G_1A, G_2A, G_3A, etc. = glucosyl-arabinose, maltosyl-
arabinose, maltotriosyl- arabinose, etc.

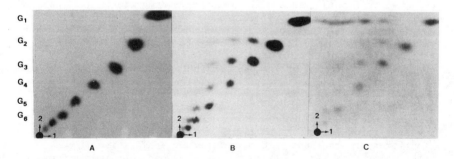

Figure 6. Oligosaccharide maps for the action of glucoamylase
on maltooligosaccharides, A = radioautogram with no enzyme, B =
radioautogram of enzymatic digest and C = paper chromatogram of
enzymic digest stained with silver nitrate and sodium hydroxide
reagents. Oligosaccharides were developed horizontally in the
first direction and vertically after enzyme treatment. G_1, G_2,
G_3, etc. = glucose, maltose, maltotriose, etc.

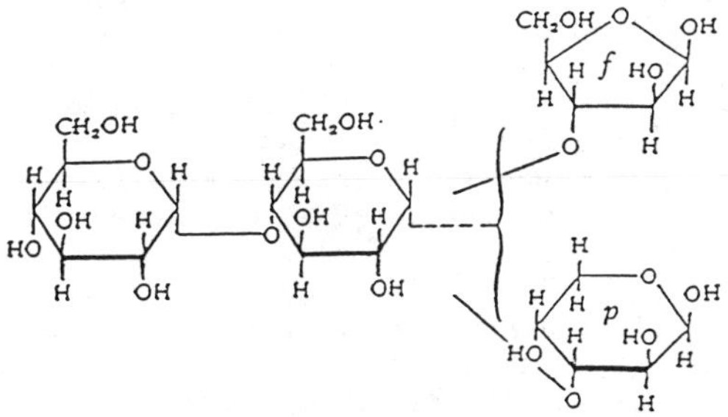

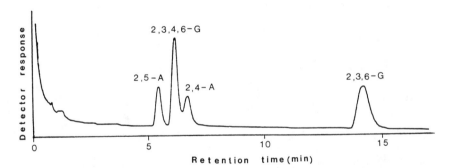

Figure 7.   Trisaccharide of glucose and arabinose showing furanose and pyranose structures for the arabinose moiety.

Figure 8.   A photograph of a GLC pattern of partially methylated alditol acetates from the trisaccharide of glucose and arabinose.   2,5-A = 1,3,4-tri-O-acetyl-2,5-di-O-methyl arabinitol; 2,3,4,6-G = 1,5-di-O-acetyl-2,3,4,6-tetra-O-methyl glucitol; 2,4-A = 1,3,5-tri-O-acetyl-2,4-di-O-methyl arabinitol; 2,3,6-G = 1,4,5-tri-O-acetyl-2,3,6-tri-O-methyl glucitol.   (Reproduced with permission from Ref. 30.   Copyright 1988 Academic Press.)

furanose and pyranose ring forms at the reducing ends of the
oligosaccharides.

Oligosaccharides containing p-aminophenyl glucose units at the
reducing ends have been synthesized with the use of macerans amylase
or fungal glucosyl transferase (54). Such oligosaccharides have
been used to prepare adsorbents for affinity columns for separating
antibodies and myeloma proteins. The isomaltose derivative was
prepared from maltose and aminophenyl glucoside with the fungal
glucosyltransferase and coupled to sepharose (56). Anti-isomaltose
antibodies and myeloma proteins were purified by affinity
chromatography on this adsorbent. Results of an experiment on the
isolation of myeloma protein specific for isomaltose units of a
dextran containing such units are shown in Figure 9. It is noted
that the serum protein was eluted with the buffer solution while the
myeloma protein fraction was eluted with the isomaltose solution.
Results of agar diffusion and gel electrophoresis tests are shown in
Figure 10.

Other derivatives of the maltooligosaccharides such as the
methyl glycosides (55), isomaltose (60) and the O-nitrophenyl
glycosides (61) have been prepared with the aid of macerans amylase.
These oligosaccharide derivatives have been very useful for
elucidating the action pattern of fungal, salivary and pancreatic
amylases on starch, glycogen and maltooligosaccharides.

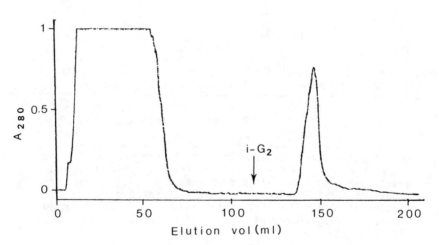

Figure 9. Elution pattern of mouse serum containing myeloma
protein W 3129 from an isomaltosyl-sepharose column: the arrow
indicates the point of elution with 1% isomaltose solution (i-
G_2). The eluates were monitored at 280 nm.

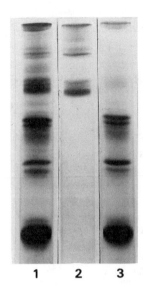

Figure 10. Gel electrophoresis patterns (A) and agar diffusion
plate (B) of proteins in serum and fractions. Number 1 is the
serum with myeloma protein. Number 2 is myeloma protein
adsorbed on an affinity column and eluted with isomaltose
solution. Number 3 is unadsorbed protein. Ag is the antigen
solution of dextran B-1355S. (Reproduced with permission from
Ref. 34. Copyright 1982 Elsevier Publishers.)

Literature Cited

1. Villiers, M. A. Compt. Rend. 1891, 112, 536-538.
2. Schardinger, F. Wien. Klin. Wochenschr. 1903, 16, 468-474.
3. Pongsawasdi, P.; Yagisawa, M. J. Ferment. Technol. 1987, 65, 463-467.
4. French, D. Advan. Carb. Chem. 1957, 12, 189-260.
5. French, D.; Knapp, D. W.; Pazur, J. H. J. Am. Chem. Soc. 1950, 72, 5148-5152.
6. Clarke, R. J.; Coates, J. H.; Lincoln, S. F. Advan. Carb. Chem. Biochem. 1988, 46, 205-249.
7. Hehre, E. J. Advan. Enzymol. 1951, 11, 297-337.
8. Cramer, F.; Saenger, W.; Spatz, H.-Ch. J. Am. Chem. Soc. 1967, 89, 14-20.
9. Tabushi, I. Acc. Chem. Res. 1982, 15, 66-72.
10. Beesley, T. E. American Laboratory 1985, 78-87.
11. Pulley, A. O.; French, D. Biochem. Biophys. Res. Commun. 1961, 5, 11-15.
12. French, D.; Pazur, J.; Levine, M. L.; Norberg, E. J. Am. Chem. Soc. 1948, 70, 3145.
13. Norberg, E.; French, D. J. Am. Chem. Soc. 1950, 72, 1202-1205.
14. French, D.; Levine, M. L.; Norberg, E.; Nordin, P.; Pazur, J. H.; Wild, G. M. J. Am. Chem. Soc. 1954, 76, 2387-2390.

15. Watanabe, T.; Akiyama, Y.; Takahashi, H.; Adachi, T.;
    Matsumoto, A.; Matsuda, K.. Carbohydr. Res. 1982, 109, 221-
    232.
16. French, D.; Levine, M. L.; Pazur, J. H.; Norberg, E. J. Am.
    Chem. Soc. 1949, 71, 353-356.
17. Saenger, W. Angew. Chem. Int. Ed. Engl. 1980, 19, 344-362.
18. Szejtli, J. Starch/Stärke 1985, 37, 382-386.
19. Armstrong, D. W.; Ward, T. J.; Armstrong, R. D.; Beesley, T. E.
    Science 232, 1132-1135.
20. Straub, T. S.; Bender, M. L. J. Am. Chem. Soc. 1972, 94, 8873-
    8881.
21. Breslow, R. Chem. Br. 1983, 126-131.
22. Mifune, A.; Shima, A. J. Synth. Org. Chem. 1977, 35, 116-130.
23. Ueno, A.; Takahashi, K.; Osa, T. J.C.S. Chem. Comm. 1980, 636,
    921-922.
24. Szejtli, J. Starch/Stärke, 1986, 38, 388-390.
25. Pazur, J.; Okada, S. J. Biol. Chem. 1966, 241, 4146-4151.
26. Okada, S.; Kitahata, S.; Higashihara, M.; Fukumoto, J. Agr.
    Biol. Chem. 1969, 33, 900-906.
27. Pazur, J. H. in Studies in Natural Products Chemistry (Atta-ur-
    Rahman, Ed.), Elsevier Publishers 1988, 321-364.
28. Okada, S; Kitahata, S.; Higashihara, M; Fukumoto, J. Agr.
    Biol. Chem. 1970, 34, 1407-1415.
29. Araya, S.; Yamada, Y. National Institute of Preventive Hygiene
    (Japan) 1975, 5-42.
30. Pazur, J. H.; Miskiel, F. J.; Liu, B. Anal. Biochem. 1988,
    174, 46-53.
31. Hinze, W. L.; Riehl, T. E.; Armstrong, D. W.; DeMond, W.; Alak,
    A.; Ward, T. Anal. Chem. 1985, 57, 237-242.
32. Konig, W. A.; Lutz, S.; Mischnick-Lubbecke, P.; Brassat, B.;
    von der Bey, E.; Wenz, G. Starch/Stärke 1988, 40; 472-476.
33. Pazur, J. H.; Tominaga, Y.; Dreher, K. L.; Forsberg, L. S.;
    Romanic, B. M. J. Carb. Nucleos. Nucleot. 1978, 5, 1-14.
34. Pazur, J. H.; Tay, M. E.; Rovnak, S. E.; Pazur, B. A. Immunol.
    Letters 1982, 5, 285-291.
35. Tilden, E. B.; Hudson, C. S. J. Am. Chem. Soc. 1939, 61, 2900-
    2902.
36. Shiosaka, M. United States Patent, 3988206, Oct. 26, 1976.
37. Kitahata, S.; Okada, S. Agr. Biol. Chem. 1974, 38, 2413-2417.
38. Horikoshi, K. Chem. Econ. Eng. Rev. 1981, 13, 7-10.
39. Bender, H. Arch. Microbiol. 1977, 111, 271-282.
40. Tilden, E. B.; Hudson, C. S. J. Bact. 1942, 43, 527-544.
41. Kitahata, S.; Tsuyama, H.; Okada, S. Agr. Biol. Chem. 1974,
    38, 387-393.
42. Kobayashi, S.; Kainuma, K.; Suzuki, S. Carbohyr. Res. 1978,
    229-238.
43. Szejtli, J. Cyclodextrins and their Inclusion Complexes,
    Akademiai Kiado, Budapest, Hungary 1982: C.A. 1982, 97, 146489.
44. Pazur, J. H. J. Am. Chem. Soc. 1955, 77, 1015-1017.
45. Freudenberg, von K.; Jacobi, R. Ann. Chemie. 1935, 518, 102-
    108.
46. French, D.; Pulley, A. O.; Whelan, W. J. Starch/Starke 1963,
    15, 280-284.
47. American Maize Products Co. Chem. Eng. News 1990, July 23, 49.

48. French, D.; Pulley, A. O.; Effenberger, J. A.; Rougvie, M. A.; Abdullah, M. Arch. Biochem. Biophys. 1965, 111, 153-160.
49. Thoma, J. A.; Wright, H. B.; French, D. Arch. Biochem. Biophys. 1959, 85, 452-460.
50. Mayer, F. C.; Larner, J. J. Am. Chem. Soc. 1959, 81, 188-193.
51. The East Publications, Inc., New York, NY (USA) 1981, 7.
52. Hamada, S.; Horikoshi, T.; Minami, T.; Okahashi, N.; Koga, T. J. Gen. Micribiol. 1989, 135, 335-344.
53. Block, R. J.; Durrum, E. L.; Zweig, G. Paper Chromatography and Paper Electrophoresis, 2nd Ed., 1958, pg. 181, Academic Press, New York.
54. Pazur, J. H.; Ando, T. Arch. Biochem. Biophys. 1961, 93, 43-49.
55. Pazur, J. H.; Marsh, J. M.; Ando, T. J. Am. Chem. Soc. 1959, 81, 2170-2172,
56. Cuatrecasas, P. J. Biol. Chem. 1970, 245, 3059-3065.
57. Affinity Chromatography, Pharmacia Fine Chemicals, Uppsala, Sweden, 1986.
58. Pazur, J. H.; Dropkin, D. J.; Dreher, K. L.; Forsberg, L. S.; Lowman, C. S. Arch. Biochem. Biophys. 1976, 176, 257-266.
59. Pazur, J. H.; Kleppe, K. J. Biol. Chem. 1962, 237, 1002-1006.
60. Summer, R.; French, D. J. Biol. Chem. 1956, 222, 469-477.
61. Wallenfels, K.; Meltzer, B.; Laule, G.; Janatsch, G. Fresenius. Z. Anal. Chem. 1980, 301, 169-170.

RECEIVED November 20, 1990

Chapter 6

Starch-Hydrolyzing Enzymes with Novel Properties

Fergus G. Priest and J. Roger Stark

Department of Biological Sciences, Heriot Watt University,
Edinburgh EH14 4AS, Scotland

Amylases for starch processing, food production and other applications are traditionally produced from a few microbial sources. The aerobic, endospore-forming bacteria of the genus *Bacillus* have long dominated the market for α-amylase along with the "liquefying" enzymes from *Bacillus amyloliquefaciens* and, more recently *B.licheniformis*. These enzymes are thermostable. Particularly the latter has a temperature optimum for activity around 90°C and can be used to hydrolyse starch at even higher temperatures (1,2). They are ideally suited for the initial size reduction of amylose and amylopectin molecules in starch during the process termed liquefaction (3). The dextrins produced from liquefaction (usually of dextrose equivalence, DE, about 15-20) are the substrates for several processes. Glucose syrups are prepared from extensive hydrolysis with the fungal enzyme glucoamylase often supplemented with the debranching enzyme pullulanase. High conversion syrups can be prepared from the liquefied starch by the use of fungal α-amylase. These α-amylases from *Aspergillus niger* and other molds are of the saccharifying type and on extensive hydrolysis of starch give rise to large quantities of glucose, maltose and maltotriose. The liquefying enzymes from most bacilli produce larger quantities of higher oligosaccharides such as maltopentaose and maltohexaose. Judicious use of fungal α-amylase and glucoamylase can give rise to syrups with various ratios of glucose and maltose and higher oligosaccharides (3). These high conversion syrups generally contain about 35-43% glucose, 30-40% maltose and 8-15% maltotriose. It is important that these syrups should have a high DE and yet be sufficiently stable not to crystallize at temperatures down to 4°C at 80-82% dry substance.

The expected demands from industry are for maltooligosaccharide preparations of various compositions and thus with various useful properties. High-maltose syrups for example are characterized by low viscosity, low hygrosopicity, resistance to crystalization and reduced in sweetness. Maltotetraose syrups on the other hand have relatively high viscosity and low colouration. To stimulate demand from the food industry, such syrups of consistent quality must be

0097–6156/91/0458–0072$06.00/0

prepared economically and this in turn has encouraged the search for new amylases with novel properties that can be used to manufacture products with novel compositions. In this contribution, some of the recent developments which centre on novel microbial amylases and their characteristics are reviewed.

Sources of Bacillus. Since *Bacillus* had proved to be such a valuable source of industrial enzymes since early this century, much screening for new enzymes concentrated initially on these bacteria (4). The results were not disappointing. The first bacterial β-amylases were discovered in *B. polymyxa* (5-7), *B. cereus* (8) and *B. megaterium* (9,10) and the genus is a rich source of pullulanases (11) with commercial production from *B. acidopullulyticus* (12). However, following the lead of the antibiotic industry, it became apparent that bacterial diversity would probably be accompanied by enzymic diversity; by screening new and often exotic bacteria novel enzymes would be discovered. Three areas have proved particularly interesting. *Pseudomonas* is a heterogeneous collection of free-living, gram negative bacteria associated with soil, water, and plants (13). Most members of the genus are non-pathogenic (some cause diseases in plants and *P. aeruginosa* is a pathogen of man and animals). Although they are not widely recognised for the secretion of extracellular enzymes, some amylolytic strains have been shown to produce interesting amylases (see below).

The actinomycetes are gram positive bacteria best known for the production of antibiotics. They are morphologically unusual since they generally grow as branched filaments and they are regarded as a distinct evolutionary group. Because of their morphology, the actinomycetes are often regarded as the bacterial equivalent of the fungi. The group encompasses more than thirty genera and several hundred species most of which are found widely distributed in soil, composts and water. They are responsible for considerable turnover of plant biomass in the environment and thus are well equipped with diverse extracellular enzymes for the degradation of plant polymers (14). Since most are non-pathogenic they are an ideal source of novel enzymes (15).

The final group of bacteria which has been shown to be a reservoir of novel, starch-degrading enzymes is the clostridia. These are the anaerobic equivalent of the bacilli and, because of the problems associated with growing bacteria in the absence of air, have been ignored until recently (16). These bacteria have been examined in detail and they will not be mentioned here (Zeikus et al, this volume).

Maltose-producing Enzymes. The common enzymes that produce maltose predominantly or exclusively from amylose are the β-amylases. These exo-acting enzymes hydrolyse amylose from the non-reducing chain end removing maltose units in the β-configuration. They are typically associated with plants such as soy bean and barley. There is a demand for maltose syrups which is low at present and is currently supplied by hydrolysing dextrins with soybean β-amylase and a debranching enzyme, either isoamylase or pullulanase (17). Bacillus β-amylases are generally too thermolabile for application in industry (optimum temperatures 40-55°C for the enzymes from *B. polymyxa*, *B. cereus* and *B. megaterium*) and the optimum pH for activity is around neutral compared to 5 to 6 for the plant enzymes. Unlike the plant enzymes the bacterial enzymes adsorb onto and partially degrade starch granules. DNA sequence comparison between soybean and *B. polymyxa* β-amylase genes shows some 30% homology indicating significant divergence as might be expected (18,19).

The need for maltose-producing enzymes with more appropriate industrial properties has been met by other sources. The enzyme from *Bacillus circulans* is a β-amylase but exhibits some characteristics that distinguish it from other bacterial β-amylases (20). The enzyme is active at 60°C and is much less susceptible to p-chloromercuribenzoate than plant β-amylases. Activity is optimal around pH 7.

An α-maltose-producing amylase has been described recently by Takasaki (21) originating from a strain of *B. megaterium.* This enzyme has all the properties of β-amylase (inhibition by sulphydryl agents such as mercuric chloride and p-chloromercuribenzoate with reactivation by cysteine) but produces maltose that is exclusively in the α-anomeric form. It may be that previous "β-amylases" described from this bacterium (9,10,23) are in fact of this type. The pH and temperature optima (7.0 and around 60°C) do not make this enzyme particularly attractive to industry. Novo Nordisk a/s manufactures an exo-attacking, α-amylase that produces maltose in the α-configuration from dextrins. This enzyme is derived from a thermophilic *Bacillus* strain (22) and is considerably more thermostable than the mesophilic *Bacillus* β-amylases.

The streptomycetes are also a rich source of amylases that produce predominantly maltose from starch. These enzymes achieve this by a mechanism that resembles that of the enzyme from the thermophilic bacillus strain mentioned above . They are endo-acting enzymes that, in common with typical α-amylases produce random oligosaccharides upon initial hydrolysis of starch. However, later in the reaction they catalyse a series of condensation and transglucosylation reactions that result in maltose production. The enzyme from *S. praecox* is very temperature labile (optimum 40°C) which severely limits its applicability for industry (24) but the enzyme from *S. hygroscopicus* has a pH optimum of 5-6 and temperature optimum of 50-55°C; they could be more useful (25). The principal primary products from amylose are maltotriose and maltotetraose. The latter is then hydrolysed to maltose and the maltotriose is converted to maltose by (i) a condensation-hydrolytic reaction and/or (ii) a synthetic step followed by a hydrolytic reaction (see Fig 1). The end result of these reactions is to produce maltose in greater than 75% yield from starch. Indeed a low viscosity syrup of (%w/v) glucose,4; maltose, 62; maltotriose, 20 and higher oligosaccharides, 14 can easily be produced using this enzyme. Since the composition of this syrup is similar to that from barley malt, application in the brewing industry is being explored.

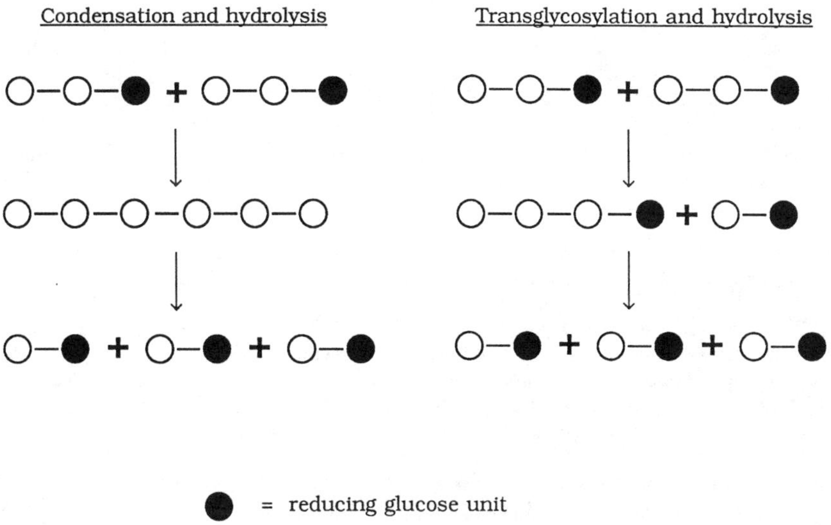

Figure 1. Breakdown of maltotriose by
Streptomyces hydroscopicus α-amylase.

Following a screening program of a diverse collection of 98 strains of *Streptomyces, S. limosus* was shown to have high amylolytic activity. Like the other streptomycete enzymes, this amylase produced large amounts of maltose from starch (26). The enzyme was not particularly thermostable and soon lost activity above 45°C. The pH optimum was near neutrality.   The α-amylase genes from *S. hygroscopicus* (27,28) and *S.limosus* (29,30) have been cloned and sequenced. There are conserved regions in these genes but the streptomycete amylases show little overall sequence homology with the *Bacillus* liquefying amylases. Indeed, they show more homology with mammalian and invertebrate amylases than with those from microorganisms or plants (29).

Maltotriose-producing Amylases. Amylases that produce maltotriose from starch as the major end-product have not been reported commonly. Indeed the only enzyme in this category is the enzyme from *Streptomyces griseus* (24). This is an exo-acting enzyme that hydrolyses its substrate from the non-reducing end to produce maltotriose in the α-configuration (31). The enzyme hydrolyses short chain amylose completely but the extent of hydrolysis with starch and waxy maize starch as substrate is 55 and 51% respectively. The optimum pH for activity is 5.6-6.0 and the enzyme is stable up to 40°C but rapidly denatured at temperatures above 45°C.

Maltotetraose-producing Amylases.   An exo-acting amylase that produces predominantly maltotetraose from starch was demonstrated in culture fluid from *Pseudomonas stutzeri* by Robyt and Ackerman (32). The enzyme releases α-maltotetraose from the non-reducing end of amylose chains.   Maltooligosaccharides longer than G4 are hydrolysed into maltotetraose and the remaining sugar. For example G7 is hydrolysed to G4+G3 and G6 into G4+G2 and maltotetraose is slowly hydrolysed to maltotriose and glucose (31). An unusual property of the enzyme is its ability to attack insoluble dyed starch which is normally resistant to attack by exo-acting amylases (33). The properties of the enzyme are not entirely conducive to industrial exploitation, the pH optimum is around pH 8 and the optimum temperature for activity is about 45°C (34).   Following an extensive screening program, a strain of *P. stutzeri* that produces a more thermostable enzyme (optimum about 50°C) with a wider pH range was isolated. This enzyme is being used for the production of maltotetraose on a large scale in a continuous bioreactor system. The syrups contain about 50% maltotetraose and have several desirable properties such as low sweetness, prevention of colouration in candy and increased humectancy in cakes. (35). Such exo-maltotetrahydrolases do not seem to be restricted to *P. stutzeri* since a similar enzyme from *Pseudomonas saccharophila* has recently been cloned in *E. coli* (36).

Maltopentaose-producing Amylase. Although the liquefying α-amylase from *B. licheniformis* produces maltopentaose as a major product (37) it seldom represents more than 33% of the products from starch after extensive hydrolysis. Nakakuki *et al.* (31) showed that the enzyme had some exo-activity on small oligosaccharides and could

hydrolyse G6 to G5+G1,   GG7 to G5+G2 and G8 to G5+G3. Nevertheless
the enzyme has the predominant characteristics of a liquefying
α-amylase. Exo-maltopentohydrolase has been demonstrated recently
in culture supernatants of a *Pseudomonas* species (38).

Maltohexaose-producing Amylases.   Both endo- and exo-amylases that
produce   maltohexaose   from   starch   have   been   identified   and
characterized. The exo-acting enzyme was first discovered as a cell
bound enzyme in *"Aerobacter aerogenes"* (now called *Klebsiella
pneumoniae)*, an organism better known for pullulanase synthesis.
The enzyme is stable at temperatures lower than 50°C and has an
optimum pH for activity around neutrality (39). *Escherichia coli,*
long thought to be devoid of amylase activity, has been shown to
synthesize a similar enzyme (40). These enzymes are periplasmic and
synthesis is induced by the presence of maltose (40).
     Endo-acting   α-amylases that produce large quantities of
maltohexaose from starch are secreted by various bacilli. *Bacillus
circulans* strains G-6 (41) and F-2 (42) were discovered
independently and shown to produce such enzymes. Temperature and pH
optima vary considerably between these enzymes, strain F-2
producing the more temperature stable (opt. 60°C) amylase with a
lower pH optimum for activity (6-6.5). This enzyme also has
considerable activity on raw starch (43).
     A strain of *B. subtilis* was isolated by Kennedy & White (44)
that   produced   an   α-amylase   that   yielded   large   amounts   of
maltohexaose from starch which would seem to be different from the
saccharifying enzyme normally associated with this species, in
particular in showing enhanced thermostability.

Enzymes with Actvity on Pullulan. Just as the distinction between
α- and β-amylases is becoming confused through the isolation of
enzymes with characteristics of both classes, so the original
classification of 1,4-α-amylases and debranching or 1,6-α-amylases
is no longer entirely relevant. Pullulan, the extracellular
polysaccharide is comprised of maltotriose units linked by 1,6-α-
bonds (see Fig 2) is the common substrate for detection of
debranching enzymes. It seems to be a suitable substrate for
various debranching enzymes and other amylases with the exception
of isoamylase which debranches amylopectin but has no activity on
pullulan (45). Dyeing pullulan with reactone red or a similar dye
aids the detection of colonies that can degrade pullulan through
the hydrolysis of the material.   The low molecular weight products
diffuse away from the colony leaving a clear zone around the colony
(46).
     Four categories of enzyme with activity on pullulan are now
recognised. (i). Glucoamylase, is an exo-acting enzyme common in
fungi that releases glucose from amylose chains and pullulan.
Activity on 1,6-α-linkages in pullulan and amylopectin by this
enzyme is considerably slower than activity on 1,4-α-bonds. (ii).
Pullulanase itself hydrolyses the 1,6-α-bonds in pullulan to
release maltotriose and also debranches amylopectin. It has very
low activity on glycogen. This enzyme is produced commercially from
B. acidopullulyticus (12) and has been studied extensively in
Klebsiella pneumoniae. (iii) The 1,4-α-linkages of pullulan

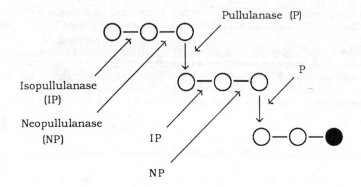

● = reducing glucose unit

Products from pullulan digestion

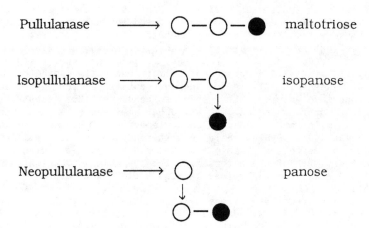

Figure 2. Hydrolysis products from pullulan.

furthest from the reducing end are hydrolysed by isopullulanase (Fig. 2) to yield isopanose. This enzyme has been demonstrated in culture filtrates of *Aspergillus niger* (47). (iv) The other 1,4-α-linkage of pullulan is suceptible to cleavage by neopullulanase with the production of panose (Fig. 2). This enzyme has a slightly confused history but it now seems that various thermophilic bacteria make a similar type of amylase (see below).   An amylase which hydrolysed pullulan to panose and had a typical saccharifying activity on starch was described from culture filtrates of *Thermoactinomyces vulgaris* by Shimizu *et al.* (48). These bacteria are morphologically similar to the true actinomycetes and have a filamentous appearance. On solid media they produce a substrate mycelium and most strains later develop a white aerial mycelium hence their name which denotes their similarity to the true actinomycetes. However, in molecular respects they resemble the thermophilic bacilli. Thermoactinomycetes share considerable ribosomal RNA homology with the bacilli and make true endospores typical of *Bacillus* endospores. The current view is therefore that they should be considered filamentous bacilli rather than actinomycetes (49). There is also confusion about the amylases synthesized by these bacteria. Original studies of *T. vulgaris* revealed a typical, saccharifying α-amylase (50). The enzyme studied by Shimizu et al. (48) also supposed to derive from *T. vulgaris* had different physical properties in addition to activity on pullulan. Moreover, according to current taxonomic criteria, *T. vulgaris* is considered to be non-amylolytic ! (49). In an attempt to clarify this situation we examined the amylases from *T. thalpophilus*, which is generally held to be the amylolytic version of *T. vulgaris* (48) and *T. putidus* a closely related species. The enzyme from *T. thalpophilus* resembled the original amylase described by Kuo and Hartman (50) and had no significant activity on pullulan (unpublished). The enzyme from *T. putidus* showed limited activity on pullulan and clear zones appeared around colonies growing in the presence of dyed pullulan. However, it was concluded that this resulted from the hydrolysis of 1,4-α-bonds in maltotetraosyl units in the pullulan. Such units occur as about 10% of most preparations of pullulan (51) and could be the sites of cleavage by an enzyme that cannot gain access to the maltotriosyl units (52). In conclusion, it would seem that the enzyme studied by Shimizu *et al.* (48) which does hydrolyse pullulan efficiently is relatively scarce and is not found in standard strains of *T. thalpophilus* or *T. putidus*.

The close relationship between thermoactinomycetes and thermophilic bacilli is also evident from the occurrence of pullulan-hydrolysing enzymes similar to that described by Shimizu *et al.* (48) for thermoactinomyces in several thermophilic *Bacillus* strains. These enzymes have now been given the name neopullulanase (53). Originally described in a strain of *B. stearothermophilus* (54), this enzyme hydrolyses pullulan (to panose predominantly), amylopectin and soluble starch as well as several low molecular weight substrates such as α- and B-cyclodextrins, phenylmaltoside and maltotriose. Optimum activity is around 55°C and pH 5.8 (55). A similar enzyme has been described by Kuriki *et al.* (56) also from

B. stearothermophilus. This enzyme also hydrolyses the 1,6-α-linkages in pullulan albeit as a slow delayed reaction (57). Such activity has also been noted in the thermoactinomycete enzyme (58) and would explain the presence of low amounts of glucose and maltose in pullulan digests recorded by Suzuki and Imai (55). There are some minor differences in the relative extents of hydrolysis of pullulan and starch by these enzymes but the general features and catalytic activities seem to be very similar.

Nevertheless, not all amylases from *B. stearothermophilus* need be of the neopullulanase type. *B. stearothermophilus* is a very heterogeneous species and comprises at least 8 distinct taxa which could be considered species (59). Many of these are amylolytic and previous studies amongst bacilli suggest that strains of different species generally secrete different amylases. Thus the original amylase prepared from *B. stearothermophilus* was a normal α-amylase (60). The maltogenic amylases (see above) have been described and a third type has been described recently. This is a pullulanase that yields maltotriose from pullulan and also hydrolyses starch in a typical α-amylolytic fashion (61). Such "amylopullulanases" are not uncommon and have also been isolated from *Thermoanaerobium* sp. (62, 63) and a strain identified as *B. subtilis* (64). It seems reasonable to assume that these different types of amylase from thermophilic bacilli are associated with different species just as different amylases are associated with the mesophilic species such as *B. subtilis, B. licheniformis* and *B. amyloliquefaciens.*

Hydrolysis of Starch Granules. There is increasing interest in amylases with high activity towards starch granules. These particles are generally resistant to amylolytic attack thus necessitating gelatinization of starch slurries by cooking at high temperature. Significant energy saving could be realized from efficient enzymic degradation of raw starch prior to, for example, fermentation with yeast to produce ethanol (65).

Microorganisms known to produce enzymes that degrade raw starch granules are generally fungi such as *Aspergillus* and *Rhizopus* in which the glucoamylases are active on this substrate. It seems that enzymes with this ability show extensive adsorbtion to starch granules probably through a COOH-terminus binding domain (67 and Svensson, this book). Bacterial enzymes that show activity on starch granules, such as the β-amylases of *B. polymyxa* and cyclodextrin glucanotransferase from *Bacillus macerans* also possess this domain. Surveys have shown that both exo- and endo-acting enzymes with activity on starch granules are fairly common particularly within the genus *Bacillus* (68, 69).

Streptomycetes are also a rich source of α-amylases with activity on starch granules (26). The enzymes from *S. hygroscopicus* (70) and *S. limosus* (26) both have high activity on granules and contain the C-terminal binding domain. Typical attack on corn starch granules is shown in Figure 3. The early stages of hydrolysis are indicated by deep holes in the granules which often reveal the layered structure of the granule. These holes presumably arise from initial binding and continued attack at one site. For this reason, in digests where the ratio of enzyme to granules is low, fully degraded granules can be observed next to

Figure 3. Scanning electron micrographs showing the action of
α-amylase on corn starch.

intact forms (Fig. 3a). Whereas, if the concentration of enzyme is higher, holes are produced uniformly (Fig. 3b). Extensive hydrolysis by the amylase from *S. limosus* gives rise to granules in the form of shells in which most of the starch has been hydrolysed (Fig. 3c).

Concluding Remarks. There is a bewildering variety of microbial amylases described in the literature and in this review we have intended to highlight the major types that have been reported recently. These classes are based at this stage on physical properties and action patterns. With the exception of the endo- versus exo- mode of action, these divisions may be somewhat artificial, particularly with regard to 1,4-α- and 1,6-α-bond specificity. It is already apparent that limited protein engineering can have a profound effect on the bond specificity of glucoamylases and enzymes with 1,4-α- and 1,6-α- activity share significant homology and should not be considered so widely different (71). As DNA sequences, particularly of the more obscure amylase genes are determined and analysed, it is certain that a deeper understanding of the classes of amylases and their evolution will unfold. This will undoubtedly be a fascinating story. Nevertheless, in the meantime there is still a future for screening for new enzymes. At least for the next few years screening is likely to remain less expensive and more successful than protein engineering for the development of a novel enzyme with specific desired properties.

Literature Cited

1. Madsen, G. B.; Norman, B. E.; Slott,S. Starch 1973, 25, 304-308.
2. Morgan, F. J.; Priest, F. G. J. Appl. Bacteriol. 1981, 50, 107-114.
3. Fogarty, W. M. In Microbiol Enzymes and Biotechnology; Fogarty, W. M., Ed.; Applied Science Publishers: London, 1983; pp. 1-92.
4. Priest, F. G. Bacteriol. Rev. 1977, 41, 711-753.
5. Robyt, J.; French, D. Arch. Biochem. Biophys. 1964, 104, 338-345.
6. Fogarty, W. M.; Griffin, P. J. J. Appl. Chem. Biotechnol. 1975, 25, 229-235.
7. Muraom, S.; Ohyama, K., Arai, M. Agric. Biol. Chem. 1979, 43, 719-724.
8   Shinke, R.; Nishira, H.; Mugibayashi, N. Agri. Biol. Chem. 1974, 38, 665-667.
9   Higashihara, M.; Okada, S. Agric. Biol. Chem. 1974, 38, 1023-1029.
10. Thomas, M.; Priest, F. G.; Stark, J. R. J. Gen. Microbiol. 1980, 118, 67-72.
11. Morgan, F. J.; Adams, K. R.; Priest, F. G. J. Appl. Bacteriol. 1979, 46, 291-294.
12. Jensen, B. F.; Norman, B. E. Proc. Biochem. 1984, 19, 129-134.

13. Palleroni, N. J. In Bergey's Manual of Systematic Bacteriology Vol. 1. Krieg, N. R.; Holt, J. G. Eds. Williams & Wilkins, Baltimore, 1984; pp 141-199.

14. Crawford, D. L. In Actinomycetes in Biotechnology Goodfellow, M.; Williams, S.T.; Mordarski, M. Eds. Academic Press: San Diego, 1988; pp 433-460.

15. Peczynska-Czoch, W.; Mordarski, M. In Actinomycetes in Biotechnology; Goodfellow, M.; Williams, S. T.; Mordarski, M. Eds.; Academic Press: San Diego, 1988; pp 219-284.

16. Saha, B. C.; Lamed, R.; Zeikus, J. G. In Biotechnology Handbooks 3 Clostridia; Minton, N. P. & Clarke, D. J. Eds.; Plenun Press, New York; 1989; pp 227-264.

17. Kennedy, J. F.; Cabalda, V. M.; White, C. A. Trends in Biotechnol. 1988, 6, 184-189.

18. Rhodes, C. J.; Strasser, J.; Frieberg, F. Nucleic Acids Res. 1987, 15, 3934.

19. Kawazu, T.; Nakanishi, Y.; Vozumi, N.; Sasaki, T.; Yamagata, H.; Tsukagoshi, N.; Udaka, S. J. Bacteriol. 1989, 171, 1564-1570.

20. Fogarty, W. M. In Current Developments in Malting, Brewing and Distilling; Priest, F. G.; Campbell, I.; Eds.; Institute of Brewing: London, 1983, 83-111.

21. Takasaki, Y. Agric. biol. Chem. 1989 53, 341-347.

22. Outtrup, H.; Norman, B. E. Starch 1984 36, 405-411.

23. Stark, J. R.; Stewart, T. B.; Priest, F. G. FEMS Microbiol. Letts. 1982, 15, 295-298.

24. Wako, K.; Takahashi, C.; Hashimoto, S.; Kaneda,J. J. Jap. Soc. Starch Science. 1987, 25, 155-159.

25. Hidaka, H.; Koase, T.; Yoshida, Y.; Niwa, T.; Shomura, T.; Niida, T. Starch 1974, 26, 413-416.

26. Fairbairn, D.; Priest, F. G.; Stark, J. R. Enzyme Microbiol. Technol. 8, 89-92.

27. McKillop, C.; Elvin, P.; Kenten, J. FEMS Microbiol. Letts. 1986, 36, 3-7.

28. Hoshuko, S.; Makabe, O.; Nojiri, C.; Katsumata, K.; Satoh, E.; Nagaoka, K. J. Bacteriol. 1987, 169, 1029-1036.

29. Long, C.; Virolle, M.-Y.; Chang, S.-Y.; Chang, S.; Bibb, M. J. J. Bacteriol. 1987, 169, 5745-5754.

30. Virolle, M.-Y.; Bibb, M. J. Molec. Microbiol. 1988, 2, 197-208.

31. Nakakuki, T.; Azuma, K.; Kainuma, K. Carbohydr. Res. 1985, 128, 297-310.

32. Robyt, J. R.; Ackerman, R. S. Arch. Biochem. Biophys. 1971, 145, 105-114.

33. Schmidt, J.; John, M. Biochim. Biophys. Acta 1979, 566, 88-99.

34. Sakano, Y.; Kashiyama, E.; Kobayashi, T. Agric. Biol. Chem. 1983, 47, 1761-1768.

35. Tsujisaka, Y. In Handbook of Amylases. Japanese Society for Starch Science; Kyoto, 1988, pp 213-215.

36. Zhou, J.; Takano, T.; Kobayashi. S. Agric. Biol. Chem. 1989, 53, 301-302.

37. Saito, N. Arch. Biochem. Biophys. 1973, 155, 292-298.

38. Okemoto, H.; Kobayashi, S.; Mommo, M.; Hashimoto, H.; Hara,
    K.; Kainuma, K. Appl. Microbiol. Biotechnol. 1986, 25, 137-
    142.
39. Palmer, T. N.; Wöber, G.; Whelan, W. J. Eur. J. Biochem.
    1973, 39, 601-612.
40. Freundlieb, S.; Boos. W. J. Biol. Chem. 1986, 261, 2946-
    2953.
41. Takasaki, Y. Agric. Biol. Chem. 1982, 46, 1539-1547.
42. Taniguchdi, H.; Chung, M. J.; Yoshigi, N.; Maruyama, Y. Agric.
    Biol. Chem. 1983 47, 511-519.
43. Kennedy, J. F.; White, C. A. Starch 1979, 71, 93-99.
44. Matsuzaki. H.; Yamane, K.; Yamaguchi, K.; Nagata, Y.; Maruo,
    B. Biochem. Biophys. Arch. 1974, 356, 235-247.
45. Harada, T. Genetic Eng. Biotech. Rev. 1984, 1, 39-64.
46. Marumo, S.; Hattori, H.; Katayama, M.; Agric. Biol. Chem.
    1985, 49, 1521-1523.
47. Sakano, Y.; Masuda, N.; Kobayashi, T. Agric. Biol. Chem.
    1971, 35, 971-973.
48. Shimizu, M.; Kanno, M.; Tamura, M.; Suekane, M. Agric. Biol.
    Chem. 1978, 42, 1681-1688.
49. Goodfellow, M.; Cross, T. In The Biology of the
    Actinomycetes; Goodfellow, M.; Mordarski, M.; Williams, S.
    T. Eds.; Academic Press, New York: 1984, pp 7-164.
50. Kuo, M. J.; Hartman, P. A. Can. J. Microbiol. 1967 13, 1157-
    1163.
51. Catley, B. J.; Whelan, W. J. Arch. Biochem. Biophys. 1971,
    143, 138-142.
52. Priest, F. G.; Stark, J. R.; Tabrizi, S. Starch 1988, 40,
    426-429.
53. Imanaka, T.; Kuriki, T. J. Bacteriol. 1989. 171, 369-374.
54. Suzuki, Y.; Chisiro, M. Eur. J. Appl. Microbiol. Biotechnol.
    1983, 17, 24-29.
55. Suzuki, Y.; Imai, T. Appl. Microbiol. Biotechnol. 21, 20-
    26.
56. Kuriki, T.; Okada, S.; Imanaka, T. J. Bacteriol. 1988, 170,
    1554-1559.
57. Imanaka, T.; Kuriki, T. J. Bacteriol. 1989, 171, 369-374.
58. Sakano, Y., Hiraiwa, S.; Fukushima, J.; Kobayashi, T. Agric.
    Biol. Chem. 1982, 46, 1121-1129.
59. Wolf, J.; Sharp, R. J. In The Aerobic Endospore-forming
    Bacteria; Classification and Identification Berkeley, R. C.
    W.; Goodfellow, M. Eds. Academic Press, London, 1981: pp
    251-296.
60. Pfeuller, S. L.; Elliot, W. H. J. Biol. Chem. 1969, 244, 48-
    54.
61. Saha, B. C.; Shen, G.-J.; Srivastava, K. C.; Le Cureux, L.
    W.; Zeikus, J. G. Enzyme Microbiol. Technol. 1989, 11, 760-
    764.
62. Coleman, R. D.; Yang S.-S.; McAlister, M. P. J. Bacteriol.
    1987, 53, 1661-1667
63. Plant, A. R.; Clemens, R. M.; Daniel, R. M.; Morgan, H. W.
    Appl.Microbiol. Biotechnol. 1987, 26, 427-433.
64. Takasaki, Y. Agric. Biol. Chem. 1987, 51, 9-16.

65. Matsumoto, N.; Fukushi, O.; Miyanaga, M.; Kakihara, K.; Nakajima, E,; Yoshizumi, H. Agric. Biol. Chem. 1982, 46, 1549-1558.
66. Veda, S.; Saha, B. C.; Koba, Y. Micrbiol. Sciences 1984, 1, 21-24.
67. Svensson, B. FEBS Letts. 1988, 230, 72-76.
68. Itkor, P.; Shida, O.; Tsukagoshi, N.; Udaka, S. Agric. Biol. Chem. 1989 53, 53-60.
69. Umesh-Kumar, S.; Rehana, F.; Nand, K. Letts. Appl. Microbiol. 8, 33-36.
70. Hidika, H.; Adachi, J. In Mechanisms of Saccharide Polymerization and Depolymerization; Marshal, J. J. Ed.; Academic Press, New York, 1980; pp 101-130.
71. Kuriki, T.; Imanaka, T. J. Gen. Microbiol. 1989, 135 1521-1528.

RECEIVED November 20, 1990

Chapter 7

Novel Thermostable Saccharidases from Thermoanaerobes

Badal C. Saha[1], Saroj P. Mathupala[2], and J. Gregory Zeikus[1-3]

[1]Michigan Biotechnology Institute, Lansing, MI 48910
[2]Department of Biochemistry and [3]Department of Microbiology and Public Health, Michigan State University, East Lansing, MI 48823

We have purified and characterized several new saccharidase activities including ß-amylase, amylopullulanase, α-glucosidase, and cyclodextrinase from thermoanaerobes. The ß-amylase was stable and active at 75°C. The amylopullulanase displayed higher affinity towards pullulan than starch; and it contained a putative single active site for cleavage of both α-1,4 and α-1,6 linkages. It produced maltotriose from pullulan, and DP2, DP3 and DP4 from starch. The α-glucosidase had an apparent MW of 162,000 and cleaved both maltosyl and isomaltosyl polymers with a decreasing affinity for longer chains. The cyclodextrinase hydrolyzed various cyclodextrins with decreasing activity rates displayed on α-CD > ß-CD > γ-CD. The unique biochemical properties and process features of these thermophilic enzymes that have potential usefulness for development of new saccharide biotechnologies are described.

Amylolytic enzymes are an important group of industrial enzymes. Three types of enzymes are involved in the production of sugars from starch: (1) endo-amylase (α-amylase), (2) exo-amylase (ß-amylase, glucoamylase) and (3) debranching enzymes (pullulanase, isoamylase). The starch bioprocessing usually involves two steps - liquefaction and saccharification. First, an aqueous slurry of starch (30-40%, DS) is gelatinized (105°C, 5-7 min) and partially hydrolyzed (95°C, 2 h) by highly thermostable α-amylase to DE (dextrose equivalent) 5-10. The optimum pH for the reaction is 6.0-6.5 and calcium is also required. Then in the saccharification step, glucoamylase and pullulanase are used (60°C, pH 4.0-4.5, 48 h) to produce more than 95-96% glucose; ß-amylase and pullulanase can also be used (55°C, pH 5.0-5.5, 48-72 h) to produce around 80-85% maltose. Thus, there is a need for thermostable saccharolytic enzymes to run the saccharification reaction at a higher temperature. Thermophiles often possess thermostable enzymes and interest in thermoanaerobic

0097–6156/91/0458–0086$06.00/0

bacteria has increased because of their unexamined potential as a source of thermostable and thermoactive enzymes including saccharidases. Thermoanaerobic bacteria may then serve as gene sources for cloning thermostable enzymes into aerobic industrial hosts (1). Our group initiated a screening program for thermostable saccharidases from diverse thermoanaerobic species.   As a result, some new organisms have been isolated from hot spring areas and some novel highly thermophilic saccharidases have been discovered. Table I summarizes our efforts on obtaining some unique saccharidases from thermoanaerobes.   *Clostridium thermo-*

Table I. Thermophilic Saccharidases from Thermoanaerobes

Source (Enzyme)	Optimum Temp. (°C)	Optimum pH	Thermal stability (up to °C)	pH Stability
Clostridium thermosulfurogenes strain 4B				
ß-amylase	75	5.5-6.0	80	3.5-6.5
α-glucosidase	-	-	-	-
Glucose isomerase	85	7.0	86	5.5-8.0
Clostridium thermohydrosulfuricum strain 39E				
Amylopullulanase	90	5.5-6.0	90	4.5-5.5
α-glucosidase	75	4.0-6.0	75	5.0-6.0
Cyclodextrinase	65	6.0	60	5.5
Glucose isomerase	85	8.0	85	6.0-8.0
Thermoanaerobacter strain B6A				
Amylopullulanase	75	4.5-5.5	70	5.0-6.0
Glucogenic amylase	70	5.0-5.5	70	4.5-6.0
ß-galactosidase	65	6.0-6.5	60	5.0-7.0
Xylanase	75	5.5	65	5.0-7.0
Cyclodextrinase	60	6.0	60	6.0
Glucose isomerase	80	7.0	85	5.5-8.0

sulfurogenes strain 4B produces an extracellular ß-amylase and intracellular glucose isomerase. *Clostridium thermohydrosulfuricum* strain 39E produces amylopullu-lanase, α-glucosidase, glucose isomerase and cyclodextrinase activities. *Thermo-anaerobacter* strain B6A produces amylopullulanase, glucogenic amylase, ß-galactosidase, cyclodextrinase, glucose isomerase and xylanase. In this chapter, we will focus our research efforts on the amylo-saccharidases from these thermo-anaerobes.

Biochemical Characteristics

ß-Amylase. ß-Amylase (EC 3.2.1.2, α-1,4-D-glucan maltohydrolase, saccharogenic amylase) is an exo-acting saccharidase which cleaves alternative α-1,4-glucosidic linkages in starch from the non-reducing end and produce ß-maltose. ß-Amylase occurs widely in many higher plants and is also produced by microorganisms. *C. thermosulfurogenes* 4B is the only anaerobe reported so far that produces extracellular ß-amylase. The enzyme is stable up to 80°C and optimally active at 75°C (2). ß-Amylase synthesis in this organism is inducible and subject to catabolic repression. A hyperproductive mutant was developed which produced 8-fold more ß-amylase in starch medium than the wild type (3). The effect of culture conditions and metabolite levels on the production of thermostable ß-amylase with the overproducing mutant of *C. thermosulfurogenes* 4B was investigated in continuous culture (4). The ß-amylase activity level reached 90 units/ml at the dilution rate of 0.07/h in 3% starch medium. Growth inhibition by acetate and low enzyme productivity at low growth rates limited the further increase in enzyme production level. Nipkow et al. (5) then developed a microfiltration cell-recycle pilot system for continuous production of ß-amylase by the thermoanaerobe. The concentration of ß-amylase rose to 220 units/ml in the reactor, which was 5.5-fold more than under comparable conditions in a chemostat.

The ß-amylase from *C. thermosulfurogenes* 4B was purified 811-fold to homogeneity from the culture broth by ultrafiltration, ethanol treatment, DEAE-Sepharose CL-6B column chromatography and gel filtration on Sephacryl S-200 (6). The purified enzyme had a specific activity of 4215 units/mg protein. It was a tetramer (MW 210,000) having an isoelectric point at pH 5.1. The enzyme displayed K_m and K_{cat} values for boiled soluble starch of 1.68 mg/ml and 400,000/min, respectively. It was antigenically distinct from sweet potato and barley ß-amylases.

The ß-amylase from *C. thermosulfurogenes* 4B readily and strongly adsorbed onto raw starch (7). pCMB treated ß-amylase lost its activity towards raw or gelatinized starch but preserved the ability to adsorb onto raw starch. The adsorbed ß-amylase was gradually released from starch in liquid phase during hydrolysis at 75°C. The degradation of raw starch by ß-amylase was greatly enhanced by the addition of pullulanase. The optimum pH for raw starch hydrolysis by ß-amylase was 4.5-5.5, whereas, that of soluble starch hydrolysis was 5.5-6.0. Raw starch adsorbed ß-amylase and soluble ß-amylase showed similar rates of hydrolysis in reaction mixtures. It was found that the adsorbed ß-amylase can be easily desorbed from raw starch by using soluble starch or maltodextrin as elutant. The soluble starch treated ß-amylase could not adsorb onto raw starch which suggests that the soluble and insoluble substrate binding sites of the ß-amylase may be the same. The ß-amylase was easily purified to homogeniety by simple raw starch adsorption-desorption techniques and octyl-Sepharose chromatography (8). A comparison of certain physicochemical characteristics of this ß-amylase with some other microbial ß-amylases is given in Table II.

A gene coding for the ß-amylase of *C. thermosulfurogenes* 4B was cloned into *Bacillus subtilis* and its nucleotide sequence was determined (12). The ß-amylase was translated from the monocistronic mRNA as a secretory precursor with a signal peptide of 32 amino acid residues. The deduced amino acid sequence of the mature ß-amylase contained 519 residues with a MW of 57,167. The amino acid sequence showed 57, 32 and 32% homology with those of *B. polymyxa*, soybean and barley ß-amylases. The hydrophobicity of several regions in the amino acid sequence of *C. thermosulfurogenes* 4B ß-amylase was found to be remarkably high as compared with that of the corresponding regions of the *B. polymyxa* ß-amylase.

Amylopullulanase. Pullulanase is a debranching enzyme which specifically attacks the α-1,6 glucosidic linkages of pullulan and starch. Pullulan is a linear polymer of about 250 maltotriosyl units linked together by α-1,6 linkages. Pullulan degrading enzymes can now be classified into five groups (Table III). Amylopullulanase, a new class of enzyme, hydrolyzes α-1,6 linkages of pullulan like normal pullulanase but unlike pullulanase which cleaves only α-1,6 linkages in starch, this enzyme cleaves α-1,4 linkages of starch (*19*).   We have suggested the name amylopullulanase. Recently, thermostable pullulanase activity has been reported in a number of microorganisms such as *C. thermohydrosulfuricum* (*21-23*), *C. thermosaccharolyticum* (*24*), *Clostridium* sp. (*25*), *Thermus* sp. (*26*), *Thermus aquaticus* (*27*), *Thermoanaerobium* Tok6-B1 (*28*), *T. brockii* (*29*), *Thermoanaerobacter* strain B6A (*30*), *T. finnii* (*23*), *Thermobacteroides ethanolicus* (*24*), *T. acetoethylicus* (*24*), *Thermoactinomyces thalpophilus* (*31*), and thermophilic *Bacillus* sp.(*32*). The pullulanase from *C. thermohydrosulfuricum* (*21, 22*), *Thermoanaerobium* Tok6-B1 (*28*), *T. brockii* (*29*), *Thermus* sp.(*26*), *Thermoanaerobacter* sp. B6A (*30*) and thermophilic *Bacillus* sp. (*32*) have already been demonstrated to be of the amylopullulanase type.

The synthesis of amylase in *C. thermohydrosulfuricum* 39E (ATCC 33223) was inducible and subject to catabolic repression (*21*). Catabolic repression resistant mutants were isolated which displayed improved starch metabolism features in terms of enhanced rates of growth, ethanol production and starch consumption (*33*). In chemostat cultures, both wild type and mutant strains produced amylopullulanase at

Table II. Comparison of Properties of ß-Amylase from Different Microbial Sources

Property	B. cereus var. mycoides (9)	B. polymyxa I, II (10)	B. megatarium (11)	C. thermo-sulfurogenes 4B (6)
Molecular weight	35,000	44,000	58,000	210,000
Optimum pH	7.0	7.5	6.5	6.0
Optimum temp (°C)	50	45	40-55	75
pH stability	6.0-9.0	4.0-9.0	5.0-7.5	3.5-5.0
Thermal stability (°C)	< 50	< 50	< 55	80
Isoelectric point (pH)	8.3	8.35, 8.59	9.1	5.9
Inhibitor pCMB	+	+	+	+
Cyclodextrin	+	+	+	+

Table III. Comparison of Substrate Product Relationship
of Different Pullulan Degrading Enzymes

| | Linkages Cleaved in | | Major End Product Formed | |
Enzyme	Pullulan	Starch	Pullulan	Starch
Pullulanase (13)	α-1,6	α-1,6	maltotriose	linear dextrins
Isopullulanase (14)	α-1,4	α-1,4	isopanose	not known
Neopullulanase (15, 16)	α-1,4	α-1,4	panose	maltose
Glucoamylase (17, 18)	α-1,4 and α-1,6	α-1,4 and α-1,6	glucose	glucose
Amylopullulanase (19, 20)	α-1,6	α-1,4	maltotriose	DP2-DP4

high levels in starch limited chemostats but not in glucose or xylose limited chemostats. The enzyme was secreted into the medium when grown under substrate (maltose) limited continuous culture (34).

The amylopullulanase from C. thermohydrosulfuricum 39E was purified from the cell-free extract by treatment with streptomycin sulfate and ammonium sulfate, and by DEAE-Sephacel, octyl-Sepharose and finally by pullulan-Sepharose column chromatography (35). The enzyme was purified 3511-fold and displayed homogeneity on SDS-PAGE. It was a monomeric glycoprotein with a MW of 136,500 and isoelectric point at pH 5.9. The purified amylopullulanase was stable and active at 90°C (Figure 1). The optimum pH for activity and pH stability ranges were 5.0-5.5 and 3.0-5.0, respectively. Table IV compares the initial reaction velocity for hydrolysis of various polysaccharides by purified amylopullulanase from C. thermohydrosulfuricum 39E with that of Bacillus pullulanase in order to assess the relative substrate specificities of these two enzymes. C. thermohydrosulfuricum 39E amylopullulanase displayed a 3-fold higher specificity towards soluble starch and amylopectin and 7-fold higher specificity towards glycogen than did Bacillus pullulanase. The apparent K_m and K_{cat} values for amylopullulanase activity on pullulan at 60°C were 0.675 mg/ml and 16240/min, respectively.

C. thermohydrosulfuricum 39E amylopullulanase cleaves a variety of linear maltooligosaccharides (20). The final reaction products obtained from various starchy polysaccharides were DP2, DP3 and DP4. The enzyme hydrolyzed low MW oligosaccharides three glucose units away from a terminal and released maltotriose as a product and did not form maltose from maltotriose. The minimum substrate requirement for its catalytic activity was a maltotriosyl unit. Kinetic analysis of the purified enzyme in a system which contained both pullulan and amylose as the two competing substrates gave positive proof for both activities belonging to the same enzyme and putatively within the same active site (20).

Coleman et al (29) cloned the amylopullulanase gene from T. brockii into Escherichia coli and B. subtilis. The cloned enzyme could cleave all of the α-1,6 glucosidic linkages (and none of the α-1,4 bonds) in pullulan, it hydrolyzed mostly

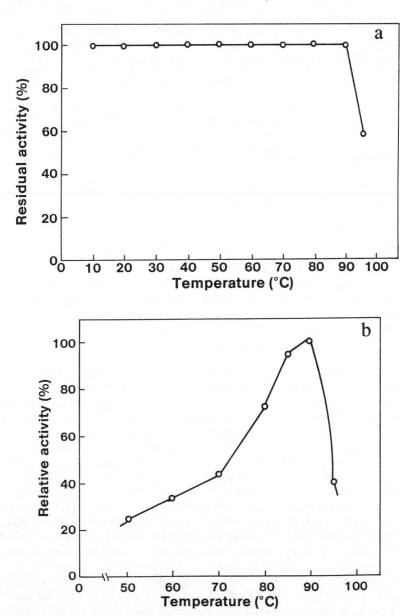

Figure 1. Effect of temperature on stability and activity of *Clostridium thermohydrosulfuricum* strain 39E amylopullulanase. (a) Thermal stability. The enzyme was placed in acetate buffer (50 mM, pH 6.0) with 5 mM $CaCl_2$ and preincubated at various temperatures for 30 min, and then residual amylopullulanase activities were assayed. (b) Effect of heat on activity. The enzyme activity was assayed at various temperatures by the standard assay method (30 min incubation). (Reproduced with permission from Ref. *35*. Copyright 1988, The Biochemical Society and Portland Press, London.)

Table IV. Comparison of Substrate Specificity of Homogeneous Amylopullulanase Purified from *Clostridium thermohydrosulfuricum* Strain 39E With That of a Commercial Pullulanase from *Bacillus* sp.[a]

Substrate	Relative Substrate Specifity (% hydrolysis)	
	C. thermohydrosulfuricum 39E Amylopullulanase	*Bacillus* sp. Pullulanase
Pullulan	100	100
Amylopectin	62	22
Glycogen (oyster)	75	10
ß-limit dextrin from glycogen (oyster)	77	33
Soluble starch	52	16

[a]Sufficient enzyme was added to cause linear release of product during the first 15 min of reaction. The substrate is expressed as the initial hydrolysis velocity as a percentage of that obtained with pullulan substrate. (Reprinted with permission from Ref. *35*. Copyright 1988, The Biochemical Society and Portland Press, London.)

α-1,4 and very few of the α-1,6 linkages in starch. The products of starch hydrolysis are various sized oligomers. Melasniemi (*22*) reported that *C. thermohydrosulfuricum* strain E101-69 produces a pullulanase and α-amylase, and suggested that the single protein had dual specificity and called it α-amylase - pullulanase. Later the enzyme was purified as two forms (*36*). Each form appeared to be dimers of two similar subunits. The two forms have similar amino acid compositions, the same N-terminal amino acid sequence (Glu-Ile-Thr-Ala-Pro-Ala-Ile) and the same isoelectric point of 4.25. Both forms are glycoproteins having rhamnose, glucose, galactose and mannose, and neutral hexose content of 11-12%. Plant et al. (*28*) purified an extracellular pullulanase from culture supernatant of *Thermoanaerobium* Tok6-B1 by ammonium sulfate precipitation, pullulan-affinity precipitation, gel filtration and ion exchange chromatography. It hydrolyzed only α-1,6 glucosidic linkages of pullulan to maltotriose but it hydrolyzed starch, amylopectin and amylose producing maltooligosaccharides DP2-DP4 as products. Maltotetraose was slowly hydrolyzed into maltose. The enzyme had a half-life of 17 min at 85°C and 5 min at 90°C. It had an optimum pH at 5.5 and the activity was promoted by Ca^{+2} ions. The enzyme had a single carboxyl group at the catalytic center of the active site (*37*). EDAC-mediated inhibition of pullulan α-1,6 linkage hydrolysis was relieved by amylose or pullulan. Similarly, both pullulan and amylose protected the activity directed at α-1,4 linkages of amylose from EDAC inhibition. When both amylose and pullulan were simultaneously present, the observed rate of product formation closely fitted a kinetic model in which both substrates were hydrolyzed at the same active site.

We have also purified to homogeneity an amylopullulanase from *Thermoanaerobacter* strain B6A (*30*). The purified enzyme is a glycoprotein having a MW of about 450,000. It is a dimer of similar subunit MW of about 220,000. The K_m for pullulan and boiled soluble starch were 0.43 and 0.37 mg/ml, respectively. The final amylose hydrolyzate contained DP2 (44%), DP3 (33%) and DP4 (17%) and higher saccharides (6%).   The enzyme activity did not show any metal ion dependence and both activities were inhibited by ß- and γ-cyclodextrins but not by α-cyclodextrin. A summary of some biochemical characteristics of amylopullulanase type enzyme from some thermoanaerobes is given in Table V.

The amylopullulanase gene from *C. thermohydrosulfuricum* strains E 101-69 and 39E has been cloned and expressed in *E. coli* (*38, 34*).

α-Glucosidase.   α-Glucosidases (EC 3.2.1.20, α-D-glucosidic glucohydrolase) hydrolyze terminal non-reducing α-1,4-linked glucose residues of various substrates, releasing α-D-glucose. These enzymes hydrolyze di- and oligosaccharides rapidly, relative to large saccharides.   α-Glucosidases possess very diverse substrate specificity. They are produced by a variety of microorganisms (*39*).   Various thermoanaerobic species were reported to produce alpha-glucosidase activity such as *C. thermohydrosulfuricum* (*40, 22*), *Clostridium* sp. (*25*), *Fervidobacterium* sp. (*41*), *Thermoanaerobium* sp. (*41*), *Thermoanaerobacter* sp. (*41*).   In the *Clostridium* isolate, the α-glucosidase was optimally active at 65°C and pH 5.0 (*25*).   The apparent K_m of α-glucosidase for maltose hydrolysis was 25 mM. Our studies with *C. thermohydrosulfuricum* 39E α-glucosidase preparation showed that the organism possessed both α-1,4 and α-1,6 glucosidase activities (*42*).   A summary of some characteristics of the α-glucosidase activity from *C. thermohydrosulfuricum* 39E is shown in Table VI. The enzyme had a half-life of 35 min at 75°C, 110 min at 70°C, and 46 h at 60°C.

Cyclodextrinase.   Cyclodextrinase (EC 3.2.1.54, cyclomaltohydrolase, decycling, CDase) is an enzyme that hydrolyzes cyclic dextrins and also linear dextrins. We have reported that *C. thermohydrosulfuricum* 39E produces cyclodextrinase activity in addition to amylopullulanase and α-glucosidase activities (*43*). We have purified the enzyme partially and characterized it. Table VII summarizes some biochemical characteristics of the CDase. It is suggested that the enzyme cleaves cyclodextrin in a "Multiple attack" pattern. It first opens up the ring of the cyclic dextrins and the linear dextrin molecules are degraded to smaller molecules. This may be the most thermostable and thermoactive CDase reported so far.   The CDase activity is inhibited by pCMB indicating a SH-group is important in the active site of the enzyme. *Thermoanaerobacter* sp. B6A also possesses CDase activity (unpublished work).

Application in Biotechnology

The ß-amylase is used for the production of maltose syrups from starch. It generally makes about 60% maltose from starch, the remainder being ß-limit dextrins. With pullulanase, it makes about 80-85% maltose. It is also used to make high conversion syrups which generally has 35-43% glucose and 30-47% maltose (fermentable sugars, about 85%).   Various maltose-containing syrups are used in the brewing, baking, soft-drinks, canning, and confectionery industries. The utility of *C. thermosulfur-*

ogenes 4B ß-amylase in making various maltose containing syrups has been demonstrated (*44-46*).

Table V. Comparison of Some Biochemical Properties of Amylopullulanase-type Enzyme from Different Sources

Property	Microbial Sources				
	1	*2*	*3*	*4*	*5*
Mol. weight (x 10^4)	136.5	370±85 330±85	450	105	120
Optimum pH	5.0-5.5	5.6	5.0	5.5	
pH stability	3.5-5.0	4.5-5.0			
Optimum temp. (°C)	90	85-90	75		
Thermal stability	90	65	70		
Isoelectric point	5.9	4.25	4.5		
Product					
Pullulan	maltotriose	same	same	same	same
Starch	DP2-DP4		DP2-DP4		DP2-DP4
Activator	-	Ca	-	Ca	
Inhibitor					
Cyclodextrin (ß- and γ-)	+	+			

1, *C. thermohydrosulfuricum* strain 39E (*20, 35*); 2, *C. thermohydrosulfuricum* strain E101 (*22, 36*); 3, *Thermoanaerobacter* strain B6A (*30*); 4, *Thermoanaerobium brockii* (*29*); 5, *Thermoanaerobium* strain Tok6-B1 (*28*).

The amylopullulanase may not be useful as a true pullulanase (debranching enzyme) for use in the production of > 95% glucose syrups from starch because so far our studies with *C. thermohydrosulfuricum* 39E and *Thermoanaerobacter* B6A amylopullulanases did not increase the glucose yield nor decrease the reaction time (unpublished result). The *Bacillus* strain 3183 amylopullulanase also did not increase the glucose yield nor decrease the reaction time (*32*).

The amylopullulanases behave likes α-amylase in their action pattern. As mentioned before, some of these enzymes are highly active and stable at 90-95°C and pH 5.0-6.0, and do not require calcium for activity. These enzymes may be used in starch liquefaction processes in place of α-amylase at low pH and low Ca^{+2}. They may be a superior alternative to the conventional α-amylase that is now used for starch liquefaction if a more highly thermostable and thermoactive amylopullulanase can be discovered by a screening program. These enzymes may also be useful for making various specialty maltodextrin syrups because of their unique product specificity.

**Table VI. Biochemical Characteristics of Thermophilic
α-Glucosidase Partially Purified from
Clostridium thermohydrosulfuricum Strain 39E[a]**

Molecular weight	162,000
Optimum pH	5.0-5.5
pH stability	5.0-6.0
Optimum temperature (°C)	75
Half-life	
at 60°C (h)	46
at 70°C (min)	110
at 75°C (min)	35
K_m (mM)	
Maltose	1.85
Isomaltose	2.95
Panose	1.72
Maltotriose	0.58
pNPG	0.31
Specificity	Hydrolyzes both α-1,4 and α-1,6 linkages
Metal ion requirement	
For stability	None
For activity	None
Inhibitor	Acarbose

[a]Unpublished work.

**Table VII. Biochemical Characteristics of Thermophilic
Cyclodextrinase from *Clostridium thermohydrosulfuricum* strain 39E**

Optimum pH	6.0
pH stabillity	5.5
Optimum temperature (°C)	65
Half-Life	
at 60°C (min)	402
at 65°C (min)	180
at 70°C (min)	75
Action pattern	Multiple attack
Metal ion requirement	
For activity	None
For stability	None
Inhibitor	pCMB

SOURCE: Adapted from ref. 43.

It was demonstrated that amylopullulanase from *C. thermohydrosulfuricum* 39E catalyzes the production of condensation products from maltose and maltotriose (*35*).

It is predicted that this enzyme may be used to produce branched cyclodextrins which are more soluble than cyclodextrins.

α-Glucosidase may be useful for making dextrose syrups. It is often considered to have transglucosidase activity. This activity is useful for the production of specialty oligosaccharides by the transglucosylation reaction. The commercial importance of cyclodextrins are growing in importance daily. These are used in various industries because of the molecules unique ability to bind specific environmental chemicals. The CDase may become useful in separation processes or in making specialty linear dextrins from cyclodextrins if the reaction can be stopped just after opening of the ring.

Conclusions

The amylo-saccharidases from thermoanaerobes because of their novel activity, extreme thermostability and thermoactivity and pH compatibility may find applications in starch conversion biotechnologies if product uses are demonstrated. These enzymes from thermoanaerobes may be cloned, overexpressed and easily purified in GRAS organisms such as *B. subtilis* (47).

Acknowledgment

This material is based upon work supported by the Kellogg Foundation and the U. S. Department of Agriculture (USDA 89-34189-4299 to Michigan Biotechnology Institute).

Literature Cited

1. Zeikus, J. G. *Enzyme Microb. Technol.* **1979**, *1*, 243-252.
2. Hyun, H. H.; Zeikus, J. G. *Appl. Environ. Microbiol.* **1985**, *49*, 1162-1167.
3. Hyun, H. H.; Zeikus, J. G. *J. Bacteriol.* **1985**, *164*, 1162-1170.
4. Nipko, A.; Shen, G-J.; Zeikus J. G. *Appl. Environ. Microbiol.* **1985**, *55*, 689-694.
5. Nipko, A.; Zeikus, J. G.; Gerhardt, P. *Biotechnol. Bioeng.* **1989**, *34*, 1075-1084.
6. Shen, G-J.; Saha, B. C.; Lee, Y-E.; Bhatnagar, L.; Zeikus, J. G. *Biochem. J.* **1988**, *254*, 835-840.
7. Saha, B. C.; Shen, G-J., Zeikus, J. G. *Enzyme Microb. Technol.* **1987**, *9*, 598-601.
8. Saha, B. C.; Lecureux, L. W.; Zeikus, J. G. *Anal. Biochem.* **1988**, *175*, 569-572.
9. Takasaki, Y. *Agric. Biol. Chem.* **1976**, *40*, 1523-1530.
10. Murao, S.; Ohyama, K.; Arai, M. *Agric. Biol. Chem.* **1979**, *43*, 719-726.
11. Higashihara, M.; Okada, S. *Agric. Biol. Chem.* **1974**, *38*, 1023-1029.
12. Kitamoto, N.; Yamagata, H.; Kato, T.; Tsukagoshi, N.; Udaka, S. *J. Bacteriol.* **1988**, *170*, 5848-5854.
13. Bender, H.; Walenfels, K. *Biochem. Z.* **1961**, *334*, 79-95.
14. Sakano, Y.; Matsuda, N.; Kobayashi, T. *Agric. Biol. Chem.* **1971**, *35*, 971-973.
15. Kuriki, T.; Okada, S.; Imanaka, T. *J. Bacteriol.* **1988**, *170*, 1554-1559.
16. Shimizu, M.; Kanno, M.; Tamura, M.; Suekane, M. *Agric. Biol. Chem.* **1978**, *42*, 1681-1688.

17. Marshall, J. J. *Starch/Starke* **1975**, *27*, 377-383.
18. Saha, B. C.; Mitsue, T.; Ueda, S. *Starch/Starke* **1979**, *31*, 307-314.
19. Saha, B. C.; Zeikus, J. G. *Trends Biotechnol.* **1989**, *7*, 234-239.
20. Mathupala, S., Saha, B. C., Zeikus, J. G. *Biochem. Biophys. Res. Commun.* **1990**, *166*, 126-132.
21. Hyun, H. H.; Zeikus, J. G. *Appl. Environ. Microbiol.* **1985**, *49*, 1168-1173.
22. Melasniemi, H. *Biochem. J.* **1987**, *246*, 193-197.
23. Antranikian, G.; Zablowski, P.; Gottschalk, G. *Appl. Microbiol. Biotechnol.* **1987**, *27*, 75-81.
24. Koch, R.; Zablowski, P.; Antranikian, G. *Appl. Microbiol. Biotechnol.* **1987**, *27*, 192-198.
25. Madi, E.; Antranikian, G.; Ohmiya, K.; Gottschalk, G. *Appl. Environ. Microbiol.* **1987**, *53*, 1161-1167.
26. Nakamura, N.; Sashihara, N.; Nagayama, H.; Horikoshi, K. *Starch/Starke* **1989**, *41*, 112-117.
27. Plant, A. R.; Morgan, H. W.; Daniel, R. M. *Enzyme Microb. Technol.* **1986**, *26*, 668-672.
28. Plant, A. R.; Clemens, R. M.; Daniel, R. M.; Morgan, H. W. *Appl. Microbiol. Biotechnol.* **1987**, *26*, 427-433.
29. Coleman, R. D.; Yang, S.-S.; McAlister, M. P. *J. Bacteriol.* **1987**, *169*, 4302-4307.
30. Saha, B. C.; Lamed, R., Lee, Y-Y., Mathupala, S. P.; Zeikus, J. G. *Appl. Environ. Microbiol.* **1990**, *56*, 881-886.
31. Odibo, F. J. C.; Obi, S. K. C. *J. Ind. Microbiol.* **1988**, *3*, 343-350.
32. Saha, B. C.; Shen, G-J.; Srivastava, K. C.; LeCureux, L. W.; Zeikus, J. G. *Enzyme Microb. Technol.* **1989**, *11*, 760-764.
33. Hyun, H. H.; Zeikus, J. G. *J. Bacteriol.* **1985**, *164*, 1146-1152.
34. Mathupala, S. P.; Saha, B. C., Zeikus, J. G. *Abstract of ASM Annual Meeting*, Anaheim, CA, May, 1990.
35. Saha, B. C.; Mathupala, S. P.; Zeikus, J. G. *Biochem. J.* **1988**, *252*, 343-348.
36. Melasniemi, H. *Biochem. J.* **1988**, *250*, 813-818.
37. Plant, A. R.; Clemens, R. M.; Morgan, H. W.; Daniel, R. M. *Biochem. J.* **1987**, *246*, 537-541.
38. Melasniemi, H.; Paloheimo, M. *J. Gen. Microbiol.* **1989**, *135*, 1755-1762.
39. Fogarty, W. M. In *Microbial Enzymes and Biotechnology* (Fogarty, W. M. ed.), Applied Science Publishers, London, **1983**, pp. 1-92.
40. Hyun, H. H.; Shen, G-J.; Zeikus, J. G. *J. Bacteriol.* **1985**, *164*, 1153-1161.
41. Plant, A. R.; Patel, B. K. C.; Morgan, H. W.; Daniel, R. M. *System. Appl. Microbiol.* **1987**, *9*, 158-162.
42. Saha, B. C.; Zeikus, J. G. *Abstract ASM Biotechnology Annual Meeting*, Orlando, FL, June, 1989.
43. Saha, B. C.; Zeikus, J. G. *Appl. Environ. Microbiol.* **1990**, *56*, (in press).
44. Zeikus, J. G.; Saha, B. C. *US patent* 4,814,267 (Mar. 21, **1989**)
45. Saha, B. C.; Zeikus, J. G. *Biotechnol. Bioeng.* **1989**, *34*, 299-303.
46. Saha, B. C.; Zeikus, J. G. *Enzyme Microb. Technol.* **1990**, *12*, 229-231.
47. Lee, C.-Y.; Bhatnagar, L.; Saha, B. C.; Lee, Y.-E.; Takagi, M.; Imanaka, T.; Bagdasarian, M.; Zeikus, J. G. *Appl. Environ. Microbiol.* **1990**, *56*, (in press).

RECEIVED September 9, 1990

Chapter 8

Strategies for the Specific Labeling of Amylodextrins

John F. Robyt

Department of Biochemistry and Biophysics, Iowa State University, Ames, IA 50011

Maltodextrins and isomaltodextrins specifically labeled in (a) the reducing-end glucopyranosyl unit, (b) the nonreducing-end glucopyranosyl unit, (c) a specific number of labeled glucopyranosyl units at one or the other end or both of the chain ends, and (d) all of the glucopyranosyl units uniformly labeled, find uses in the study of the function of the dextrins and especially in the study of the mechanisms of enzymes that interact with the dextrins and related substrates. Branched maltodextrins and cyclodextrins can also be specifically labeled. All of the methods that will be discussed involve the use of specific enzymes that are commercially available or can be readily prepared in the laboratory.

Synthesis of Uniformly labeled Maltodextrins

C-14-Uniformly labeled glycogen can be readily synthesized using uniformly labeled C-14-sucrose and *Neisseria perflava* amylosucrase *(1,2)*. The labeled glycogen can be converted into linear α-1→4 linked maltodextrins by the action of *Pseudomonas amyloderma* isoamylase that specifically hydrolyzes the α-1→6 branch linkages of glycogen to give a mixture of uniformly labeled maltodextrins of degree of polymerization (D.P) 10 and greater. These can be converted into specific maltodextrins of D.P. 2 to 6 by the action of specific amylases. For example, maltose can be obtained from the action of β-amylases; maltotriose from the action of G3-amylase from *Streptomyces griseus* *(3)*; maltotetraose from the action of G4-amylase from *Pseudomonas stutzeri* *(4)*; maltopentaose from the action of G5-amylase from *Pseudomonas sp.* *(5)*; and maltohexaose from the action of G6-amylase from *Aerobacter aerogenes* *(6)*. A mixture of maltose — maltoheptaose can be obtained by the action of *B. amyloliquefaciens* α-amylase in which different amounts of maltodextrins can be obtained, depending on the amount of enzyme used and the length of time of the reaction (7). Uniformly labeled cyclodextrins can be synthesized by the action of *B. macerans* cyclodextrin glucanosyl transferase acting on C-14 labeled amylo-dextrins obtained from the action of *N. perflava* amylosucrase and C-14 sucrose. See Table I for a list of sources of the enzymes.

Recently, we synthesized maltotetraose specifically labeled with C-13 at carbon-1 of each glucopyranosyl residue by converting 1-C-13-enriched D-glucose into 1-C-13-α-glucopyranosyl fluoride, which is a substrate for *N. perflava* amylosucrases to give 1-C-13-glycogen. The 1-C-13-glycogen was reacted with isoamylase to cleave the α-1→6 branch linkages, followed by reaction with *P. stutzeri* G4 amylase to give the 1-C-13-labeled maltotetraose (8) (Figure 1). This can be considered a special type of uniformly labeled malto-dextrin in which each of the glucopyranosyl residues are labeled at C-1 with C-13.

Synthesis of Uniformly Labeled Isomaltodextrins

A series of C-14-uniformly labeled isomaltodextrins can be obtained by the action of *Leuconostoc mesenteroides* B-512F dextransucrase and U-C-14-sucrose and U-C-14-glucose. The glucose acts as an acceptor to which dextransucrase catalyzes the transfer of the glucopyranosyl moiety of sucrose to the C-6-OH of the glucose acceptor (9,13); this product, isomaltose can then act as an acceptor to give isomalto-triose, which in turn is an acceptor to give isomaltotetrose, isomaltopentaose, etc. The number and amount of each isomalto-dextrin dependent on the relative concentration ratio of sucrose and glucose (9). Using an equimolar amount of sucrose and glucose gives isomaltodextrins down to D.P. 7. The amounts of the isomal-todextrins, however, decrease as the size of the dextrins become larger (9).

Synthesis of Reducing-end Labeled Maltodextrins

Reducing-end labeled maltodextrins can be prepared by the acceptor or coupling reaction catalyzed by *B. macerans* cyclodextrin glu-canosyl transferase reaction between nonlabeled cyclomaltohexaose with C-14 D-glucose (10). The first labeled product is malto-heptaose, which is specifically labeled in the reducing-end glucopyranosyl unit. This reducing-end labeled maltoheptaose then undergoes a series of disproportionation reactions in which initially two maltoheptaose molecules react. For example, one of the disproportionation reactions could give maltose and malto-dodecaose, a second could give maltotriose and maltoundecaose, etc. The products of the disproportionation reactions of maltoheptaose themselves can undergo disproportionation reactions (Figure 2). All of these reactions give a homologous series of maltodextrins, specifically and exclusively labeled in the reducing-end gluco-pyranosyl residue (10). Several other acceptors other than D-glucose could also be used, e.g., maltose, sucrose, α-methyl-D-glucopyranoside, isomaltose, and maltooligosaccharides to give homologous series in which the acceptor is located at the reducing-

Table I   Sources of Enzymes Used to Prepare Specifically
Labeled Amylodextrins

Neissera perflava amylosucrase	Prepared from culture, ref. 1
Bacillus amyloliquefaciens α-amylase	Sigma Chemical Co., St. Louis, MO
Streptomyces griseus G3-amylase	Prepared from culture, ref. 3
Pseudomonas stutzeri G4-amylase	Prepared from culture, ref. 4
Pseudomonas sp. G5-amylase	Prepared from culture, ref. 24
Aerobacter aerogenes G6-amylase	Prepared from culture, ref. 6
Pseudomonas amyloderma isoamylase	Sigma Chemical Company
B. macerans cyclodextrin glucanosyl transferase	Amano International Enzyme Company, Inc., Troy, VA
Leuconostoc mesenteroides B-512F dextransucrase	Sigma Chemical Company
Pullulanase	Amano International Enzyme Company, Inc.
Muscle phosphorylase	Sigma Chemical Company
β-amylase	Sigma Chemical Company

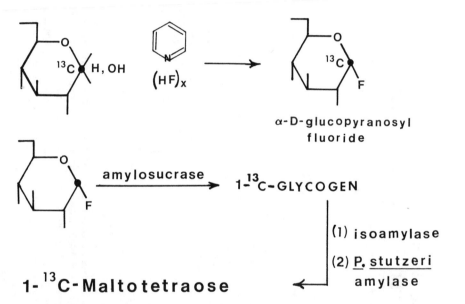

Figure 1. Synthesis of 1-C-13-maltotetraose, using amylo-sucrase, isoamylase, and *Pseudomonas stutzeri* G4-amylase.

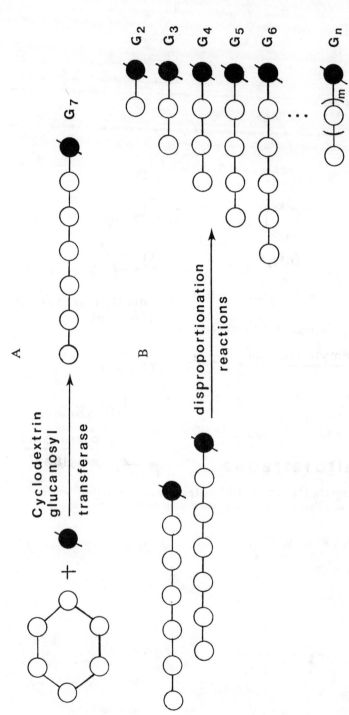

Figure 2. Synthesis of reducing-end labeled maltodextrins by reaction of *Bacillus macerans* cyclodextrin glucanosyl transferase with cyclomaltohexaose and U-C-14-D-glucose. Reaction takes place in two steps: (A) the coupling reaction to give reducing-end labeled maltoheptaose and (B) disproportionation reactions, first between two maltoheptaose molecules and then between the various resulting disproportionation products. The circle represents a glucopyranosyl unit and a circle with a slash represents a reducing glucopyranose unit; black are labeled and white are unlabeled.

end or the potential reducing-end of the resulting maltodextrin chain (*11*) (Figure 3).

Synthesis of Reducing-end Labeled Isomaltodextrins

C-14 Reducing-end labeled isomaltodextrins can be synthesized by using nonlabeled sucrose and C-14 labeled D-glucose or C-14 labeled α-methyl-D-glucopyranoside with *L. mesenteroides* B-512F dextran-sucrase (*9,12,13*). The labeled acceptor will be specifically located at the reducing or potential reducing-end (*13*) (Figure 4). As with the *B. macerans* cyclodextrin glucanosyl transferase, several different acceptors can be used. Over thirty different acceptors have been identified (*14*). Not all of them, however, react with equal efficiency. The best acceptor is maltose, followed by isomaltose, nigerose, α-methyl-D-glucopyranoside, 1,5-anhydro-D-glucitol, and D-glucose with efficiencies relative to maltose of 89, 58, 52, 30, and 17 percent, respectively (*9*). When the ratio of maltose to sucrose is relative high in the dextransucrase acceptor reaction, the branched trisaccharide, panose, is the major product (*12*) (Figure 5). Thus, by using U-C-14 sucrose and nonlabeled maltose, nonreducing-end labeled panose is formed. Using different types of acceptors, either labeled or nonlabeled, different series of isomaltodextrins are produced with the different kinds of acceptors located at the reducing end of the chains.

Synthesis of Labeled Branched Maltodextrins

When other maltodextrins, such as maltotriose, are used as acceptors in the dextransucrase-sucrose reaction, two acceptor products are formed in which the glucopyranosyl moiety of sucrose is transferred to the C-6-OH group of the nonreducing-end glucose residue and to the reducing glucose residue of maltotriose. This gives 6^3-α-D-glucopyranosyl maltotriose and 6^1-α-D-glucopyranosyl maltotriose (*15*) (Figure 5). The former has the structure of the smallest α-amylase limit dextrin (B4) formed in the hydrolysis of amylopectin by most α-amylases (*16,17*). Thus, when the maltotriose is non-labeled and the sucrose is labeled, a nonreducing-end labeled B4 is synthesized. An interesting double-labeled saccharide would result with label in both the reducing-end and the nonreducing-end when both reducing-end labeled maltotriose and labeled sucrose are used (reaction B of Figure 6). Similar kinds of reactions will occur when maltotetraose is the acceptor to give 6^4-α-D-glucopyranosyl maltotetraose and 6^1-α-D-glucopyranosyl maltotetraose (reaction C of Figure 5).

Synthesis of Nonreducing-end Labeled Maltodextrins

Nonreducing-end labeled maltodextrins of D.P. 5 and larger can be synthesized using the reaction of nonlabeled maltotetraose acceptor with labeled α-D-glucopyranosyl-1-phosphate, catalyzed by muscle phosphorylase, which will give nonreducing-end labeled maltopentaose (Figure 7). Maltotetraose is the smallest acceptor that can be used and hence, the smallest possible nonreducing-end labeled malto-dextrin is maltopentaose. Maltoheptaose was used as an acceptor

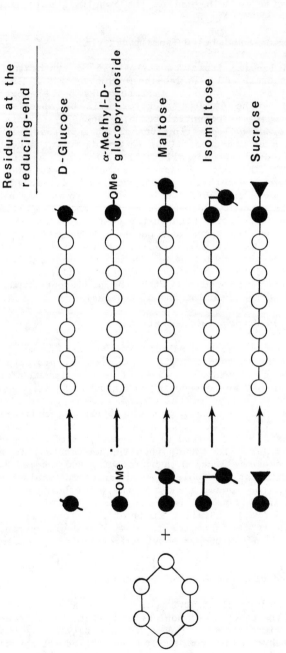

Figure 3. Synthesis of reducing-end labeled maltodextrins by reaction of *B. macerans* cyclodextrin glucanosyl transferase with cyclomaltohexaose and various kinds of acceptors. Symbols are the same as in Figure 2.

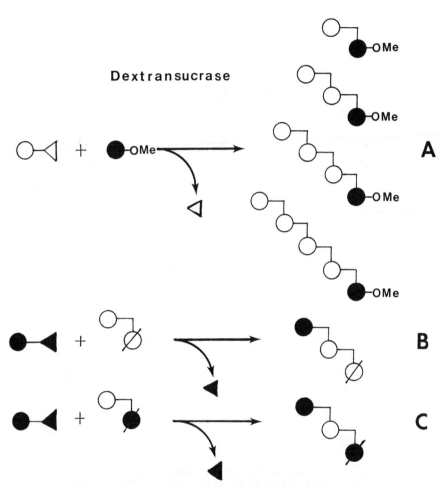

Figure 4. Synthesis of (A) reducing-end labeled isomaltodextrins by reaction of dextransucrase with nonlabeled sucrose
and labeled sucrose and labeled α-methyl-D-glucopyranoside;
synthesis of (B) nonreducing-end isomaltodextrins by reaction
with labeled sucrose and nonlabeled isomaltose; and synthesis
of (C) dual labeled isomaltodextrins by reaction with labeled
sucrose and reducing-end labeled isomaltose.   The circle
represents a glucopyranosyl unit and a circle with a slash
represents a reducing glucopyranose unit; a triangle represents
the fructose unit; black are labeled and white are unlabeled.

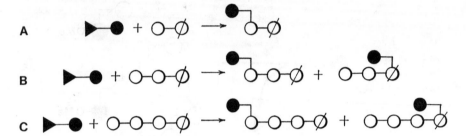

Figure 5. Synthesis of nonreducing-end labeled branched maltodextrins by reaction of dextransucrase with labeled sucrose
and nonlabeled maltodextrins: reaction with (A) maltose,
(B) maltotriose, and (C) maltotetraose. Symbols are the same
as in Figure 4.

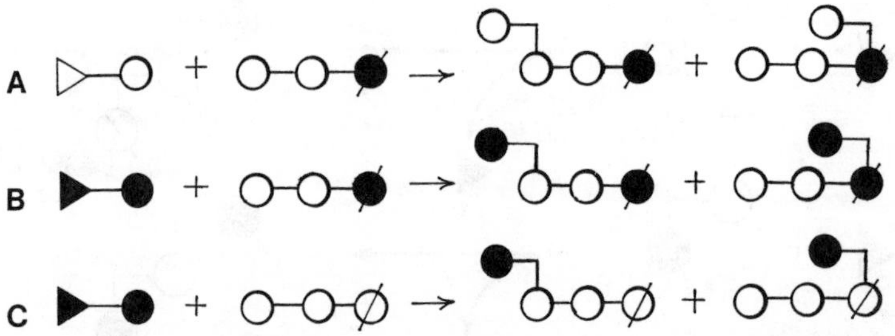

Figure 6. Synthesis of variously labeled branched maltotetra-
saccharides by reaction of dextransucrase with (A) nonlabeled
sucrose and reducing-end labeled maltotriose, (B) labeled
sucrose and reducing-end labeled maltotriose, and (C) labeled
sucrose and nonlabeled maltotriose. Symbols are the same as
in Figure 4.

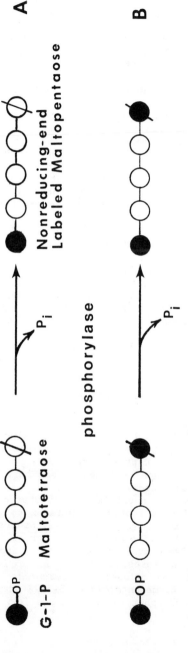

Figure 7. Synthesis of the smallest nonreducing-end labeled maltodextrin, maltopentaose, (A) by reaction of phosphorylase with nonlabeled maltotetraose and labeled α-glucopyranosyl-1-phosphate; and dual labeled maltopentaose (B) by reaction with reducing-end labeled maltotetraose and labeled α-gluco-pyranosyl-1-phosphate. Symbols are the same as in Figure 2.

to give nonreducing-end labeled maltooctaose (18). When equimolar amounts of maltoheptaose and C-14-α-glucopyranosyl-1-phosphate were used, nonreducing-end labeled maltooctaose was the major product with a small amount of labeled maltononaose, which was labeled in the last two glucopyranosyl residues at the nonreducing-end. Dual labeled maltodextrins can be formed with the glucose residues labeled at both the nonreducing-end and the reducing-end by reaction of muscle phosphorylase with reducing-end labeled maltodextrin and labeled α-glucopyranosyl-1-phosphate (Figure 7).

Synthesis of Nonreducing-end Labeled Isomaltodextrins

Nonreducing-end isomaltodextrins can be synthesized by dextran-sucrase, sucrase, starting with nonlabeled isomaltodextrin acceptor and labeled sucrose. For example, starting with nonlabeled iso-maltose and labeled sucrose, nonreducing-end labeled isomaltotriose would be formed, and starting with nonlabeled isomaltotriose and labeled sucrose, nonreducing-end isomaltotetraose would result. Dual labeled isomaltodextrins can also be synthesized, starting with reducing-end labeled isomaltodextrin acceptors and labeled sucrose (Figure 4).

Synthesis of Labeled Maltosyl Branched-cyclomaltodextrins

Labeled maltosyl branched-cyclomaltodextrins can be synthesized by the action of isoamylase (19-20) or pullulanase (22,23) with various types of labeled maltose and cyclodextrin. Three types of labeled maltosyl cyclodextrins can be obtained: (a) reducing-end labeled maltose attached to cyclodextrin; (b) uniformly labeled maltose attached to cyclodextrin; and (c) nonlabeled maltose attached to uniformly labeled cyclodextrin (Figure 8).

Summary

We, thus, have shown how a wide variety of amylodextrins can be specifically labeled in different ways by using twelve different kinds of enzymes with different kinds of labeled glucosyl donors and different kinds of labeled acceptors or by using combinations of different enzymes in sequence or by using combinations of labeled glucosyl donors and labeled acceptors together to give dual labeled products. The type of isotope that mostly has been discussed is C-14, but other types could equally be used, such as C-13 or H-3. As a summary, let us list the specific enzymes that can be used in conjunction with isotopically labeled substrates to give specifi-cally labeled amylodextrins: N. perflava amylosucrase; specific exo-acting amylases, such as, β-amylase, S. griseus G3-amylase, P. stutzeri G4-amylase, Pseudomonas sp. G5-amylase, and A. aerogenes G6-amylase; endo-acting amylases, such as, B. amyloliquefaciens amylase; P. amyloderma isoamylase; B. macerans cyclodextrin glucanosyl-transferase; L. mesenteroides B-512F dextransucrase; pullulanase; and muscle phosphorylase. Sources for these enzymes are given in Table 1.

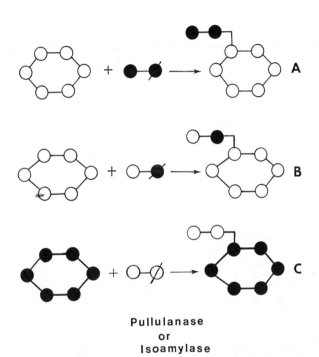

**Pullulanase
or
Isoamylase**

Figure 8. Synthesis of labeled branched cyclomaltohexaose by reaction of pullulanase or isoamylase with cyclomaltohexaose and (A) uniformly labeled maltose, (B) reducing-end labeled maltose, and (C) nonlabeled maltose and uniformly labeled cyclomaltohexaose. Symbols are the same as in Figure 2.

Literature Cited

1. Okada, G.; Hehre, E. J., *J. Biol. Chem.* **1974**,*249*,126-35.
2. Tao, B. Y.; Reilly, P. J.; Robyt, J. F., *Carbohydr. Res.* **1988**, *181*,163-74.
3. Wako, K.; Takahashi, E.; Hashimoto, S.; Kanaeda, J., *Denpun Kagaku* **1978**,*25*(1978),155-60.
4. Robyt, J. F.; Ackerman, R. J., *Arch. Biochem. Biophys.* **1971**,*145*, 105-12.
5. Kobayashi, S.; Okemoto, H.; Hara, K.; Hashimoto H., *Agric. Biol. Chem.* **1990**,*54*,147-56.
6. Kainuma, K.; Wako, K.; Kobayashi, S.; Nogami, A.; Suzuki, S. *Biochim. Biophys. Acta* **1975**,*410*,333-41.
7. Robyt, J. F.; French, D., *Arch. Biochem. Biophys.* **1963**,*100*,451-67.
8. Tao, B. Y.; Reilly, P. J.; Robyt, J. F., *Biochim. Biophys. Acta* **1989**,*995*,214-20.
9. Robyt, J. F.; Eklund, S. H., *Carbohydr. Res.* **1983**,*121*,279-86.
10. Pazur, J. H., *J. Amer. Chem. Soc.* **1955**,*77*,1015-8.
11. Norberg, E.; French, D., *J. Amer. Chem. Soc.* **1950**,*72*,1202-4.
12. Jones, R. W.; Jeanes, A.; Stringer, C. S.; Tsuchiya, H. M., *J. Amer. Chem. Soc.* **1956**,*78*,2499-502.
13. Robyt, J. F.; Walseth, T. F., *Carbohydr. Res.* **1978**,*61*,433-45.
14. Robyt, J. F.; Eklund, S. H., *Bioorganic Chem.* **1982**,*11*,115-32.
15. Fu, D.; Robyt, J. F., *Arch. Biochem. Biophys.* in press **1990**.
16. Kainuma, K.; French, D., *FEBS Letters* **1969**,*5*,257-60.
17. Kainuma, K.; French, D., *FEBS Letters* **1970**,*6*,182-5.
18. Robyt, J. F.; French, D., *J. Biol. Chem.* **1970**,*245*,3917-27.
19. Abe, J.; Mizowaki, N.; Hizukuri, S.; Koizumi, K.; Utamura, T., *Carbohydr. Res.* **1986**,*154*,81-6.
20. Kitahata, S.; Yoshimura, Y.; Okada, S., *Carbohydr. Res.* **1987**,*159*, 303-8.
21. Hizukuri, S.; Abe, J.; Koizumi, K.; Okada, Y.; Kubota, Y.; Sakai, S.; Mandai, T., *Carbohydr. Res.* **1989**,*185*,191-8.
22. Abdullah, M.; French, D., *Arch. Biochem. Biophys.* **1970**,*137*,483-5.
23. Shiraishi, T.; Kusano, S.; Tsumuraya, Y.; Sakano, Y. *Agric. Biol. Chem.* **1989**,*53*,2181-8.
24. Okemoto, H.; Kobayashi, S.; Momma, M.; Hashimoto, H.; Hara, K.; Kainuma, K.; *Appl. Microbiol. Biotechnol.* **1986**,*25*,137-42.
25. Fu, D.; Robyt, J.F., *Prep. Biochem.* **1990**,*20*,93-106.

RECEIVED October 3, 1990

Chapter 9

Maltohexaose-Producing Amylase of *Bacillus circulans* F–2

Hajime Taniguchi

National Food Research Institute, Kannondai 2–1–2, Tsukuba,
Ibaraki 305, Japan

Bacillus circulans F-2 produces an alpha-
amylase which has a potent starch granule-
digesting activity and produces maltohexaose
from soluble starch. Action pattern of this
amylase was examined in detail. When non-
reducing end-labelled amylopectin was subjected to enzymatic hydrolysis, radioactivity was released rapidly. During enzymatic
hydrolysis of amylopectin, intermediate
size products were not detected on gel filtration. Relationship between decrease in
iodine staining and increase in reducing
value during the enzymatic hydrolysis of
amylopectin was plotted. The amylase
gave a steadily declining line which is
very close to that obtained with beta-
amylase while alpha-amylase gave a rapidly
declining one. The amylase produced labelled glucose, maltose and maltotriose from
reducing end-labelled maltoheptaose, -octa-
ose and -nonaose, respectively. The anomeric
configuration of produced maltohexaose and
maltotetraose was found to be alpha. This
amylase was classified as a member of unique
alpha-amylases which act in an exo-manner
but produce saccharides with alpha configuration.

Bacillus circulans F-2, isolated from potato tubers, is
a bacterium that can grow on potato starch granules as
the sole carbon source(1, 2). The amylase of this bacterium has been studied extensively and it was found to
have the following 3 unique properties. (a) The purified
enzyme has a potent potato starch-digesting activity(3).
(b) Enzyme production is induced only when the bacterium

is grown on starch granules, potato starch being most
effective. (c) The purified enzyme produced maltohexaose
exclusively from soluble starch in an early stage of
hydrolysis.   The second point has been studied in de-
tail(4-6); it was found that this bacterium has a catabo-
lite repression system which is extremely sensitive to
glucose.
        In this paper, the third property of this enzyme,
its action pattern on soluble substrates such as amylopectin and maltooligosaccharides, was studied in detail.

Methods

Cultivation of the bacterium, purification of amylase,
enzyme assay method, basic analytical methods such as
determination of protein and sugars, polyacrylamide gel
electrophoreses with or without sodium dodecyl sulfate
(SDS) and thin layer chromatographic method were described in previous papers(1, 3, 7).
        High performance liquid chromatography(HPLC) was
carried out using Shimadzu Liquid Chromatograph LC-4A
equipped with Senshu-Pak 5 NH$_2$ column(4.6 mm x 25 cm) and
Ermer RI Detector ERC-7510.   Sixty % acetonitrile in
water was used as solvent at a flow rate of 1 mL/min at
room temperature.   Sugars as low as 0.1 µg/injection
could be detected quantitatively with this system.
        Gel filtration of the reaction products was carried
out on a Sephadex G-75 column (1.8 x 90 cm) using 0.1 M
NaCl as a solvent.   Two mL fractions were collected and
the amount of sugars in each fraction was determined by
the phenol-sulfuric acid method(8).
        Forty mg of amylopectin were incubated with 3 U of
the enzyme in 8 mL of 50 mM phosphate buffer(pH 6.5) at
30 °C.   At time intervals, 2 mL portions were withdrawn
and subjected to gel filtration on a Sephdex G-75 column.
Fifty uL portions were withdrawn concomitantly for
determination of reducing sugars by the method of Somogyi
(9) and Nelson(10).
        Reaction mixtures containing 20 mg of amylopectin
and 0.4 U of amylase in 20 mM phosphate buffer(pH 6.5) in
a total volume of 2 mL were incubated at 30 °C with samples withdrawn at time intervals.   A 40 uL sample was
added to 3 mL of 0.005 % I$_2$- 0.05 % KI solution and the
absorbance at 660 nm was obtained.   Reducing value was
determined using a 50 µL sample.
        ^{14}C-Amylopectin was prepared by incubating amylopec-
tin with ^{14}C-glucose-1-phosphate(277 mC$_i$/mmole) in the
presence of rabbit muscle phosphorylase b in 10 mM α-
glycerophosphate buffer (pH 6.5) and 10 mM AMP.   Labelled amylopectin(3 mg, 20,000 cpm) was incubated with
the enzyme(0.02 U) in 600 µL of 0.1 M phosphate buffer(pH
6.5) at 30 °C.   At intervals, portions were withdrawn and
spotted on Toyo Roshi No 50 paper.   The paper was irrigated with 65 % aqueous n-propanol at 40 °C by an ascend-
ing method and then autoradiogrammed using Fuji X-ray

film Kx(Fuji Film Co.).   Radioactive spots were cut out
and their activity was counted by a Packard Tri-Carb 3315
scintillation spectrometer using toluene cocktail.
Portions of the reaction mixture were withdrawn at the
same intervals for the determination of reducing value.
    Reducing end labelled maltooligosaccharides were
synthesized by iucubating 50 μC_i (0.5 mg) of ^{14}C-glucose
with 2 mg of α-cyclodextrin in the presence of 1.76 U of
<u>Bacillus</u> <u>macerans</u> cyclodextrin glucanotransferase in 600
μL of 0.1 M acetate buffer(pH 5.6) containing 1 mM $CaCl_2$
for 5 hrs at 50 °C.   Twenty μL of this reaction mixture
was spotted on paper, irrigated in the first direction,
and dried.   The chromatogram was sprayed with purified
enzyme solution (0.7 U/mL), incubated at 40 °C for 30 min
and then irrigated again in the second direction.   An
autoradiogram was obtained as described above.
    The anomeric form of reaction products was deter-
mined by a JASCO DIP-140 Digital Polarimeter equipped
with a microcell(3.5 x 100 mm, 2 mL).   Enzymes(4,5 U)
were incubated with 2 % (w/w) of short chain amylose
(Hayashibara, D.P.=17) or maltohexaose in 50 mM phosphate
buffer(pH 6.5) at room temperature and the $[\alpha]_D$ was
followed at 1 min intervals.   At the points indicated by
an arrow(Fig. 11), 2 mg of sodium carbonate were added to
facilitate mutarotation.

<u>Results and   Discussion</u>

<u>Puriification and general properties of Bacillus circu-
lans F-2 amylase. (3)</u>   The bacterium produced several
amylase components on the medium containing potato starch
granules.   When the culture supernatant was subjected
to polyacrylamide gel electrophoresis followed by activity staining, four groups of activity bands were detected
as shown in Figure 1.   Group I amylase was purified
recently and found to be very unique(7).   It hydrolyzes
both α-1,4-glucosidic linkage in soluble starch and
α-1,6- glucosidic linkage in pullulan, approximately at
the same rate.   Amylases of groups II - IV showed a
different action pattern, maltohexaose(G_6) being the only
product in an early stage of hydrolysis.   As the hydrolysis reaction proceeds, maltotetraose(G_4) and maltose
(G_2) were produced in addition to G_6.   However, no other
oligosaccharides were produced during this period. The
amylase component of group II and its properties are
detailed in the following section.
    Group II amylase was purified by fractionation with
ammonium sulfate, adsorption and desorption from starch
granules, gel filtration on Bio-Gel P-100 and ion exchange chromatography on Whatman DE-32.   Results of
purification are summarized in Table I.   Most of the
heterogenous proteins were removed during adsorption and
desorption from starch granules.   The final preparation
was homogeneous on SDS polyacrylamide gel electrophoresis
as shown in Figure 2.   It has a specific activity of

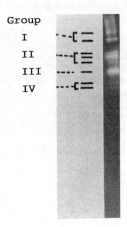

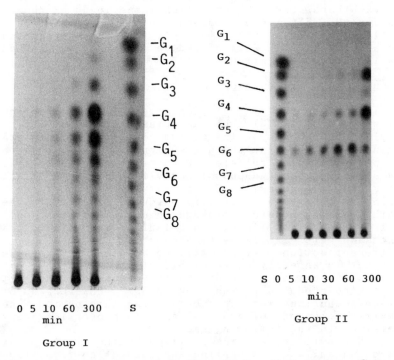

Figure 1. Action pattern of *Bacillus circulans* F–2 amylase. Culture supernatant was applied to PAGE and stained for amylase activity with starch–iodine method. Action patterns of groups I and II toward soluble starch were studied by thin-layer chromatography. (Reproduced with permission from ref. 7. Copyright 1989 Elsevier.)

Table I.   Purification of <u>Bacillus circulans</u>
Group II Amylase(2)

Purification Step	Total Protein (mg)	Total Activity (U)	Specific Activity (U/mg)	Yield (%)
Culture Sup	9,200	21,800	2.37	100
30% Ammo.Sulf.	3,390	22,400	6.60	102.8
Starch Eluate	220	8,290	37.6	38.0
Bio-Gel P-100	174	7,500	43.0	34.5
Whatman DE-32	101	5,560	54.8	25.5

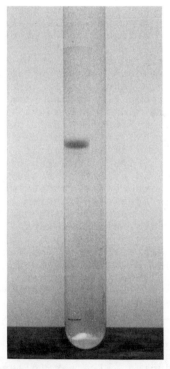

Figure 2. SDS–polyacrylamide gel electrophoresis of the purified amylase. (Reproduced with permission from ref. 3. Copyright 1983 Japan Society for Bioscience, Biotechnology, and Agrochemistry.)

54.8 U/mg protein and was obtained with 25.5 % yield
based on culture supernatant.
        General basic properties of the purified enzyme
are summarized in Table II.   Its molecular weight calculated from SDS polyacrylamide gel electrophoresis is
93,000 dalton, considerably higher than those of other
amylases.   The enzyme migrated as two close bands on the
electrofocusing gel. The pI values of these bands were
calculated to be 4.88 and 4.93.   The maximum activity
of the enzyme is expressed at pH 6.0 - 6.5 and at 60°C.
The enzyme is quite stable at alkaline ranges up to pH 12
while it is unstable at < pH 5.   The enzyme requires
Ca^{2+} for its full activity.   It was found to have a
remarkably higher digestive activity toward potato starch
granules compared to those of porcine pancreatic alpha-
amylase and Streptococcus bovis alpha-amylase as shown in
Figure 3.

Action on soluble polysaccharides.   The purified enzyme
was incubated with amylopectin and the amount of various
oligosaccharides formed was determined by HPLC.   Results
are shown in Figure 4.   It is clear that only G_6 was
produced initially, with G_4 and G_2 produced slowly and
concomitantly on further incubation.   However, no other
oligosaccharide was produced until the extent of hydrolysis (expressed as glucose equivalent of reducing sugars)
exceeded 20 %.   This result is in accordance with the
action pattern of group II amylase shown in Figure 1.
        Amylopectin was hydrolyzed to various extent from 0
(top) to 30.5 %(bottom) and the reaction mixtures were
analyzed by gel filtration on Sephadex G-75 to determine
if hydrolysis products of intermediate size are produced.
Results are shown in Figure 5. Sugar content in the
eluate was determioned by phenol-sulfuric acid method and
ordinate indicates absorbance at 490 nm.   Hydrolysis %
is expressed as just described above .   Figures in the
parenthses indicate % amount of sugars recovered at the
bed volume to the total eluted sugars. The peak appeared
at the void volume contained only amylopectin and partially hydrolyzed amylopectin.   The later peaks at the
bed volume contained only G_6, G_4 and G_2.   No other peak
was detected until hydrolysis reached 19.6 %,   when 61.6
% of added amylopectin was recovered as a mixture of the
three oligosaccharides at the bed volume.   This result
strongly suggests that the enzyme attacks amylopectin
molecules in an exo-manner.
        The enzyme was incubated with amylopectin and the
relationship between a decrease in blue iodine value and
an increase in reducing value was followed as shown in
Figure 6.   Barley malt beta-amylase, a typical exo-
amylase, gave a line which declined slowly whereas Bacillus liquefaciens amylase, a typical endo-amylase, gave a
line which declined quite rapidly.   The purified amylase
gave a line close to that of beta-amylase, again suggesting an exo-type hydrolytic action.

Table II.   General Properties of The Puriified Amylase(3)

Molecular Weight	93,000 (SDS-PAGE)
Ioelectric Point	pH 4.88 and 4.93
Optimum pH	6.0-6.5
Stable pH	6.5-12.0
Optimum Temp.	60 C
Stable Temp.	100% at 35 C, 0 % at 55 C
Inhibitor	Hg^{2+}, Ag^+, Cu^{2+}, Pb^{2+}
Activator	Ca^{2+}

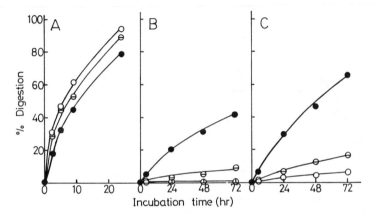

Figure 3. Digestion of starch granules by the purified amylase. ●, the purified amylase; ○, porcine pancreatic amylase; and ◒, *Streptococcus bovis* amylase. A, corn starch; B and C, potato starch. The amount of added amylases was increased six-fold in C. (Reproduced with permission from ref. 3. Copyright 1983 Japan Society for Bioscience, Biotechnology, and Agrochemistry.)

Time Course of Hydrolysis of Amylopectin

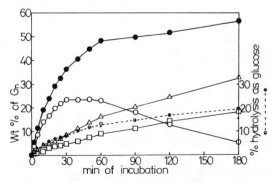

Figure 4. Time course of hydrolysis of amylopectin by the purified amylase. ○, G_6; △, G_4; □, G_2; ●, $G_6 + G_4 + G_2$.

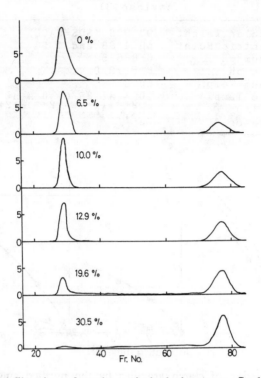

Figure 5. Gel filtration of amylopectin hydrolyzates on Sephadex G–75.

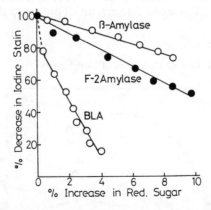

Figure 6. Relationship between reducing value and iodine stain during hydrolysis of amylopectin by various amylases. BLA is *Bacillus* liquefying amylase and F–2Amylase is the purified amylase.

Amylopectin with a ^{14}C labelled nonreducing end glucose moiety was used as a substrate of the enzyme to determine exo preference. As shown in Figure 7, radioactivity in the amylopectin disappeared quite rapidly, before less than 10 % of amylopectin molecule was hydrolyzed as apparent glucose. The labelled hydrolysis products were mainly G_6 and G_4 as shown in Figure 8. These results indicate that the enzyme attacks amylopectin preferentially from its nonreducing end.

Action on oligosaccharides. Action of the purified enzyme on a series of reducing end-labelled maltooligosaccharides was examined by two-dimensional paper chromatography. The autoradiogram shown in Figure 9A indicates formation of labelled glucose from malto-pentaose and -heptaose; labelled maltose from malto-hexaose and -octaose; labelled maltotriose from maltononaose; and so on. These results are schematically illustrated in Figure 9B. It is clear that the enzyme cleaves the sixth glucosidic linkage in oligsacccharides with D.P. higher than 6, and fourth glucosidic linkage in maltopentaose and maltohexaose.

Kinetic parameters of this enzyme toward maltooligosaccharides were obtained using HPLC. K_m, V_{max} and related perameters are summarized in Table III. It is apparent that the enzyme has negligible activity toward malto-triose and -tetraose as compared with the next higher oligosaccharides. Furthermore, in terms of both K_m and V_{max}, the enzyme has a slightly lower activity toward maltotetraose compared to that for malto-octaose and -nonaose. This difference is more striking when K_o/K_m values for these oligosaccharides are compared. This parameter reflects the actual enzymatic activity toward the substrate at lower concentrations. This result, together with the data shown in Figures 1, 4 and 8, is in good accordance with the interpretation that amylopectin is initially hydrolyzed to maltohexaose. When the amount of this saccharide reached to a certain concentration, it is then split into maltotetraose and maltose.

The action pattern of this enzyme can be summarized as in Figure 10. The mode of action on amylopectin is an attack from its nonreducing end releasing maltohexaose successively, which, in turn, is hydrolyzed into maltotetraose and maltose on further incubation.

Anomeric form of the hydrolyzed products. The action pattern of the purified amylase was found to be an exo-type as shown above. Exo-amylases such as glucoamylase and beta-amylase are known to produce saccharides with beta configuration. However, results indicate that both maltohexaose produced from amylose and maltotetraose produced from maltohexaose are in alpha configuration as shown in Figure 11. Therefore, this enzyme must be considered to belong a unique group(11-15) of alpha-

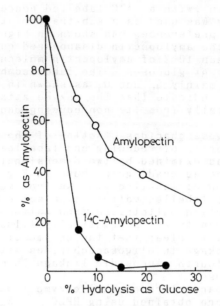

Figure 7. Hydrolysis of nonreducing end-labeled amylopectin with the purified amylase.

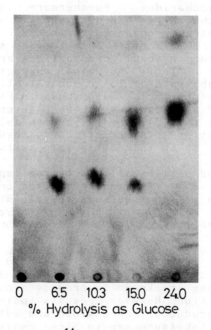

Figure 8. Autoradiogram of ^{14}C-amylopectin hydrolyzed by the purified amylase.

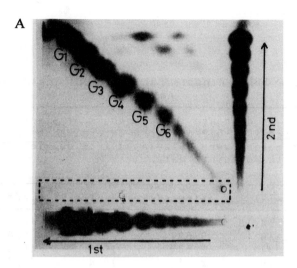

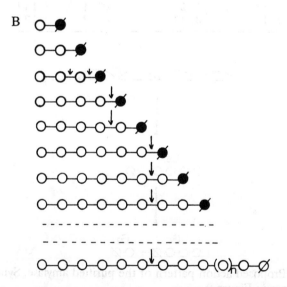

Figure 9. Autoradiogram of reducing end-labeled maltooligosaccharides hydrolyzed by the purified amylase. A, autoradiogram (Reproduced with permission from ref. 2. Copyright 1982 Japanese Society of Starch Science.). B, action pattern. ○, glucose residue; ●, labeled glucose residue; ∅, reducing end glucose residue; and ↓, cleavage point.

Table III.   Kinetic Parameters of The Purified
Amylase

Rates of decrease in the amount of starting
saccharides(0.1-5mg/mL) were determined by HPLC and
the parameters were calculated from double reciprocal
plot of the obtained data.

D.P.	Km(M)	V(mole/min)	$K_o(V/e_o)$	k_o/Km
3	7.14×10^{-3}	0.0037×10^{-6}	0.023×10^2	0.0032×10^5
4	1.16	0.027	0.17	0.015
5	0.34	2.53	15.5	46.5
6	0.19	2.94	18.4	96.8
7	0.22	3.59	22.4	97.4
8	0.18	4.22	26.4	147
9	0.16	3.41	21.3	133

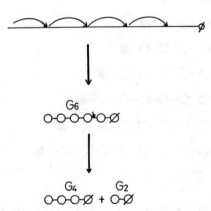

Figure 10. Proposed action pattern of the purified amylase. Symbols are the
same as those in Figure 9.

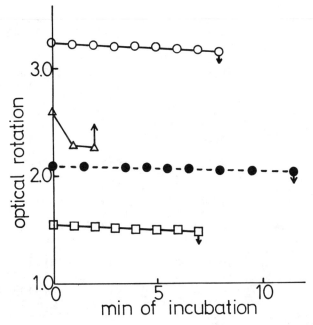

Figure 11. Mutarotation of the reaction products formed by the action of various amylases. ○, the purified amylase on 2% amylose; ●, the purified amylase on 2% maltohexaose; Δ, sweet potato beta-amylase on 2% amylose; and □, *Bacillus subtilis* alpha-amylase on 2% amylose.

amylases which attack amylaceous polysaccharides such as amylopectin in an exo-manner.

Acknowledgments

I am grateful to Dr. Shoichi Kobayashi of National Food Research Institute for generous gift of the purified cyclodextrin glucanotransferase from Bacillus macerans, and to Nihon Shokuhinn Kako Co. Ltd., for generous gifts of maltooligosaccharides.

Literature Cited

1. Taniguchi, H.; Odashima, F.; Igarashi, M.; Maruyama, Y.; Nakamura, M. Agric. Biol. Chem. 1982, 46, 2107-15.
2. Taniguchi, H.; Chung, M.; Maruyama, Y.; Nakamura, M. J. Jap. Soc. Starch Sci. 1982, 29, 107-16.
3. Taniguchi, H.; Chang, M.; Yoshigi, N.; Maruyama, Y. Agric. Biol. Chem. 1983, 47, 511-9.
4. Sata, H.; Taniguchi, H.; Maruyama, Y. Agric. Biol. Chem. 1986, 50, 2803-9.
5. Sata, H.; Taniguchi, H.; Maruyama, Y. Agric. Biol. Chem. 1987, 51, 1521-7.
6. Sata, H.; Taniguchi, H.; Maruyama, Y. Agric. Biol. Chem. 1987, 51, 3275-80.
7. Sata, H.; Umeda, M.; Kim, C.; Taniguchi, H.; Maruyama, Y. Biochim. Biophys. Acta 1989, 991, 388-94.
8. Dubois, M.; Gilles, K.A.; Hamilton, J.K.; Rebers, P.A.; Smith, F. Anal. Chem. 1956, 28, 350-6.
9. Somogyi, M. J. Biol. Chem. 1952, 195, 19-23.
10. Nelson, N. J. Biol. Chem. 1944, 153, 375-80.
11. Hidaka, H.; Adachi, T.; Yoshida, K.; Niwa, T. J. Jap. Soc. Starch Sci. 1978, 25, 148-54.
12. Wako, K.; Takahashi, C.; Hashimoto, S.; Kaneeda, J. Jap. Soc. Starch Sci. 1978, 25, 154-61.
13. Robyt, J.F.; Ackerman, R.J. Arch. Biochem. Biophys. 1971, 145, 105-14.
14. Kobayashi, S.; Okemoto, H.; Hara, K.; Hashimoto, H.; Agric. Biol. Chem., 1990, 54 147-56.
15. Kainuma, K.; Wako, K.; Kobayashi, S.; Nogami, A.; Suzuki, S. Biochim. Biophys. Acta. 1975, 410, 333-46.

RECEIVED November 9, 1990

Chapter 10

Properties of CGTases from Three Types of *Bacillus* and Production of Cyclodextrins by the Enzymes

Michikatsu Sato and Yoshiaki Yagi

Central Research Laboratories, Sanraku Inc., 9–1, Johnan 4-Chome, Fujisawa 251, Japan

Comparative studies of CGTases from Bacillus ohbensis, B. macerans and B. circulans and an enhanced production method of gamma-cyclodextrin (CD) are reported. Each CGTase was purified to a homogeneity. The CGTase from B. ohbensis was successfully crystallized. The enzyme from B. ohbensis had the smallest molecular weight (35,000) and had the lowest affinity to starch among the enzymes examined. The CGTase had similar enzymatic properties as the other enzymes, but it exhibited high temperature and broad pH stability in the presence of substrates. CGTase from B. macerans had advantages for the production of alpha-CD but the reaction conditions had to be strictly controlled. CGTase from B. ohbensis can be used favorably for the production of beta- and gamma-CD. We elucidated an efficient gamma-CD production system by addition of compounds with high affinity to the CD. Of the compounds tested, glycyrrhizic acid was the most effective for production of gamma-CD. But the addition had little effect on gamma-CD production using CGTase from B. macerans and B. circulans.

Cyclodextrin (CD) forming enzyme, cyclomaltodextrin glucanotransferase (CGTase, alpha-1,4-glucan 4-glycosyltransferase, EC.3.2.1.19) is an enzyme which produces CDs from alpha-1,4-glucan, such as maltodextrin, amylose, starches, etc. There have been many reports on CGTase producing bacterial strains (1~8). The enzymes are classified into two groups according to their major product; one is a group which mainly produces alpha-CD, and the other is one which mainly produces beta-CD. Primary examples of the alpha-CD producing group and beta-CD group are the CGTase from B. macerans (1) and the enzyme from B. circulans (4) respectively. We isolated a strain producing CGTase from a soil sample in Ohba district in Fujisawa city, Japan (5) and found that the enzyme formed a large

0097–6156/91/0458–0125$06.00/0

amount of beta-CD and did not produce alpha-CD under conventional reaction conditions for the enzyme. Therefore, we examined the properties of the enzyme and compared them with those of other CGTases of Bacillus origin. CD producing strains, B. macerans IAM 1227 and B. circulans IFO 3329 from our laboratory's culture collection, were studied and compared to the CGTases from B. ohbensis, B. macerans and B. circulans (9). Recently, we devised a new gamma-CD production system by adding a well fitted guest compound for gamma-CD in the reaction mixtures (10). We report here comparative studies of the CGTases and a new method for enhanced production of gamma-CD by the enzyme from B. ohbensis.

Purification and Properties of 3 Kinds of CGTases

CGTase from B. macerans is known as a CD-producing enzyme (1) and French et al. (11) reported detailed research on the enzyme in 1957. The CGTase predominantly produces alpha-CD and there was no report on other types of CGTases until 1972. Okada et al.(3) first reported the CGTase of B. megaterium, a beta-CD type CGTase. Since then many CGTases of Bacillus origin have been reported, and we also were able to isolate a beta-CD type CGTase producing bacterium (5). Since the enzyme did not form alpha-CD under normal enzyme reaction conditions, we examined the properties of the enzyme and compared them with the other CGTases from different Bacillus strains.

Purification of CGTases from B. ohbensis   The CGTase from B. ohbensis was purified to homogeneity (12) by acetone precipitation, starch adsorption (13) with ammonium sulfate, DEAE-cellulose column chromatography and crystallization (14). The specific activity was elevated 15-fold from the acetone precipitation and the hexagonal crystal with a protruding center was formed (14).

Purification of CGTase from B. macerans   B. macerans IAM 1227 was cultured and the supernatant of the cultured broth was adsorbed onto corn starch and active fractions were recovered by eluting with warm water (50°C). The starch adsorption was effective for decoloring and increasing the specific activity of the CGTase. The active fractions were combined, subjected to twice repeated DEAE-cellulose column chromatography and Sephadex G-100 column chromatography. The activity was increased 150-fold from the culture broth and the enzyme was purified as an almost single protein (9).

Purification of CGTase from B. circulans   B. circulans IFO 3329 was cultured and the CGTase was purified by starch adsorption. The eluate from the starch was concentrated by ultrafiltration. The enzyme was further purified by ammonium sulfate precipitation, DEAE-column chromatography, Sephadex G-100 column chromatography and gel isoelectric focusing. In the isoelectric focusing electrophoresis, 2 active bands appeared. The major band's isoelectric point (Ip) was 8.8 and that of the other band was 8.5. The enzyme of the major band was used for further experiments. The Ip 8.8 enzyme was purified 2,240-fold from the crude enzyme preparation (9).

Comparison of Properties of the CGTases

Enzymatic properties of the respective CGTases such as optimum pH, pH stability, optimum temperature, thermal stability, behavior on starch adsorption, molecular weight by SDS-PAGE and isoelectric point were compared. The result of the starch adsorption is shown in Table I (9).

Table I.   Adsorption of CGTase on Starch

$(NH_4)_2SO_4$ conc. (% saturation)	B. ohbensis adsorption(%)		B. macerans adsorption(%)		B. circulans adsorption(%)	
	+	−	+	−	+	−
0	3.4	92	61	36	53	23
10	64	18	60	30	52	20
20	74	8.6	63	24	53	17

+: percent adsorbed; −: percent not adsorbed.

The enzyme from B. ohbensis had less affinity for starch than the other enzymes. The CGTase was only adsorbed by adding ammonium sulfate to 20 % saturation. The enzyme from B. macerans had a high affinity for starch, and it did not need any addition of ammonium sulfate. The CGTase from B. circulans showed moderate affinity and there was no dependence on ammonium sulfate.

The properties of three enzymes are summarized in Table II (9).

Table II.   Properties of CGTases

	B. ohbensis	B. macerans IAM 1227	B. circulans IFO 3329
Main product	beta-CD	alpha-CD	beta>alpha-CD
Optimum pH	5.5	5.5	6.0
pH stability	6.5-9.5	6.5-8.5	6.0-9.5
Optimum temp.(°C)	60	50	55
Thermal stability(°C)*	55	50	55
Molecular weight	35,000	150,000	(200,000)
Isoelectric point	<4	4.2	8.5, 8.8
Affinity to starch	−	+++	++

*The temperature which showed the residual activity (>80 %) after exposure for 60 min at each temperature at pH 6.0.

Enzymatic activities of the enzymes were similar, but the molecular weight, isoelectric point and affinity to starch were different. Of greatest note is the molecular weight; the size of the enzyme from B. ohbensis is the smallest that has ever been reported from

genus Bacillus.   The molecular weight was measured by SDS-PAGE
and confirmed by ultra-centrifugation.   We also carried out gel-
permeation column chromatography to elucidate further information
on the molecular weight, but the value showed about 70,000~80,000.
The reason is not known, but it may have been by aggregation of
the enzyme molecules.
      Amino acid compositions of the CGTases from 3 kinds of Bacillus
were examined (9).   None of the three enzymes contained cysteine,
and the amounts of neutral and basic amino acids were similar.   A
relatively large amount of acidic amino acids was contained in the
CGTase from B. ohbensis, which may be related to the acidity of
the enzyme.

Cyclodextrin Formation by the CGTases

We performed comparative studies of the 3 kinds of enzymes on CD
production.   The substrate specificities, influence of incubation
time, pH, temperature and quantity of the CGTase produced under
various conditions were examined.

Substrate Specificity   The substrate specificity of the CGTases
was examined.   Incubation conditions were as follows: substrate
concentration, 1 % (w/v); incubation temperature, 50°C; incubation
time, 20 hrs.   The results are shown in Fig. 1.   The CGTase from
B. ohbensis produced the largest amount of beta-CD of all the
enzymes.   The enzyme from B. macerans was a typical alpha-CD type
CGTase; alpha-CD was the dominant product in any of its substrates,
and the amount of alpha-CD was the largest in the CGTases examined.
CGTase from B. circulans was a beta-CD type enzyme but it also
formed a relatively large amount of alpha-CD.   The properties of
CGTase from the strain IFO 3329 was different from that reported
by Okada et al. (4).   The CGTase from B. ohbensis did not form any
amount of alpha-CD in the reaction conditions and the CGTase was
found to be unique of all the beta-CD type enzymes which have been
reported.   The enzyme from B. ohbensis produced the largest amount
of gamma-CD of the enzymes, and the amount (8~10 % to substrate)
may be the largest among the CGTases reported (15).

Time Course of CDs Formation by the CGTases   We examined the
time course of CD formation by the enzymes.   The incubation conditions were as follows: substrate, 5 % (w/v) of soluble starch;
incubation pH, 6.0; incubation temperature, 50°C.   The results are
shown in Fig. 2.   The CGTase from B. ohbensis produced beta-CD
from the beginning of enzyme reaction and gamma-CD was formed later,
but alpha-CD was not produced throughout the incubation period.
The CGTase from B. macerans began to produce alpha-CD at first,
but the formation ceased after 10 hours of incubation.   Beta-CD
formation was a little delayed but it continued even after 24
hours of incubation.   Therefore, the ratio of alpha-CD/beta-CD
decreased in the later phase of enzyme reaction.   In the case of
B. circulans enzyme, beta-CD was dominantly produced and alpha-CD
was also produced from the early stage of the enzyme reaction.

Effect of pH on CD Formation by the CGTases   The enzyme reaction
conditions were the same as the previous section.   The reaction pH

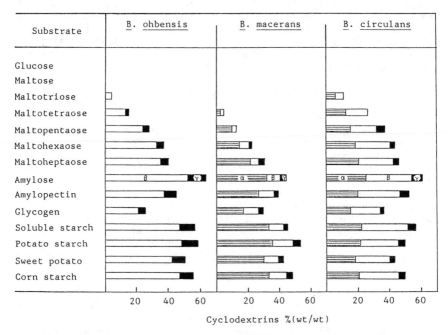

Figure 1.   Substrate specificity of CGTases on CD formation.
Incubation conditions were described in the text.   (Reproduced
with permission from Ref. 9.   Copyright 1986 M. Nakamura.)

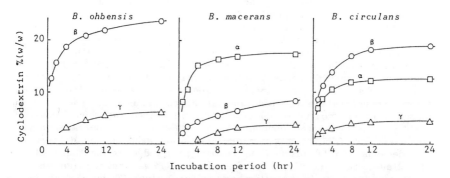

Figure 2.   Time course of CD formation.   Incubation conditions
were described in the text.   (Reproduced with permission from
Ref. 9.   Copyright 1986 M. Nakamura.)

was adjusted with 100 mM of various buffers: pH 4 to 8 with MacIlvaine buffer, pH 8 to 11 with borate-Na_2CO_3 buffer and pH 11 to 12
with Na_2HPO_4-NaOH buffer. The results are shown in Fig. 3. The
enzyme from B. ohbensis produced CDs in a wide range of pH (pH 5
to 10). As for the CGTase from B. macerans, CD production rate
was changed by the incubation pH. In the acidic pH range (pH 5~6),
the ratio of alpha-CD was high while in the neutral to alkaline
range, it became low. The CGTase from B. circulans had a narrow
pH activity range and was active only from pH 5 to 7.

Effect of Incubation Temperature on CDs Formation   The incubation
pH was adjusted to pH 6.0 with 100 mM MacIlvaine buffer and other
incubation conditions were the same as the previous experiment. The
results are shown in Fig. 4.   The enzyme reaction proceeded best
in the range of 40°C to 70°C for the CGTase from B. ohbensis.   In
the enzyme from B. circulans, the reaction proceeded best from
40°C to 60°C.   But the CGTase from B. macerans had a very narrow
temperature activity range around 50°C.

Effect of Substrate Concentration on CD Formation   Soluble starch
was used as the substrate.   The enzyme reaction was carried out at
pH 6.0 and 50°C and for 24 hrs.   The results are shown in Fig. 5.
Generally, the yield decreased with increased substrate concentration.   The decrease was significant in the enzyme from B. macerans
and it had to be incubated with relatively low concentration of
starch.

Effect of CGTase Concentration on CD Formation   The incubation
conditions were as follows: substrate concentration, 5 %; incubation
pH, 6.0; incubation time, 24 hrs.   One unit of enzyme was defined
as the amount of enzyme which produces 1 mg of CD/hr at 50°C, pH
6.0 with 5 % of soluble starch as substrate.   Enzyme units were
determined by HPLC using the method of Sato et al. (16).   The
results are shown in Fig. 6.   As the amount of enzyme increased,
the amount of CD produced increased and then decreased.   With B.
ohbensis, the optimum amount of enzyme for production of beta- and
gamma-CD was approximately the same, but production of alpha-CD
did not start until this optimal amount of enzyme was used.   As
the amount of enzyme increased, the production of alpha-CD was
increasing as the amount of beta- and gamma-CD was decreasing.   An
increase and then decrease in CD production with B. circulans enzyme
was also found as enzyme concentration was increased.   The optimal
enzyme concentration for all CDs was approximately the same.   With
B. macerans enzyme, CD production again increased and then decreased
with increasing enzyme concentration.   However, the optimal enzyme
concentration for the maximum production of each CD was different.
The decrease in CD production seemed to be due to another action
of CGTase, such as disproportionation.   If we employ high enzyme
concentration, it is possible to produce alpha-CD by CGTase from
B. ohbensis.   In the B. macerans enzyme, beta-CD ratio increased
with the increased concentration of the enzyme, so, the proper
amount of enzyme must be selected to produce alpha-CD.   As for the
CGTase from B. circulans, the CD production ratio was not changed
with the various enzyme concentrations.
     The enzyme from B. macerans is the most advantageous for

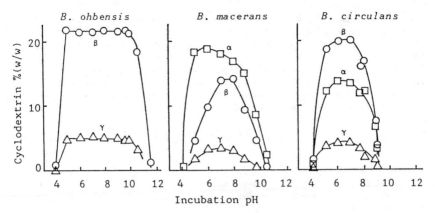

Figure 3. Effect of pH on CD formation. Incubation conditions were described in the text. (Reproduced with permission from Ref. 9. Copyright 1986 M. Nakamura.)

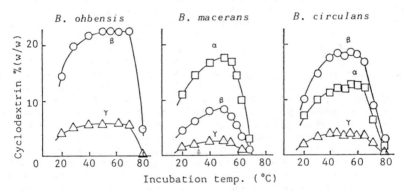

Figure 4. Effect of temperature on CD formation. Incubation conditions were described in the text. (Reproduced with permission from Ref. 9. Copyright 1986 M. Nakamura.)

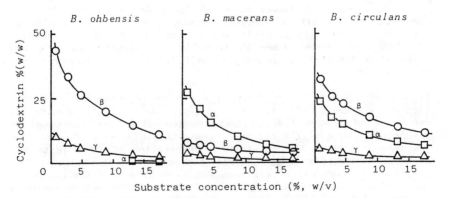

Figure 5. Effect of substrate concentration on CD formation. Incubation conditions were described in the text.

producing alpha-CD, but the incubation conditions must be relatively strictly controlled.   CGTase from B. ohbensis is now being used for the industrial production of beta-CD and a wide range of temperature and pH can be used for the production of beta-CD.   For the production of gamma-CD, the enzyme from B. ohbensis is also favorable and is being used because the enzyme does not produce alpha-CD under normal enzyme reaction conditions, allowing easy elimination of beta-CD from the reaction broth.

Enhanced Formation of Gamma-CD by the Addition of High Affinity Guest Compound

It is very costly to produce gamma-CD because of the small amount produced by the enzymes. Here, we report a new system for producing gamma-CD by employing a guest compound which has a high affinity for gamma-CD.

Screening of Guest Compounds with According to Their Affinity to Gamma-CD   In studies to mask the bitterness of stevioside, a natural sweetener, gamma-CD was found to be effective to eliminate this bitterness and to complex the stevioside very well (17). Fig. 7 shows a phase solubility diagram of CD-stevioside system by the method of Higuchi et al. (18).   Only gamma-CD could form an inclusion complex very well with the guest compound.   The stability constant was 11,200/M.   From the experimental results, we examined the effect of the addition of stevioside on gamma-CD formation. The incubation conditions were as follows: additional amount of stevioside, 0.2~3.0 % (w/v); substrate, 5 % (w/v) of soluble starch; incubation pH, 6.5 with 100 mM MacIlvaine buffer; incubation temperature, 50°C; incubation time, 20 hrs.   As was had expected, gamma-CD increased proportionally with the increased concentration of stevioside from 0.2 to 1.0 % (Fig. 8).   As there had been no report on enhanced gamma-CD production by addition of a selective guest compound, we screened the compounds by its affinity to gamma-CD.   We especially focused on the guest compounds which were natural products because they may be safer than chemically synthesized compounds.   The screening system was as follows: concentration of guest compound to be added, 1 % (w/v); enzyme used, CGTase from B. ohbensis; substrate, 5 % (w/v) of soluble starch; incubation pH, 7.5; incubation temperature, 50°C; incubation time, 20 hrs. Some examples of effective guest compounds are listed in Fig. 9. Glycyrrhizic and glycyrrhetic acids were found to be superior compounds to enhance gamma-CD formation.   Glycyrrhizic acid was selected as the guest compound for further studies to enhance gamma-CD formation.

Effects of Glycyrrhizic Acid on the Formation of Gamma-CD   Incubation pH, substrate concentration and glycyrrhizic acid concentration were studied.   When the substrate concentration was increased in the presence of glycyrrhizic acid, gamma-CD yield per g of substrate was constant up to the substrate concentration of 20 % (data not shown).   Incubation pH affected the gamma-CD formation.   The addition of glycyrrhizic acid was not very effective for enhancing the gamma-CD production below pH 6.0, because the guest compound is not water soluble in acidic conditions.   The optimum pH for

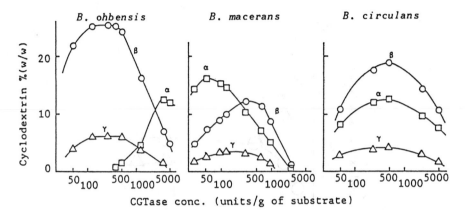

Figure 6. Effect of CGTase concentration on CD formation. Incubation conditions were described in the text. (Reproduced with permission from Ref. 9. Copyright 1986 M. Nakamura.)

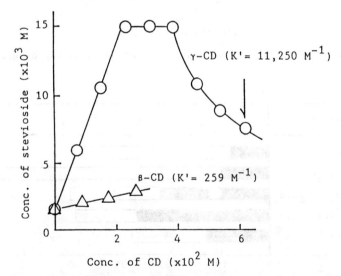

Figure 7. Phase solubility diagram of stevioside-CD system in water at 25°C.

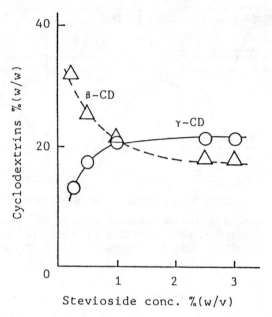

Figure 8. Effect of stevioside addition on formation of gamma-CD.

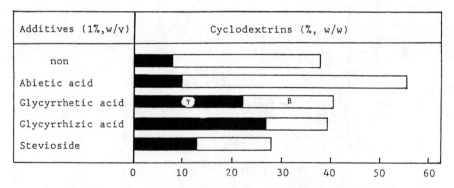

Figure 9. Enhancement of gamma-CD formation with various additives.

enhanced production of gamma-CD with glycyrrhizic acid was around pH 7.5 (data not shown). As we increased the guest compound concentration, gamma-CD increased almost linearly to the conversion yield of 40 % per substrate incubated (Fig.10). The incubation conditions were as follows: substrate, 5 % of soluble starch; incubation pH, 7.5; incubation temperature, 50°C; incubation period, 20 hrs. Fig. 11 shows the HPLC pattern of the reaction broth with 2.5 % of glycyrrhizic acid. The HPLC conditions were as follows: column, Shodex Ionpak S-614 (Showa Denko Co., Japan); detector, Shodex RI SE-11 (Showa Denko Co., Ltd.); eluent, CH_3CN/H_2O (65/35); flow rate, 1.5 ml/min; column temperature, 60°C; injection volume, 10 µl. Gamma-CD was almost the only product in the incubation broth.

Comparative Studies of CGTases with Guest Compounds for Enhanced Production of Gamma-CD   We compared the effect of glycyrrhizic acid, glycyrrhetic acid and stevioside on gamma-CD formation using CGTases from B. ohbensis, B. macerans and B. circulans.   The reaction conditions were as follows: substrate, 5 % (w/v) of soluble starch; incubation pH, 6.5; incubation temperature, 50°C; incubation period, 18 hrs. The results are shown in Fig. 12. Interestingly, only the CGTase from B. ohbensis formed a large amount of gamma-CD. The reason is not known, but we guess that it may be related to the relatively high gamma-CD production ability of the B. ohbensis enzyme in the CGTases tested.

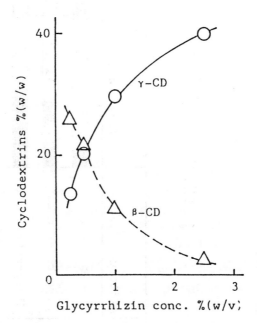

Figure 10. Effect of glycyrrhizic acid addition on gamma-CD formation.

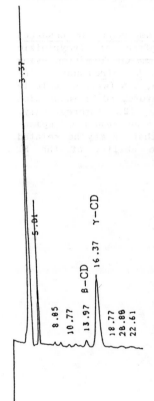

Figure 11. HPLC profile of broth incubated with 2.5 % (w/v) of glycyrrhizic acid.

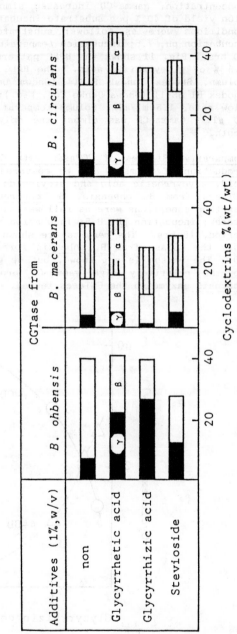

Figure 12. Formation of CDs with additives by CGTases from Bacillus.

Literature Cited

1. Schardinger, F. Wiener Klinishe Wochenshrift 1904, 17, 207.
2. Tilden, E. B.; Hudson, C. S. J. Bacteriol. 1942, 43, 527.
3. Okada, S.; Tsuyama, N.; Kitahata, S. Proc. Symp. on Amylase (in Japanese), 1972, 7, 61.
4. Okada, S.; Kitahata, S. Proc. Symp. on Amylase (in Japanese), 1973, 8, 21.
5. Iguchi, H.; Ichikawa, F.; Yagi, Y.; Takamatsu, A. Abstract Papers, Annual Mtg. of Agric. Chem. Soc. Japan, 1973, p 46.
6. Horikoshi, K.; Akiba, T. Alkalophilic Microorganisms, A New Microbial World, Japan Scientific Soc. Press, Tokyo, 1982.
7. Shiosaka, M.; Bun-ya, H. Proc. Symp. on Amylase (in Japanese), 1973, 8, 43.
8. Bender, H. Arch. Microbiol. 1977, 111, 271.
9. Yagi, Y.; Sato, M.; Ishikura, T. J. Jpn. Soc. Starch Sci. 1986, 33, 144-151.
10. Sato, M.; Nagano, H.; Yagi, Y.; Ishikura, T. Jpn. Patent Kokai, 60-227693 (1985).
11. French, D. Adv. Carbohydr. Chem. 1957, 12, 189.
12. Sato, M.; Nagano, H.; Yagi, Y.; Ishikura, T. Abstract Papers, Annual Mtg. of Agric. Chem. Soc. Japan, 1984, p 503.
13. Kobayashi, S; Kainuma, K.; Suzuki, S. Carbohydr. Res. 1978, 61, 229.
14. Nagano, H.; Sato, M.; Yagi, Y.; Ishikura, T. Abstract Papers, Annual Mtg. of Agric. Chem. Soc. Japan, 1985, p 703.
15. Kato, T.; Horikoshi, K. J. Jpn. Soc. Starch Sci. 1986, 33, 137-143.
16. Sato, M.; Nagano, H.; Yagi, Y.; Ishikura, T. Agric. Biol. Chem. 1985, 49, 1189-1191.
17. Yagi, Y.; Sato, M. Jpn. Patent Kokai, 60-98957 (1985).
18. Higuchi, T.; Connors, K.A. Adv. Anal. Chem. Instr. 1965, 4, 117.

RECEIVED October 19, 1990

ANALYSIS AND CHARACTERIZATION

Chapter 11

Analysis of Amylodextrins

V. M. B. Cabalda[1], J. F. Kennedy[1], and K. Jumel[2]

[1]Research Laboratory for the Chemistry of Bioactive Carbohydrates and
Proteins, School of Chemistry, University of Birmingham,
Birmingham B15 2TT, England
[2]Chembiotech Ltd., Institute of Research and Development, University
of Birmingham Research Park, Vincent Drive, Edgbaston,
Birmingham B15 2SQ, England

Available chromatographic, spectroscopic and
chemical methods for the characterization
and quantitative determination of amylodextrin products are reviewed, discussed and
evaluated. The analyses of amylodextrins are
of vital importance to researchers involved,
for example, in the characterization
of novel starch degrading enzymes and
starches of novel rheological properties
and the rationalization of structure/
function relationships of amylases.

Amylodextrins are several hydrolytic products of starch,
amylose or amylopectin via chemical and/or enzymatic
processes. They consist of glucose units of varying chain
lengths linked by (1--4)- and/or (1--6)-alpha-glycosidic
linkages, with maltose and isomaltose being the shortest.
In addition, amylodextrins could be linear, branched or
cyclic and are available as mixtures (e.g. corn syrup,
glucose syrup) and, in some cases (especially for low
molecular weight oligosaccharides) in purified forms (e.g.
maltose). Some hydrolysates of other polysaccharides, e.g.
glycogen and pullulan, are similar to some starch
amylodextrins.
  One of the purposes of biotechnology is the
exploitation of abundant agricultural crops by
enzymatically modifying them to meet the ever-increasing
human daily needs as well as many luxuries. Amylodextrins
are used in varying ways by the food industry as fat
substitutes, sweeteners, viscosity modifiers, gelling
agents, flavour encapsulating agents, etc.; the
fermentation industries use them as feedstocks; the mining
industries use them for drilling operations; and the
pharmaceutical industries use them as synthons for
synthesizing medicinal drugs. Hence, continuing interest
in the optimization of their production processes and

0097–6156/91/0458–0140$08.75/0

their further applications lead to: a) active investigations on the rationalization of structure/function
relationship in amylases, b) active search for novel
sources of starch degrading enzymes with new activities
and specificities and c) search for new sources of starch
of novel rheological properties. The latter is also to
enable, in particular, developing countries to grow their
own "acclimatized" starch producing crops to assist them
in their quest for self-reliance through biotechnological
means.

Interest in the genetic engineering of both plants
and micro-organisms for the production of tailor made
amylose, amylopectin and/or starches has also been
reported. Furthermore, investigations on the enzymatic
modification of starch and its major components, for
example the introduction of additional branches composed
of glucose and/or other monosaccharides and/or uronic
acids as well as amino acids or peptides, to produce
carbohydrates of possibly comparable functionality to
galactomannans, pectin, gum arabic, etc., has been
initiated. Also studies on the metabolic fate of
carbohydrates in food, the biosynthesis of starch, the
fine structure of starch from different sources, the
effect of electrolytes on the gelatinization of starch and
the development of enzymic methods for starch analyses are
still active.

Analysis of amylodextrins is a necessity for
scientists engaged in the above areas of research. This
paper presents, discusses and reviews available
chromatographic, chemical and spectroscopic methods for
the analyses of amylodextrins. There exists a number of
reviews (1-3) of carbohydrate analyses to which the reader
is referred for further reading and more detailed
description.

Analyses of amylodextrins could be subdivided into
two types, namely, structural analyses and rheological
measurements. The former is more of interest to the basic
researcher whilst both are of use to the applications
researcher. Rheological measurements will not be discussed
in this review and the reader is referred to works by
Collyer and Clegg (4), Mitchell (5) and Robinson (6) for
information on that particular subject.

STRUCTURAL ANALYSIS

Analysis of amylodextrins is of vital importance to the
researcher investigating activity mechanisms of
carbohydrate degrading enzymes and to the researcher
undertaking structural studies of unmodified and modified
polysaccharides and the carbohydrate moiety of
proteoglycans and glycoproteins. Structural
characterization of amylodextrins produced after starch
hydrolysis by newly discovered/created carbohydrolases
helps in studies regarding their activity patterns and
specificity; structures of novel polysacharides, etc.

could be elucidated after hydrolyzing these compounds with enzymes of known activity and specificity with subsequent analysis of the oligosaccharide distribution and identification of the resulting amylodextrin products (if any). Some of the contributions in this book exemplify such investigations.

Moreover, since the application of a particular starch hydrolysate product depends on its molecular weight (distribution) and/or oligosaccharide composition, analyses of these materials are a necessity to the applications chemist whilst the health and nutrition conscious public demands quantitative determination of carbohydrates in food.

Structural characterization involves:

    a) determination of molecular weight distribution or degree of polymerization especially for dextrins of low degree   of hydrolysis,

    b) determination of oligosacharide composition,

    c) glucosidic linkage and sequence determination,

    d) determination of anomeric configuration, and

    e) determination of reducing property,

for which chromatographic methods, spectroscopic methods and chemical methods have played important roles in their investigations.

CHROMATOGRAPHIC METHODS

Hydrolysis of starch or its major components often produces amylodextrins of a wide range of molecular weight. Fortunately, development in recent years allows, in some cases, the efficient separation of individual carbohydrates with their simultaneous characterization in terms of molecular weight (and, therefore, chain length or degree of polymerization), quantitation, the mode of linkage (i.e. linear, branching or cyclic) and anomeric configuration.

Amylodextrins could be separated in varying efficiencies using any of the different chromatographic techniques namely gas liquid chromatography, liquid chromatography, planar chromatography and supercritical fluid chromatography. The reader is referred to the book of Snyder and Kirkland (7) for the basic principles of chromatography, and to an excellent review by Churms (8) on recent developments in the chromatographic analyses of carbohydrates.

Gas Liquid Chromatography (GLC)

GLC is applicable only for the analysis of volatile compounds. Amylodextrins are non-volatile due to the presence of polar hydroxyl groups and need, therefore, to

be converted to their volatile derivatives such as ethers
(silyl, methyl, etc.) and esters (acetyl,
trifluoroacetyl,etc.).

Structural elucidation using GLC. As a preliminary step
to the structural analysis (separation, isolation and
purification) GLC is not a preferred method. The need for
a derivatization step is time-consuming, quantitative
recovery is not guaranteed (9,10) and it is only possible
to analyse oligosaccharides up to DP6-7 (3,11,12), see
Figure 1A. In addition, all detector components of GLC
are destructive. On the other hand, determination of the
monosaccharide composition of a polysaccharide after
complete hydrolysis is possible with this method (2).
    However, as a subsequent step to a nondestructive
separation and purification procedure (e.g. HPLC) of the
amylodextrin products after selective and partial
hydrolysis of starch, GLC coupled with mass spectroscopy
(MS) is a very powerful tool for structural elucidation of
the original carbohydrate molecule. For example (2),
individual amylodextrin components are first methylated
(13) or ethylated to form ethers at the free hydroxyl
groups. The derivatized carbohydrate is subsequently
hydrolyzed to the corresponding monosaccharides and free
hydroxyl groups form where there were glycosidic linkages
(and ring linkages) previously. Reduction by sodium
borohydride (or borodeuteride for deuterium tagging of C_1
or C_2 to differentiate aldoses from ketoses) is then
carried out followed by acetylation or silylation of the
reduced monosaccharides and finally analyses of the
methylated alditol acetates by GLC-MS to ascertain
original linkages (2,14). Figure 1B shows a GLC
separation of partially methylated alditol acetates.
    Periodate oxidation coupled with GLC is also used
for structural characterization (2,15-17), monosaccharide
sequence and glycosidic linkage determination (see below).
The presence of branching (e.g. in starch) poses some
problems (18) however, these have been circumvented
(17). In some cases, partial cleavage of gycosidic
linkages is carried out by methanolysis or acetolysis to
give products derivatized at the anomeric carbon previously involved in the glycosidic linkage and the originally
free anomeric carbon.

Routine Quantitative Analysis. Due to the relative
difficulty in alkylating sugars, methylation or ethylation
are reserved for structural elucidation. For routine
analysis, such as the quantitative determination of low
molecular weight sugars in food stuffs, the hydroxyl
groups are silylated (19-21), acetylated (22) or
trifluoroacetylated (12,23). In some cases, the need for
simple chromatograms to obtain non-equivocal qualitative
identification and quantitative measurement requires prior
specific derivatization of the aldehyde group. Simplified
chromatograms could be achieved by reduction of the

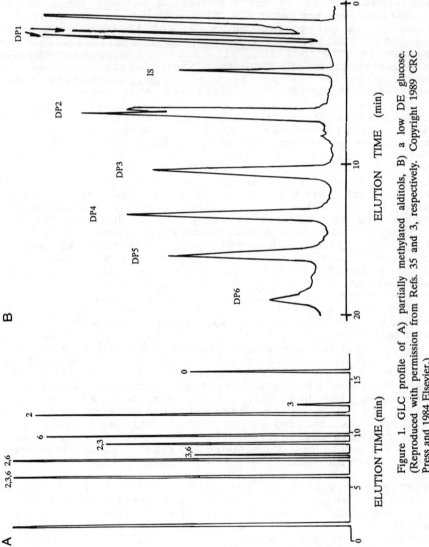

Figure 1. GLC profile of A) partially methylated alditols, B) a low DE glucose. (Reproduced with permission from Refs. 35 and 3, respectively. Copyright 1989 CRC Press and 1984 Elsevier.)

aldehyde with sodium borohydride to the corresponding alditol (24), conversion of the aldehyde into an oxime or a substituted oxime such as O-methyl-oxime (25,26), conversion of the aldehyde into an oxime with subsequent dehydration of the oxime into a nitrile (27,28), or conversion of the aldehyde to diethyl dithioacetals (29). These procedures result in the non-production of multiple peaks (alpha- and beta-anomers, pyranoside and furanoside rings) by each sugar due to the elimination of stereochemistry at the anomeric carbon and the opening of sugar rings. On the other hand, quantitation of the anomeric forms of sugar solutions could be determined using strictly controlled conditions (30).

Determination of the Degree of Polymerization. Another use of GLC with regard to amylodextrin analysis is the determination of the degree of polymerization of high molecular weight amylodextrins based on the derivatization of the anomeric carbon not linked through glycosidic bond prior to acid hydrolysis. A simple method is to reduce the end group to give the corresponding alditol (31). The relative amount as well as the structure of the original end groups, and therefore the degree of polymerization, can be calculated either by prior separation, e.g. by selective absorption on ion-exchange resin (31), of the alditol and aldoses, or simultaneous analysis of different derivatives of alditols and aldoses (28). Periodate oxidation could also be employed for determination of the degree of polymerization (32).

Separation of D- and L-Enantiomers. In addition, the absolute configuration of component monosaccharides (i.e. whether D- or L-enantiomers) could be determined by using chiral stationary phases (33) and/or by reacting the monosaccharide with a reagent which will introduce chiral centres (34).

Procedures for etherification and esterification of carbohydrates for GLC analysis, advantages and disadvantages of the different methods of hydroxyl and aldehyde group derivatization, columns used for the separation of the various derivatives, detection methods for GLC, mass spectroscopy and fast atom bombardment (FAB) as well as outlines of some strategies for structural analysis of carbohydrates are described, discussed and reviewed in an excellent book on the analyses of carbohydrates by GLC (35).

Recent developments in GLC include modification of existing methods to produce volatile compounds using less time consuming, less laborious techniques (36) whilst permitting: the use of lesser amounts of carbohydrate (37), use and development of capillary columns to give more efficient and faster resolution of sugars (36,38), compilation of retention time data of specific carbohydrate derivatives using columns of varying polarity as an aid to the identification of unknown carbohydrates

(39,40), development of chiral derivatization techniques (34) and chiral columns (33), and use of fast atom bombardment (FAB)-mass spectroscopy (41). Churms (8) gives a more extensive review on recent advances in GLC.

Liquid Chromatography

Liquid chromatography was, is and always will be an essential technique for the purification, separation and isolation of amylodextrins. Liquid chromatography is herein classified into low pressure column chromatography and high pressure liquid chromatography (HPLC). The basic distinction between the two is the size of the column packing with 25 um as the typical division.

Low Pressure Column Chromatography

Low pressure column chromatography is characterized by the use of compressible column packings (>25 um) which require the use of low pumping pressures, e.g. hydrostatic pressure or peristaltic pumps, and extended analysis times in the region of 2-18 hours. Detection is frequently performed using automated assay systems, e.g. an L-cysteine sulphuric acid system, or by manual assay following collection of eluant fractions (1).

Gel Permeation Chromatography.
Gel permeation chromatography (GPC) or size exclusion chromatography (SEC) is defined as the separation of compounds according to their molecular size, or more correctly according to their hydrodynamic volume, which in most cases is directly related to the molecular weight. Even with the advancement of HPLC techniques, low pressure GPC is still frequently used by most research workers. The slower flow rates of low pressure GPC provide better resolutions so that distinct peaks corresponding to various components of amylodextrin mixtures can be seen and therefore can be better isolated from each other. This technique provides a highly workable method for preparative isolation of individual amylodextrins for further analysis or for purification of amylodextrins of defined structures for chromatographic standards. Samples are run under mild conditions in aqueous media and samples are completely recoverable. Low pressure GPC separates amylodextrins up to DP 20 .

Furthermore, it has recently been shown that this technique can be used for identifying a range of oligosaccharides containing more than one type of glycosidic linkage (42-44), see Figure 2A and 2B. Thus, it is possible to identify particular oligosaccharides of defined DP since the greater the number of (1--6)-alpha-D-glucosidic linkages makes the oligosaccharide elute earlier than an oligosaccharide of the same DP but with more (1--4)-alpha-D-glucosidic linkages. Cyclomalto-oligosaccharides are apparently 5% smaller than malto-

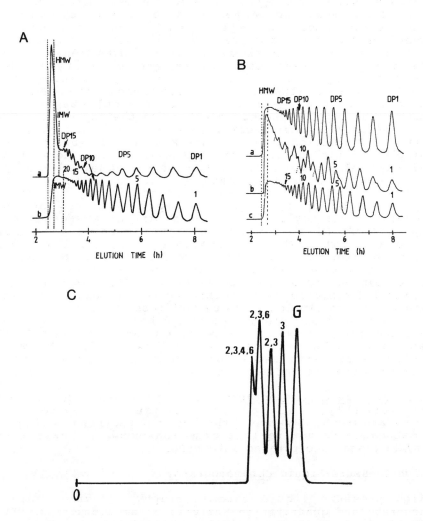

Figure 2. Low pressure GPC profiles of A) cellulose (a) and amylose (b); B) dextran (a), pullulan (b), and amylose (c); C) methyl derivatives of glucose. (Reproduced with permission from Refs. 42–44, respectively. Copyright 1989 VCH, 1988 VCH, and 1974 Springer, respectively.)

oligosaccharides of the same DP whilst cello-oligosaccharides appear to be 34% larger than the corresponding malto-oligosaccharides. In addition, this technique has been shown to separate mono-, di-, tri- and tetra-methyl-glucoses (Figure 2C). The presence of methyl groups increases the apparent size of the molecule by almost one glucose unit per methyl group and therefore could be used for structural analysis (45).

Sephadex and Biogel materials (tridimensional cross-linked gels based on dextran and polyacrylamide, respectively) have found useful application in amylodextrin analyses. It is recommended that eluants containing, for example, 0.1M NaCl should be used to prevent adsorption phenomena affecting the separation. The use of polyacrylamide-based gels is also recommended for optimum resolution at high temperatures (e.g. 60C) or when samples of biological origin are used in order to prevent elution of extraneous carbohydrate material due to disruption of the gel matrix. Such a system can be operated with very specific automated assay detection systems for up to 1 year or more with no loss of resolution for amylodextrins containing oligosaccharides up to DP 13-15. High pressure size exclusion chromatography (HPSEC) techniques with e.g. low angle light scattering for detection would be more appropriate for the determination of average molecular weight distribution for amylodextrins with DP>15. Microheterogeneity and the effects of branching in the polysaccharide reduce the resolution of higher oligosaccharides.

Although low pressure GPC had been recommended for routine analysis (46) it is being replaced increasingly by more rapid HPLC methods. However, low pressure GPC will still remain invaluable for academic researchers with their limited financial resources.

Ion-Exchange Chromatography. Low pressure GPC has almost universally replaced low pressure ion-exchange chromatography. However, the basic principles, which in most cases do not reflect true ion-exchange, have been refined and incorporated into HPLC.

High Pressure Liquid Chromatography

High pressure liquid chromatography provides rapid, accurate and quantitative analysis of amylodextrins. HPLC is typified by small particle sizes (<25 um), narrow bore columns, high inlet pressures and short analysis times.

Normal Phase Chromatography. Normal phase mode involves the polar interaction of sample molecules with a polar stationary phase, commonly silica-based packings with an aminopropyl bonded phase. Alternatively, pure silica gel modified in situ by equilibration (and simultaneous regeneration) with a mobile phase containing a polyfunctional amine that is adsorbed could be used (47).

The development of smaller (3 um) packings has now permitted resolution of linear D-gluco-oligosaccharides to DP 30-35 within 35 minutes using isocratic elution (48) whilst the application of two novel column packings, a polyamine resin bonded to silica gel and a carbamoyl resin bonded to silica gel, have resolved malto-oligosaccharides up to DP 25 in 40 minutes (49).These three packings have also been successfully applied to the HPLC analysis of branched cyclodextrins containing up to 13 D-glucosyl residues (49). For very high molecular weight amylodextrins normal phase, as well as reverse phase, chromatography can not give satisfactory analysis due to prolonged retention times and solubility problems in acetonitrile-water eluants. For full analysis, separation by HPSEC or HPIEX is necessary.

In situ modification of 5 um silica by continuous elution of 50% aqueous acetonitrile containing 0.01% of 1,4-diaminobutane, subsequent to equilibration of the column with acetonitrile and water in the same proportion but using 0.1% amine modifier, achieved satisfactory resolution of malto-oligosaccharides up to DP 25 in 45 minutes (47). Whilst diamine modifiers give optimum resolution between oligosaccharides of different DP, polyamine modifiers give improved selectivity and separation of oligosaccharides of the same DP (3).

Amine-modified silica columns have several advantages over amine-bonded silica: exceptionally long life, high stability over a wide range of pH and solvent composition, high capacity for carbohydrate solutes and relatively low cost. Prolonged use of bonded amine columns is accompanied by loss of performance due to the tendency of glycosylamine formation between the reducing sugars and the amino groups (Schiff's base reaction) of the stationary phase. Disadvantages of the amine-modified columns are the variability in retention times due to non-reproducibility of the amine loading of the silica and fluctuating baselines due to variable amine delivery. In addition, problems are encountered when using UV detection due to the presence of amine in the eluent. All silica-based packings dissolve to a small extent in water-rich eluents. Figures 3A and 3B show chromatographic profiles of corn syrup fractionated by commonly available amino-bonded column and amino-modified silica column.

Reverse Phase Chromatography. Microparticulate silica columns, and, more recently, more stable polymer matrices (e.g. vinyl alcohol copolymer gels (50) and polystyrene divinyl benzene resins) are modified with relatively non-polar hydrocarbon chains (usually C_{18}) to produce reverse phase (RP) columns.

Successful fractionations of a series of maltodextrins (up to DP 23), cyclodextrins and isomaltodexrins have been achieved using RP columns (50,51), of which water was an eluant. For these particular examples, a reverse phase partition mechanism

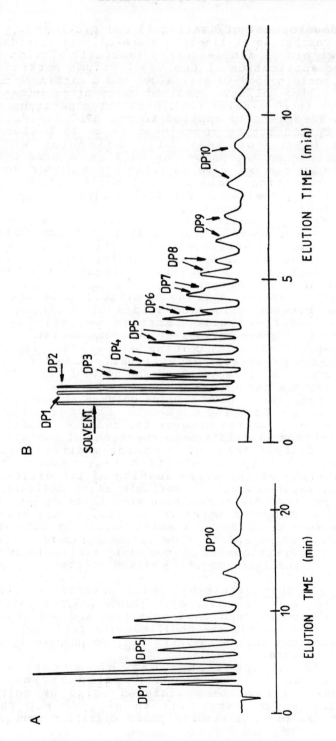

Figure 3. HPLC profiles of glucose syrups using A, amino-bonded and B, reverse-phase columns. *Continued on next page.*

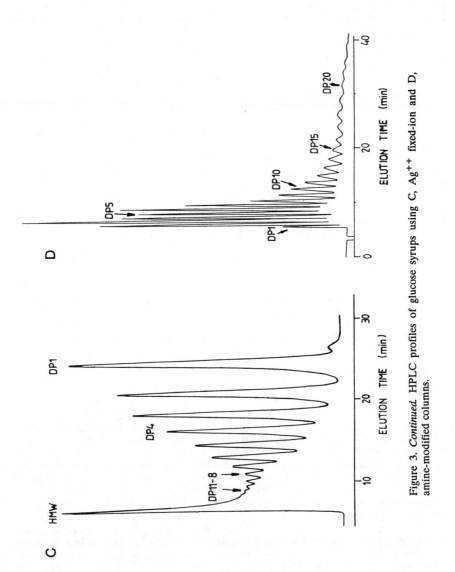

Figure 3. *Continued.* HPLC profiles of glucose syrups using C, Ag^{++} fixed-ion and D, amine-modified columns.

does not govern the separation, which is probably a form of hydrophobic chromatography. Van der Waals type interactions between the oligosaccharides and the bonded C_{18} chains are involved. Earlier trials to separate oligosaccharides by this mechanism were not sucessful. Double peaks owing to the separation of anomers of lower oligosaccharides and broad peaks due to incomplete resolution of the anomeric forms of higher oligosaccharides were encountered. Increasing separation temperature to $60^{\circ}C$ (52), use of eluants which accelerates mutarotation (53) or the improvement via modification of hydrophobic interaction (54) (e.g. use of non-ionic detergent in both stationary phase and mobile phase or addition of tetramethylurea to the eluant) have been successfully used to achieve separation of single peaks up to DP 6-13. Cyclodextrins and branched cyclodextrins which have no anomeric centres are also well resolved by RP columns (55). Resolution of branched cyclodextrins on C_{18} is now considered superior to normal phase partition (49). Some advantages of using RP columns for oligosaccharide separations are their wide availability, versatility, ease of regeneration and, as in the above cases, the use of water as eluant. Figure 3C shows an HPLC chromatogram of corn syrup separated by a commercially available RP column.

True reverse phase partition where resolution depends on preferential non-polar, hydrophobic interaction/affinity of the sample molecules for either the stationary phase or the mobile phase occurs only when analysing derivatized oligosaccharides. Application of this mechanism has great potential as an alternative to GLC for structural analysis. Such has been demonstrated by Albersheim and co-workers (56-58) who fractionated alkylated, reduced oligosaccharides containing up to six sugar residues with varying degrees of branching employing aqueous acetonitrile eluants. Detection and partial identification was carried out by MS, that is, with the chemical ionization chamber of the MS interfacing the column and the eluant serving as the CI reactant gas (57,58). The alkylated oligosaccharide-alditol were then isolated and further examined by GLC coupled with EI-MS. This technique for structural analyses should further be developed. Development should not only be seen through more sophisticated HPLC-MS interfaces but also on derivatization techniques to produce compounds amenable to reverse phase or normal phase chromatography which contain strong ultraviolet chromophores to facilitate detection (e.g. use of 4-aminobenzoic acid ethyl ester, ABEE). The Albersheim method for structural analysis makes possible the sequencing of complex carbohydrates with samples on the milligram scale, which is less sensitive than GLC techniques. Acetylated malto-oligosaccharides up to DP 35 had been separated by Wells and Lester in 1979 (59).

Development in polymer based octadecyl columns should increase potentials for preparative RP-HPLC for large scale fractionation of oligosaccharides of defined structures. Cross-linking of polymers could be manipulated either to produce low capacity and/or high capacity columns (e.g. RP-polymer based columns for hydrophobic protein fractionation.).

Reverse phase   columns are not applicable to high molecular weight amylodextrins due to prolonged retention times and supplementary HPSEC is therefore needed for full analysis.

High Pressure Size Exclusion Chromatography (HPSEC). Principles underlying this technique are the same as that of low pressure GPC. Crosslinking of the stationary gel matrix is controlled such that the distribution of pore sizes within the gel matrix can fractionate molecules according to their hydrodynamic volume over a considerable range of elution volumes. V_e is directly proportional to log (molecular weight). High molecular weight species which could not penetrate the pores are excluded and elute in a volume equal to the void volume (V_o) of the column. Very small molecules penetrate the gel freely and are eluted in a volume (V_t) equal to the sum of V_o and the volume of solvent within the gel matrix available to small molecules (V_p, pore volume). Molecules of intermediate size can penetrate some of the pores of the gel matrix and the degree of penetration is reflected by V_e which lies within the range V_o and V_t.

The development of non-compressible matrices for GPC analysis of water soluble materials has not reached the degree of sophisticaton available for the organic eluant compatible matrices. Nevertheless, some advances have been made; for example, microparticulate epichlorohydrin crosslinked agarose (60,61), water compatible hydroxylated polyether based semi-rigid matrixes (62,63), porous silica deactivated by chemically bonded polyether, or hydroxylic phases have been developed. Currently available materials have fractionation ranges which extend down to DP 10-12 whilst unmodified silica with 6nm pore sizes can extend the fractionation range down to DP 5-6. Consequently, size exclusion chromatography is primarily used for the determination of molecular weight distribution and average molecular weight of very high molecular weight amylodextrins (DP >50) (64), amylodextrins which normally could not be analyzed by normal or reverse phase chromatography, low pressure GPC, and by using fixed ion and pellicular anion exchangers, (Figures 4A and 4B show the efficiency of a newly developed amino-bonded column in fractionating amylose hydrolysates with an average DP of 17). HPSEC is required for molecular weight characterizat-ion (65) of low DE amylodextrins (Figure 4A). Applications of such (e.g. fat replacer, thickeners, glues and in the after care of patients requiring high energy but low fluid volume and electrolyte content diets) are on the increase.

Polymer based packings for SEC have lower selectivity compared to silica based matrices. However this is offset

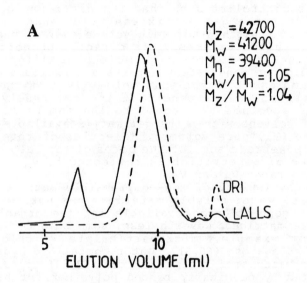

Figure 4A. HPSEC profile of a high-molecular-weight amylodextrin product using RI/LALLS. (Reproduced from Ref. 65. Copyright 1991 American Chemical Society.)

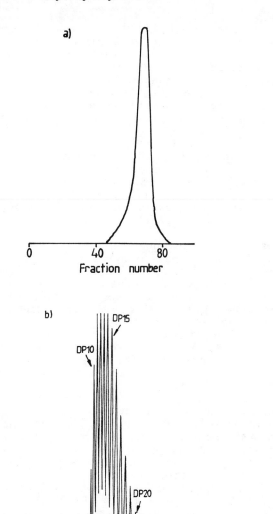

Figure 4B. Elution profile of a short chain amylose (c. DP 17) on a, SEC and b, amino-bonded columns. (Reproduced with permission from Ref. 48. Copyright 1985 Elsevier.)

by their greater stability towards alkaline pH (water insoluble high molecular weight amylodextrins could be solubilized at alkaline pH as well as in dimethylsulphoxide/water mixtures). Furthermore, modified silica still exhibits adsorption effects, and elution with ionic buffers (0.1M) is recommended within the range pH 2-7. Silica based matrices also tend to dissolve in aqueous eluents and polymer based SEC packings are preferred.

Detection by low angle laser light scattering has been recommended for the determination of average molecular weights and molecular weight distributions of polysaccharides (64). Indirect estimation of molecular weight values from the retention time (or elution volume) is less accurate even when the same kind of standard polymer is used (64).

High Pressure Ion-Exchange Chromatography (HPIEX). Until recently, the principal mechanism for the separation of carbohydrates was actually not true ion-exchange. Resolution in widely applied fixed ion resin HPLC columns involves ion or size exclusion, hydrophobic adsorption or ligand exchange. Fixed-ion resins are usually composed of the sulfonic acid-type polystyrene divinylbenzene resins, having a degree of cross-linking between the styrene polymers of 4-8%, being loaded with a particular cationic counterion. Counterions affect desired separation whilst remaining permanently (in theory) on the column throughout its useful life. There are several cationic forms available, Ca, Pb, H, K and Ag . The choice of cation is important. Ca^{++} on 4% crosslinked resin separates malto-oligosaccharides to DP 6-8 (66,67) and the more stable Ag^{++} on 4% crosslinked resin fractionates malto-oligosaccharides to DP 8-12 (68) (Figure 3D) depending on the size (or number) of the chromatographic columns and time of analysis. More recently, 2% cross-linked resin in the H^+ form has been demonstrated to resolve malto-oligosaccharides up to DP 12-14 (69,70) provided that flow rates, and hence back pressure, are kept low (not more than 0.35 ml/min at 60C and 0.5 ml/min at 85C).

Fixed ion columns are always run at elevated temperatures to eliminate formation of doublets or broad peaks owing to resolution or partial resolution of anomers and to accelerate the slow, diffusion-controlled partitioning process. Except for the H^+ form which uses about 0.005M aqueous sulphuric acid, columns using the other counter-ions use water as eluant. Use of water as eluant is a major advantage especially for preparative purposes (71), albeit the H^+ form could be directly applied without prior clean-up procedures to the analysis of the products of acid and enzymic hydrolysis of starch and components. Samples should be free from protein and fat to prevent deposition on the column causing loss of efficiency and resolution and should be neutral (except for H^+ form) and deionized before injection to prevent the stripping off or exchange of the fixed cation. Regeneration of the cations with their respective nitrates should take place regularly to prevent and correct column deterioration.

Size exclusion is the principal mechanism operating with these columns since large molecules are eluted prior to smaller molecules whilst selectivity is primarily dependant on the ligand (fixed ion) interaction/exchange (72) (i.e. of the water molecules surrounding the fixed ion). Degree of crosslinking determines the length of separation time whilst increasing resolving power. For further reading, the review of Pirisino (73) on fixed ion resin columns and amino bonded columns is recommended as well as the book by Mikes (74) on HPLC techniques.

The fixed ion resin technique, in particular Ca^{++} and Ag^{++} forms of resin, is replacing traditional GPC for routine corn syrup analysis in the food industry. The major drawback of these columns includes the compressibility of the gel matrix, extended analysis time, efficiency losses of the order of 50% for a doubling of the flowrate, the need for high temperature (85C) operation, and the need for specialized regeneration and re-packing of contaminated columns. However these drawbacks are offset by the ability to obtain an almost total analysis of material applied to the column and the use of water as the only eluant.

Cyclodextrins are also resolved on the Ca^{++} form, at 90C, eluting in the order of alpha < gamma < beta at retention times later than those of the corresponding linear D-gluco-oligosaccharides (75).

Another application of ion exchange columns in amylodextrin analysis is the separation of oligosaccharides as negatively charged borate complexes using a true anion-exchange mechanism (76,77).

More recently, the availability of high performance anion exchange columns stable at high pH conditions and the development of the pulsed amperometric detector (PAD) (78,79) have introduced a most powerful method for the HPLC of higher oligosaccharides. Such a system was introduced by Rochline and Pohl (80), and commercialized into the Dionex system.

Prior to this technique, solubilization problems of higher oligosaccharides have restricted its full analysis using amine-bonded, reverse phase and fixed ion columns. Albeit, high molecular oligosaccharides are soluble in DMSO and this solvent could be used on some SEC columns. Good resolution of high oligosaccharides is not obtained by this technique and only molecular weight distribution and average molecular weight can be measured.

Most carbohydrates have pKa values in the range 12-13 and hence are ionizable at pH >13. Not only does high pH solubilize amylodextrins but it also enhances the occurence of carbohydrate oxidation, the basis for PAD detection. Thus, liquid chromatography-electochemical detector applications typically employ mobile phases containing 0.001M to 0.15M OH$^-$. The addition of acetate decreases the capacity factors of all solutes and this is utilized to further the application to higher oligosaccharides and resolution of isomeric disaccharides. Using a linear gradient, from 0 to 600 mM sodium acetate in 100 mM sodium hydroxide over 30 min, followed by isocratic elution at the final concentration for 5 min, resolution of malto-oligosaccharides up to DP 43 (81) has been achieved. Sensitivity of the detector was increased 3-fold for the oligomers of DP 25-35, and 30-fold to detect those above DP 35. More recently, malto-oligosaccharides from amylose hydrolysates and from waxy maize after debranching with isoamylase were resolved up to DP 50 and DP 90 (see Figure 5A), respectively (82). Hydrolyzates of other polysaccharides were also resolved to DP >50 (82) whilst branched cyclodextrins containing up to twelve D-glucosyl residues were analyzed (83).

Anion Pellicular Columns. Columns employed by the Dionex system make use of electrostatically latex-coated pellicular ion exchangers. These materials consist of 3 regions: a) an inert PS-DVB core, b) a shallow sulphonated layer on the surface of the inert core, and c) a monolayer coating of colloidal ion-exchange particles which are permanently attached to the oppositely charged functionalized surface of the inert core. The latex coating consists of fully functionalized anion exchange latex particles made from vinylbenzylchloride (VBC) polymer crosslinked with DVB and fully funtionalized with an appropriate tertiary amine for desired anion exchange selectivity (84,85). Use of pellicular packings guarantees a much shorter analyte diffusion path increasing considerably column efficiencies.

The detection system employed, a pulsed amperometric detector, permits remarkable sensitivity (100 ppb), and provides the most sensitive workable commercially available detector yet developed for HPLC of underivatized carbohydrates. The drawback is that it requires strongly alkaline conditions for optimum carbohydrate oxidation (and detection). Eluants therefore contain high concentrations of non-volatile salts (typically sodium acetate and sodium hydroxide) and further structural elucidation or identification by e.g. mass spectroscopy and/or NMR requires prior desalting. The use of an anionic micromembrane suppressor downstream of the detector, thus converting the sodium hydroxide and sodium acetate to water and acetic acid, respectively, has been found satisfactory for ^{1}H NMR at 500 MHZ (86). However, with the inherent insensitivity of NMR and the low capacity of pellicular HPAEC columns, preparation for more sensitive analytical methods, e.g. MS, is desirable. Derivatization of fractionated oligosaccharides (either by methylation techniques or reductive coupling of 4-amino-benzoic acid ethyl ester, ABEE) and subsequent

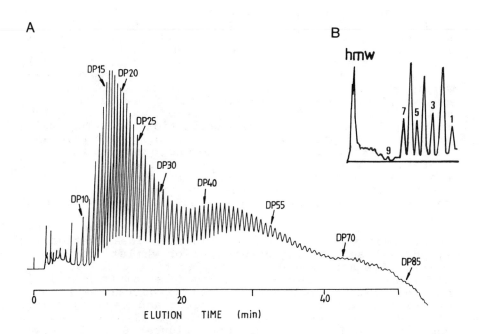

Figure 5. A) HPAEC profile of amylomaize hydrolysate. B) TLC profile of products from alpha-amylolysis of starch granules. (Reproduced with permission from Refs. 82 and 90, respectively. Copyright 1989 Elsevier and 1985 Elsevier.)

purification via reverse phase chromatography prior to
FAB-MS is highly successful, most especially with the use
of ABEE (87).

Affinity Chromatography. The interaction of concanavalin
A with alpha-D-glucans has been used in the determination
of the fine structure of amylopectin and glycogen(88),
where correlation between the external chain length of the
glucan and the hydrodynamic volume has been established.

Planar Chromatography

Planar chromatography of carbohydrates has not developed
to the same extent in recent years as have the column
methods. Paper and thin layer chromatography, in some
instances, still present a more convenient method than
HPLC for the separation of oligosaccharides. In general,
TLC is superior to PC with regard to resolution,
sensitivity and especially the time required for analysis.
However, until recently, quantitative fractionation of
malto-oligosaccharides up to DP 25 was achieved by
multiple descending PC whilst applications of HPTLC using
multiple development were limited to oligosaccharides up
to DP 12. Resolution of malto-oligosaccharides was
therefore extended by Koizumi et al (48) to give good
resolution up to DP >20 on silica gel TLC plates and Si
50000 HPTLC plates. Branched cyclodextrins are also
resolvable on Si 50000 HPTLC plates and NH_2-bonded HPTLC
plates (89).
 · TLC is a convenient technique; samples can be analyzed simultaneously and it provides an ideal method for
rapid amylase activity determination whilst providing a
qualitative assessment of its specificity and activity
mechanism. Application of scanning densitometric analysis
on visualized chromatograms permits documentation and
quantification of results. Macgregor and Macgregor have
studied the activity mechanism of the alpha-amylase in
cereals using TLC techniques (90), Figure 5B. Recent developments on detection up to nanogram amounts, and
development of HPTLC and NH_2-bonded HPTLC plates which
provide better resolution of amylodextrins up to DPs
comparable to HPLC might see a resurgence of the use of
TLC on such analyses. However, variation on a single
chromatogram as well as between chromatograms could occur
due to slight differences in temperatures and duration of
heating and coverage of spray reagent, and a lot of skill
is involved in performing TLC techniques.

Supercritical Fluid Chromatography

 Utilization of SFC for the separation of suitably
derivatized oligosaccharides,e.g. trimethylsilylated and
permethylated amylodextrins from corn syrup has been
demonstrated (91,92). Resolution of the trimethylsilylated oligomers up to DP 18 and permethylated oligomers up

to DP 15 have been achieved, at temperatures of 90C and 120C, respectively, using fused silica columns with bonded-methylpolysiloxan phases. The pressure of CO_2 mobile phase was programmed from 115 to 400 atm over 75 mins.

SFC offers a less laborious technique with better resolving power for amylodextrin analysis than GLC. Application of SFC is expected to increase, particularly in combination with CI mass spectrometry, with which it is now interfaced.

SPECTROSCOPIC METHODS

Mass Spectrometry (MS). MS is one of the key techniques used in structure determination of carbohydrates and analyses via electron impact (E.I.) and chemical ionization (C.I.) methods are performed routinely on low molecular weight permethylated or peracetylated carbohydrates. Recently, MS procedures have found wider application in the structure elucidation of less volatile higher molecular weight oligosaccharides as a result of instrumental developments (in particular, desorption methods of ionization, based on fast atom bombardment (FAB) (93), field desorption (FD), laser desorption (LD), plasma desorption (PD), and secondary ion (SI) mass spectrometry) and improvements in derivatization techniques. For example, a series of malto-oligosaccharides, starch and other glycans have been examined with LD FD-MS (94,95) whilst FAB techniques have been employed for studies of cello- and malto-oligosaccharides (96) and branched cyclodextrins (97).

MS techniques can give some sequence information, as fragmentation occurs in an ordered fashion (albeit studies by Bosso et al (96) on cello- and malto-oligosaccharides did not observe any fragmentation pertinent to the carbohydrate sequence) whilst an indication of the positions of linkage can also be obtained. Furthermore, molecular weight could be estimated on a few picomoles of the native material and characterization of homo-oligosaccharides, even as a mixture has been reported (96). At present, only derivatized oligosaccharides less than DP 10 could be analyzed by MS techniques.

FAB spectra are relatively easy to acquire and it is likely that FAB-MS with complementary E.I.MS and C.I.MS of permethylated oligosaccharides and the partially methylated alditols of their constituent monosaccharides (separated with capillary GLC) is the most efficient stratagem for MS analysis of oligosaccharides.

Nuclear Magnetic Resonance. Considerable progress has been achieved in the NMR spectroscopic characterization of carbohydrates in the solid and solution state. ^{13}C nmr has already proved to be a powerful tool for the structural elucidation of polysaccharides, providing information on their composition, sequence, overall conformation and anomeric configuration (98,2). In ^{13}C nmr measurements, it is necessary to make comparisons with standard compounds. Such measurements and comparisons have shown that chemical shifts in monosaccharides are essentially the same as the shifts obtained with a polysaccharide of the same units. Any substituent on a sugar causes a change (usually an increase) in the chemical shift of the carbon directly involved in the linkage.These patterns of chemical shift differences between monosaccharides and polysaccharides indicate the positions of the glycosidic linkages; also similarities in chemical shifts, especially on selective carbon atoms known to be sensitive to change in anomeric configuration, can be employed for the determination of linkage configuration. ^{13}C nmr could only be performed on more than 5 mg of a material whilst high resolution proton (^{1}H) nmr could give diagnostic spectra on as little as 10 nanomoles of material. ^{1}H nmr does not provide as much structural information due to incomplete separation of the proton resonance obtained for carbohydrates. It can, however, be used to quantitate individual oligosaccharide contents in situations where the unique selectivity of the method overcomes interference problems encountered when using other spectroscopic methods of great sensitivity but inferior selectivity.

More recent innovations include for example a series of proton and carbon two-dimensional (2D) NMR techniques, multistep relayed correlation spectroscopy and homonuclear ^{1}H-^{1}H Hartmann-Hahn correlation spectroscopy (99-101) which are providing extremely valuable information about the normally crowded regions of conventional one dimensional (1D) carbohydrate spectra. Examples of current application of NMR in amylodextrin analysis involves studies of their conformation (102,103) the effect of hydration (104), and the formation of inclusion complexes (105,106). Gidley (107) employed NMR techniques for the analysis of branching and reducing residues in starch and Birch and Kheiri (108) determined the degree of hydrolysis and therefore the dextrose equivalent of some glucose syrups. Furthermore, with the advent of Fourier transform techniques development is underway to prepare on line HPLC detectors (109) which will provide valuable structural information on components in mixtures rather than the more normal determination of net properties of a component. This non-destructive detector will allow it use in conjunction with other detectors or fraction collectors if preparative scale chromatography is used. The reader is referred to reviews by Coxon (110) and Rathbone (111) on NMR methodology and interpretation of spectra.

Other Spectroscopic Methods. Electron spin resonance (ESR), infrared (IR) and raman spectroscopy (112,113), fluorescence spectroscopy and photoelectron spectroscopy could also be employed in the elucidation of solution properties and the structure of starch, its major components and starch hydrolysates (114).

CHEMICAL METHODS

Colorimetric Methods. Colorimetric methods are mainly
used for the gross determination of total carbohydrate
content or total reducing sugar content although specific
assay methods have been developed for the quantitation of
particular oligosaccharides from a mixture by the use of
specific enzymes. Both total carbohydrate and reducing
sugar contents are important features of oligosaccharide
analysis since it is possible to determine the size of an
oligosaccharide, or, for a mixture of related oligosaccharides, a mean value for the ratio of total residues to
terminal reducing residues. For example, in the food
industry, glucose syrups are categorized and characterized
in terms of their dextrose (i.e. D-glucose) equivalent or
DE value. Recognition of the shortcomings of defining
mixtures of oligosaccharides in terms of only the DE value
resulted in the (increasing) characterization of corn and
glucose syrups based on their oligosaccharide composition
by chromatographic techniques. However, chemical methods
(in particular the Lane and Eynon method) still remain
the principal method type for corn and glucose syrup
characterization. Methods for total sugar assay based on
the hydrolytic action of concentrated sulphuric acid with
subsequent dehydration of the monosaccharide released to
give derivatives of furfural and which react with certain
compounds to give coloured products and methods for the
determination of the reducing properties of sugars are
described and reviewed by White and Kennedy (1) and
Chaplin (115).
     One of the major uses of colorimetric methods is in
the monitoring of chromatographic columns. All the low
pressure GPC columns can be connected to automated colorimetric assay systems based on any of the total sugar assay
methods. There is a potential in the adoption of such
systems (analogous to flow injection analysis) into HPLC
systems for the specific and more sensitive (compared to
RI and UV) detection of underivatized glucose oligomers.

Structure Elucidation. The use of classical analytical
methods based on chemical modifications and degradation,
such as Smith degradation (periodate oxidation) and methylation analysis, is continually being extended by
combination with modern spectroscopic techniques e.g. NMR
and MS.

Methylation Analysis. Methylation analysis, as described
previously, involves complete methylation of a
polysaccharide, the hydrolysis of the methylated product
to a mixture of partially methylated monosaccharides, the
reduction of the methylated monosaccharides, the
acetylation of the reduced products and, finally, analysis
of the partially methylated alditol acetates by GLC and
MS. Methylation analysis gives information on the carbons

involved in glycosidic linkages but not the anomeric configuration of the glycosidic linkages nor the sequence of
the monosaccharide residues.

Periodate Oxidation. Information regarding structures and
monosacharide sequence is based on the formation of
degradative products of particular proportions
corresponding to the specific glycosidic linkage involved
and the monosaccharide involved (17). Briefly, carbohydrate residues containing glycol groups are oxidized to
dialdehydes but if residues contain hydroxyl groups on
three adjacent carbons formic acid is produced. Also,
oxidation of a primary hydroxyl group adjacent to a secondary hydroxyl group yields formaldehayde. Reaction of
polysaccharides with periodic acid is followed by
measurement of the amounts of reagent consumed and of the
formic acid formed. Alternatively, oxidative products
after reduction and mild acid hydrolysis are analysed by
GLC, usually as their trimethylsilyl ethers. Additional
structural information, i.e. monosaccharide sequence, can
be obtained by subjecting the oxidized and reduced
carbohydrate to a second periodate oxidation (Smith
degradation) (116). Another analytical scheme based on
periodate oxidation is the Barry degradation. Periodate
oxidized polysaccharides are reacted with phenylhydrazine
in dilute acetic acid. Phenylhydrazones are formed with
the aldehyde groups generated by the oxidation and
released by hydrolysis. Some polysaccharides can be
degraded with sequential removal of monosaccharide
residues.

Acetolysis. Acetolysis of polysaccharides results in the
complete acetylation of the free hydroxyl groups of the
polysaccharide and the selective cleavage of glycosidic
bonds. There exist differences in the rate constants for
the cleavage of glycosidic bonds under acetolysis
conditions and therefore it is possible to fragment polysaccharides preferentially at specific glycosidic bonds.
Acetolysis can be used as a complementary method for acid
hydrolysis. The two procedures will yield different
fragments and comparison of the fragments can give structural information.

Alkaline Degradation. This method of analysis of
polysaccharides provides little information about overall
structure of polysaccharides due to the complex nature of
the reactions involved which includes isomerization,
oxidation, reduction, molecular rearrangements of reducing
residues and the fragmentation of polysaccharide chains.

ENZYMIC METHODS

Starch degrading enzymes are used for the specific
quantitative measurement of starch, amylose, amylopectin
and/or starch hydrolysates (amylodextrins) in mixtures.

Also, they are used for the structural analysis of
carbohydrates (117). The information obtained is not
limited to that obtained by analysis of the hydrolysis
fragments because the specificity of enzyme action, a
specificity based on type of monosaccharide and type of
linkage, leads to significant data being obtained, by a
process of elimination, from enzyme resistant structures
and partially hydrolyzed structures. The enzymes which
hydrolyse polysaccharides are divided into two groups,
endo- and exo-polysaccharide hydrolases. Endo-
polysaccharide hydrolases are specific for linkage and
monosaccharide residue and cause fragmentation of
homopolysacharides to give a homologous series of oligosaccharides. Exo-polysaccharide hydrolases are specific for
monosaccharide unit and stereochemistry at C_1. They
cleave polysaccharides by sequential removal of residues
from one end of the molecule, usually the non-reducing
end.
   The following are starch degrading enzymes (118)
usually used for structural elucidation of starches and
amylodextrins:
   a) alpha-amylase   - (endohydrolase;   EC 3.2.1.1),
   1,4-alpha-D-glucan glucanohydrolase,
   b) beta-amylase   - (exohydrolase;   EC 3.2.1.2),
   1,4-alpha-D-glucan maltohydrolase,
   c) glucoamylase   - (exohydrolase;   EC 3.2.1.3.),
   1,4-alpha-D-glucan glucanohydrolase,
   d) isoamylase -   (endohydrolase;   EC 3.2.1.7),
   glycogen-6-glucanohydrolase,
   e) pullulanase   - (endohydrolase;   EC 3.2.1.4.1),
   pullulan-6-glucanohydrolase.
For example, debranching enzymes (pullulanase and
isoamylase) are employed for the determination of chain
length and degrees of branching in starch, amylopectin and
amylose. Enzymes used for structural analysis must be very
pure and have established specificities.

OTHER METHODS

Electrophoresis. This method provides very useful
complementary information to chromatographic techniques
because it utilizes different criteria for separation,
namely molecular charge, size and shape. The use of high
voltage paper electrophoresis has been applied to the
separation of monosaccharides and oligosaccharides. These
are neutral compounds which can form electrically charged
complexes with electrolytes such as sodium borate, arse-
nite, and molybdate. The relative mobilities of the
carbohydrates can be varied by changing the complexing
agent used, where steric factors often determine the
formation of different complexes. Choice of electrolytes
will often lead to identification of the carbohydrate and
its structure and bonding.

Membrane Techniques. Reliable preparative purification of

maltodextrins (DP 3-10) and starch hydrolysates have been achieved by membrane techniques (119,120). Such applications are foreseen to increase as membrane technology progresses.

LITERATURE CITED

1. White, C.A.; Kennedy, J.F. In Carbohydrate Analysis: a Practical Approach; Chaplin M.F.; Kennedy, J.F.,Eds.; IRL Press: Washington, D.C., 1986; Chapter 2.
2. Pazur, J.H. In Carbohydrate Analysis: a Practical Approach; Chaplin M.F.; Kennedy, J.F.,Eds.; IRL Press: Washington, D.C., 1986; Chapter 3.
3. Folkes, D.J.; Brookes, A. In Glucose Syrups: Science and Technology; Dziedzic, S.Z.; Kearsley, M.W., Eds.; Elsevier Applied Science Publishers:London, 1984; Chapter 6.
4. Collyer, A.A.; Clegg, D.W. Rheological Measurement; Elsevier Applied Science: London, 1988.
5. Mitchell, J.R. In Polysaccharides in Food; Blanshard, J.M.V.; Mitchell, J.R., Eds.; Butterworths: London, 1979; Chapter 4.
6. Robinson, G., European Food and Drink Review 1990, 3, 51.
7. Snyder, L.R.; Kirkland, J.J. Introduction to Modern Liquid Chromatography; John Wiley and Sons, Inc.: New York, 1979.
8. Churms, S.C. J. Chromatogr. 1990, 500, 555.
9. Furneaux, R.H. Carb. Res. 1983, 113, 241.
10.Sweet, D.P.; Shapiro, R.H.; Albersheim, P. J. Chromatogr. 1975, 40, 217.
11.Nilsson, B.; Zopf, D. Methods Enzymol. 1982, 83, 46.
12.Englmaier, P. Carb. Res. 1985, 40, 217.
13.Hakamori, S. J. Biochem (Tokyo) 1964, 55, 205.
14.Mononen, I.; Finne, J.; Karkkainen, J. Carb. Res. 1978, 60, 371.
15.Baird, J.K.; Holroyde, M.J. ; Ellwood, D.C. Carb. Res. 1973, 27, 464.
16.Turner, S.H.; Cherniak, R. Carb. Res. 1981, 95, 137.
17.Englmaier, P; Honda, S.; Biermann, C.J. In CRC Analysis of Carbohydrates by GLC and MS; CRC Press, Inc.: Florida,1989; Chapter 12.
18.Ishak, M.F.; Painter, T. Carb. Res. 1974, 32, 227.
19.Sweeley, C.C.; Bentley, R.; Makita, M.; Wells, W.W. J. Am. Chem. Soc. 1963, 85, 2497.
20.Beadle, J.B. J. Agric. Food Chem. 1969, 17, 904.
21.Sosulski, F.W.; Elkowicz, L.; Reichert, R.D. J. Food Sci. 1982, 47, 498.
22.Gunner, S.W.; Jones, J.K.N.; Perry, M.P. Can. J. Chem. 1961, 39, 1892.
23.Vilkas, M.; Hiu, I.J.; Boussac,G.; Bonnard, M.C. Tetrahedron Lett. 1966, 14, 1441.
24.Whiton, R.S.; Lau, P.; Morgan, S.L.; Gilbert, J.; Fox, A. J. Chromatogr. 1985, 347, 109.

25.Neeser, J.-R.; Schweizer, T.F. Anal. Biochem. 1984, 142, 58.
26.Schweer, H. J. Chromatogr. 1982, 236, 361.
27.McGinnis, G.D. Carb. Res. 1982, 108, 284.
28.Morrison, I.M. J. Chromatogr. 1975, 108, 361.
29.Honda, S.; Yamauchi, N.; Kakehi, K. J. Chromatogr. 1979, 169, 287.
30.Bentley, R.; Botlock, N. Anal. Biochem. 1967, 20, 312.
31.Tanka, M. Carb. Res. 1981, 88, 1.
32.Whelan, W.J. In Methods in Carbohydrate Chemistry; Vol. IV.; Whistler,R.L. Ed.; Academic Press: New York, 1964; pp 72-78.
33.Konig, W.A.; Benecke, I. J. Chromatogr. 1983, 269, 19.
34.Little, M.R. Carb. Res. 1982, 105, 1.
35.Biermann, C.J. ; McGinnis, G.D. CRC Analysis of Carbohydrates by GLC and MS; CRC Press Inc.: Florida, 1989.
36.Harris, P.J.; Henry, R.J.; Blakeney, A.B.; Stone, B.A. Carb. Res. 1984, 127, 59.
37.Waeghe, T.J.; Darvill, A.G.; McNeil, M.; Albersheim,P. Carb. Res. 1983, 123, 281.
38.Lomax, J.A.; Conchie, J. J. Chromatogr. 1982, 236, 385.
39.Lomax, J.A.; Gordon, A.H.; Chesson, A. Carb. Res. 1985, 138, 177.
40.Seymour, F.R.; Chen, E.C.M.; Stouffer, J.E. Carb. Res. 1980, 83, 201.
41.Barber, M.; Bordoli, R.S.; Sedgwick, R.D.; Tyler, A.N. J. Chem. Soc. Chem. Commun. 1981, 325.
42.Kennedy, J.F.; Stevenson, D.L.; White, C.A. Starch/Starke 1989, 41, 72.
43.Kennedy, J.F.; Stevenson, D.L.; White, C.A. Starch/Starke 1988, 40, 396.
44.Schmidt, F.; Enevoldsen, B.S. Carlsberg Res. Commun. 1974, 40, 91.
45.Enevoldsen,B.S. Abstracts of Papers for Xth International Symposium on Carbohydrate Chemistry, 1980.
46.Kennedy, J.F.; Noy, R.F.; Stead, J.A.; White, C.A. Starch/Starke 1985, 37, 343.
47.White, C.A.; Corran, P.H.; Kennedy,J.F. Carb. Res. 1980, 87, 165.
48.Koizumi, K.; Utamura, T.; Okada, Y. J. Chromatogr. 1985, 321, 145.
49.Koizumi, K.; Utamura, T.; Kubota, Y.; Hizukuri, S. J. Chromatogr. 1987, 409, 396.
50.Koizumi, K.; Utamura, T. J. Chromatogr. 1988, 436, 328.
51.McGinnis, G.D.; Prince, S.; Lowrimore,J. J. Carb. Chem. 1986, 5, 83.
52.Vratny, P.; Coupek, J.; Vozka, S.; Hostomska, Z. J. Chromatogr. 1983, 254, 143.
53.Verhaar, L.A.Th.; Kuster, B.F.M.; Claessens, H.A. J. Chromatogr. 1984, 284, 1.

54.Cheetham, N.W.H.; Teng, G. J. Chromatogr. 1984, 336, 161.
55.Koizumi, K.; Kubota, Y.; Okada, Y.; Utamura, T; Hizukuri, S.; Abe, J.-I. J. Chromatogr. 1988, 437, 47.
56.Valent, B.S.; Darvill, M.; McNeil, M.; Robertsen, B.K.; Albersheim, P. Carb. Res. 1980, 79, 165.
57.Aman, P.; McNeil, M.; Franzen, L.-E.; Albersheim, P. Carb. Res. 1981, 95, 263.
58.McNeil, M.; Darvill, A.G.; Aman, P.; Franzen, L.-E.; Albersheim, P. Methods Enzymol. 1982, 83, 3.
59.Wells, G.B.; Lester, R.L. Anal. Biochem. 1979, 97, 184.
60.Andersson, T.; Carlsson, M.; Hagel, L.; Pernemalm, P.-A; Janson, J.-C. J. Chromatogr. 1985, 326, 33.
61.Praznik, W.; Beck, R.H.F.; Eigner, W.-D. J. Chromatogr. 1987, 387, 467.
62.Alsop, R.M.; Vlachogiannas, G.J. J. Chromatogr. 1982, 246, 227.
63.Cullen, M.P.; Turner, C.; Haycock, G.B. J. Chromatogr. 1985, 337, 29.
64.Hizukuri, S.; Takagi,T. Carb. Res. 1984, 134, 1.
65.Jumel, K.; Melo, E.H.M.; Cabalda, V.M.; Kennedy, J.F. ACS Symposium Series. In press.
66.Fitt, L.E.; Hassler, S.; Just, D.E. J Chromatogr. 1986, 187, 381.
67.Schobell, H.D.; Brobst, K.M. J. Chromatogr. 1981, 212, 51.
68.Hicks, K.B. Adv. Carb. Chem. Biochem. 1988, 46, 17.
69.Derler, H.; Hormeyer, F.; Bonn, G. J. Chromatogr. 1988, 440, 281.
70.Hicks, K.B.; Hotchkiss, A.T.Jr. J. Chromatogr. 1988, 441, 382.
71.Hicks, K.B.; Sondey, S.M. J. Chromatogr. 1987, 389, 183.
72.Goulding, R.W. J. Chromatogr. 1975, 103, 229.
73.Pirisino, J.F. In Food Constituents and Food Residues; Lawrence,J.F., Ed.; Marcel Dekker: New York, 1984.
74.Mikes, O. High-Performance Liquid Chromatography of Biopolymers and Biooligomers. Part B: Separation of Individual Compound Classes; Elsevier: Amsterdam, 1988; Chapter 11.
75.Brunt, K. J. Chromatogr. 1982, 246, 145.
76.Mopper, K.; Dawson, R.; Liebezeit, G.; Hansen, H.-P. Anal. Chem. 1980, 52, 2018.
77.Honda, S.; Matsuda, Y.; Takahashi, M.; Kakehi, K. Anal. Chem. 1980, 52, 1079.
78.Hughes, S.; Johnson,D.C. Anal. Chim. Acta 1981, 132, 11.
79.Neuberger,G.G.; Johnson, D.C. Anal. Chem. 1987, 59, 150.
80.Rocklin, R.D.; Pohl, C.A. J. Liq. Chromatogr. 1983, 6, 1577.
81.Olechno, J.D.; Carter, S.R.; Edwards, W.T.; Gillen, D.G. Am. Biotechnol. Lab. 1987, 5, 38.

11. CABALDA ET AL. *Analysis of Amylodextrins* 169

82.Koizumi, K.; Kubota, Y; Tanimoto, T.; Okada, Y. J.
Chromatogr. 1989, 464, 365
83.Koizumi, K.; Kubota, Y.; Tanimoto, T.; Okada, Y. J.
Chromatogr. 1988, 454, 3289.
84.Stillian, J.R.; Pohl, C.A. J. Chromatogr. 1990, 499,
249
85.Slingsby, R.W.; Pohl, C.A. J. Chromatogr. 1988, 458,
41.
86.Spellman, M.W.; Basa, L.J.; Leonard, C.K.; Chakel,
J.A.; O'Connor, J.V.; Wilson, S.; van Halbeek, H. J.
Biol. Chem. 1989, 264, 14100.
87.Basa, L.J.; Spellman, M.W. J. Chromatogr. 1990, 499,
221.
88.Collona, P.; Biton, V.; Mercier, C. Carb. Res. 1985,
137, 151.
89.Koizuma, K.; Utamura, T.; Kuroyanagi, T.; Hizukuri, S.;
Abe, J.-I. J. Chromatogr. 1986, 360, 397.
90.Macgregor, E.A.; Macgregor, A.W. In New Approaches to
Research on Cereal Carbohydrates; Hill, R.D.; Munck,
L., Eds.; Elsevier: Amsterdam, 1985; pp 149-160.
91.Chester, T.L.; Innis, D.P. J. High Resolut.
Chromatogr. Chromatogra. Commun. 1986, 9, 209.
92.Reinhold, V.N.; Steeley, D.M.; Kuei, J.; Her, G.R.
Anal. Chem. 1988, 60, 2719.
93.Dell, A.; Morris, H.R.; Egge,H.; von Nicolai, H.;
Strecker, G. Carb. Res. 1983, 115, 41.
94.Coates, M.I.; Wilkins, C.L. Biomed. Mass. Spectrom.
1985, 12, 424.
95.Coates, M.I.; Wilkins, C.L. Anal. Chem. 1987, 59, 197.
96.Bosso, C.; Defaye, J.; Heyraud, A.; Ulrich, J. Carb.
Res. 1984, 125, 309.
97.Koizumi, K.; Utamura, T.; Sato, M.; Yagi, Y. Carb.
Res. 1986, 153, 55.
98.Jennings, H.J.; Smith, I.C.P. In Methods in
Carbohydrate Chemistry, Vol VIII; Whistler, R.L.;
BeMiller, J.N., Eds.; Academic Press: New York, 1980;
Chapter 12.
99.Bax, A. Two Dimensional Nuclear Magnetic Resonance in
Liquids; Reidel: Boston, 1982.
100.Homans, S.W.; Dwek, R.A.; Rademacher, T.W. Biochem
1987, 26, 6571.
101.Koerner, T.A.W.; Prestegard, J.H.; Yu, R.K. Methods
Enzymol. 1987, 138, 38.
102.Hewitt, J.M.; Linder, M.; Perez, S.; Buleon, A. Carb.
Res. 1986, 154, 1.
103.Veregin, R.P.; Fyfe, C.A.; Marchessault, R.H.; Taylor,
M.G. Carb. Res. 1987, 160, 41.
104.German, M.L.; Blumenfeld, A.L.; Yuryev, V.P.;
Tolstoguzov, V.B. Carb. Polym. 1989, 11, 139.
105.Inoue, Y.; Kuan, F.-H.; Chujo, R. Carb. Res. 1987,
159, 1.
106.Hall, L.D.; Lim, T.K. J. Am. Chem. Soc. 1986, 106,
1858.
107.Gidley, M.J. Carb. Res. 1985, 139, 85.
108.Birch, G.G.; Kheiri, M.S.A. Carb. Res. 1971, 16, 215.

109.Dorn, H.C.   Analyt. Chem. 1984, 56, 747A.
110.Coxon, B.   In Developments in Food Carbohydrates-2;
    Lee, C.K., Ed.; Applied Science: London, 1980; p. 351.
111.Rathbone, E.B.   In Analysis of Food Carbohydate;
    Birch,G.G., Ed.; Elsevier Applied Science: London,
    1985; p. 149.
112.Wilson, R.H.; Kalichevsky, M.T.; Ring, S.G.; Belton,
    P.S. Carb Res. 1987, 166, 162.
113.Bulkin, B.J.; Kwak, Y.; Dea, I.C.M. Carb. Res. 1987,
    160, 95.
114.Yalpani, M.   Polysaccharides Syntheses, Modifications
    and Structure/Property Relations; Elsevier: Amsterdam,
    1988.
115.Chaplin, M.F.   In Carbohydrate Analysis: A Practical
    Approach; Chaplin, M.F.; Kennedy, J.F., Eds.; IRL
    Press Ltd: Oxford, 1986; Chapter 1.
116.Lindberg, B.; Lonngren, J.; Ruden, U.; Nimmich, W.
    Acta Chem. Scand. 1973, 27, 3787.
117.Kennedy, J.F.; White, C.A.W. Bioactive Carbohydrates
    in Chemistry, Biochemistry and Biology; Ellis Horwood
    Ltd.: Chichester, 1983. Chapter 4.
118.Fogarty, W.M.   In Microbial Enzymes and Biotechnology;
    Fogarty, W.M., Ed.; Applied Science Publishers:
    London 1983; Chapter 1.
119.Vercellotti, J.R.; Vercellotti, S.V. In New Develop-
    ments in Industrial Polysaccharides; Crescenzi,V.;
    Dea, I.C.M.; Stivala, S.S., Eds.; Gordon and Breach:
    New York, 1985; p. 125.
120.Klein, E.; Feldhoff, P.; Turnham, T.; Wendt, R.P. J.
    Membr. Sci. 1983, 15, 15.

RECEIVED October 3, 1990

Chapter 12

Use of Multidetection for Chromatographic Characterization of Dextrins and Starch

A. Heyraud and M. Rinaudo

Centre de Recherches sur les Macromolécules Végétales, Centre National de la Recherche Scientifique, B.P. 53 X, 38041 Grenoble cedex, France

Optical rotation and refractometric detectors were combined to characterize a series of dextrins by gel permeation chromatography and high performance liquid chromatography (HPLC). Gel permeation fractionates according to the hydrodynamic volume when adsorption is avoided, and for this reason separates isomaltodextrins from linear maltodextrins. Specific rotation power [α] is directly obtained and confirms the chemical structure. Elution of cyclodextrins is also tested and discussed. HPLC reverse phase chromatography separates the anomers as shown by optical rotation and allows good resolution in the range of low DP.
Light scattering and refractometric detectors were used to analyze starch with different amylose contents. Fractionations were also performed and iodine complexation tested in relation with the molecular structure. Gel permeation chromatography in DMSO was used to determine the molecular weight distribution of different starch samples without any calibration.

Chromatography is one of the best techniques to characterize oligomers and polymers for analytical purpose. The fractionation involves different mechanisms from which steric exclusion chromatography (GPC or SEC) is the most important for characterization of polymers. In SEC the elution is controlled by the hydrodynamic

0097–6156/91/0458–0171$06.00/0
© 1991 American Chemical Society

volume of the molecules as expressed by the product $[\eta]M$ where $[\eta]$ is the intrinsic viscosity and M is the molecular weight. High performance liquid chromatography (HPLC) in direct or reverse phase with a fractionation based on specific interactions with the stationary phase is certainly the most interesting for oligomers, even if sometimes it becomes necessary to combine GPC and HPLC techniques to solve an analysis of a complex mixture. A review on the chromatography of oligosaccharides has been published previously (1).

In the following, the chromatographic separation of starch and oligomers related to the $\alpha(1 \rightarrow 4)$ and $\alpha(1 \rightarrow 6)$ D-glucose series will be presented and discussed.

Separation of Oligosaccharides

SEC. Previously, gel permeation chromatography of cello and maltodextrins was investigated (2-4). It was shown that Biogel P_2 gave good separation based on the hydrodynamic volume of the carbohydrate, provided column temperature was over 65°C. Elution at 65°C suppressed the loose adsorption which exists on this matrix. A plot of $[\eta]$ M versus Ve gave a unique curve for the two series of dextrins which corresponds to a pure steric exclusion separation process. Assuming a difference between the hydrodynamic volume of a branched dextrin and that of a linear maltodextrin of the same number of monomers, a separation may be possible, at least in the range of low degrees of polymerization. Figure 1 contains chromatograms obtained for a series of maltodextrins and isomaltodextrins. Partition coefficients (K_d) given in Table I are calculated from their elution volumes (Ve) and plotted in Figure 2. The totally included volume (V_T) was obtained from the elution of D-glucose.

As usual, a linear dependency is obtained for log K_d plotted as a function of the degree of polymerization (DP). Universal calibration for the oligosaccharides was established (Figure 3) for the maltodextrins by plotting log $[\eta]$ M against Kd (3).

In Table II, the partition coefficients are given for the first oligomers ($\alpha 1 \rightarrow 6$; $\alpha 1 \rightarrow 4$). Panose is considered as a model for the branched trisaccharide. Differences in the Kd values are significant to show the role of the $\alpha(1 \rightarrow 6)$ linkage on the hydrodynamic volume of the corresponding oligomers. The use of optical rotation (i.e. $[\alpha]_D$) is also convenient to confirm the chemical structure of the oligomers. Panose has K_d and $[\alpha]_D$ value intermediate between those of DP 3, $\alpha(1 \rightarrow 6)$ and $\alpha(1 \rightarrow 4)$ linear oligomers. The optical rotation will be discussed shortly.

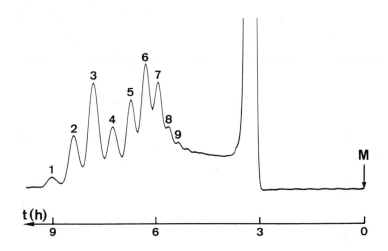

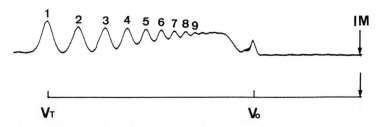

Figure 1. Maltodextrins (M) and isomaltodextrins (IM) separation by steric exclusion chromatography on a Biogel P$_2$ column (conditions given in Table I).

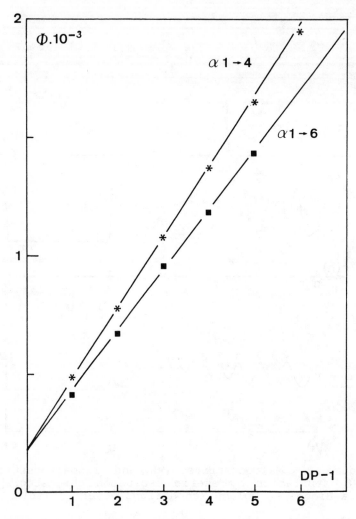

Figure 2. Partition coefficient as a function of (DP-1) for $\alpha(1 \rightarrow 6)$, $\alpha(1 \rightarrow 4)$ oligomers and cyclodextrins (CD).

Table I. Characteristics of the $\alpha(1 \to 4)$ and $\alpha(1 \to 6)$
oligomers

	DP	Kd	$[\alpha]_D$
$\alpha\ 1 \to 4$	1	1	52.7
	2	0.901	140.7
	3	0.800	153.5
	4	0.710	162
	5	0.623	164.2
	6	0.540	166.3
	7	0.487	168.4
	8	0.435	179
	9	0.382	–
	10	0.339	–
	11	0.302	–
$\alpha\ 1 \to 6$	2	0.847	121.5
	3	0.717	133
	4	0.610	145
	5	0.519	143
	6	0.448	145
	7	0.381	–
	8	0.327	–
	9	0.282	–
	10	0.245	–

Column : Biogel P_2, h = 205 cm, ϕ = 1.5 cm
Flow rate : 30 ml/h ; temperature : 65°C ; eluent : H_2O.

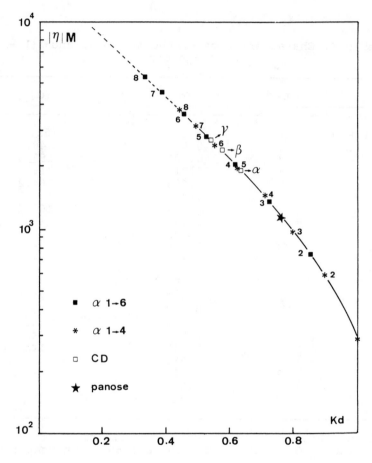

Figure 3. Universal calibration curve obtained by SEC.

Table II. Kd for lower DP oligosaccharides produced
during starch hydrolysis

DP	Kd	$[\alpha]_D$
2 (α 1 → 4)	0.901	140.7
2 (α 1 → 6)	0.855	121.5
3 (α 1 → 4) linear	0.800	153.5
Panose	0.762	145.4
3 (α 1 → 6) linear	0.717	133

The chromatogram of α, β, γ cyclodextrins is shown in Figure 4. Values of Kd are given in Table III and compared with the values for the comparable linear oligomers.

Table III. Characteristics of the cyclic and
corresponding linear oligomers

Oligomers	Kd	$[\alpha]_D$	$[\eta]$ ml/g
α	0.628	–	2.01
DP6	0.549	166.3	2.55
β	0.570	–	2.12
DP7	0.487	168.4	2.74
γ	0.534	–	2.09
DP8	0.435	179	2.85

Same conditions as Table I.

From these values, it is apparent that the hydrodynamic volumes of the cyclic oligomers are lower than the corresponding linear oligomers. In Table III, the values of K_d, $[\alpha]_D$ and the intrinsic viscosity $[\eta]$ are determined from universal calibration. The values of $[\eta]$ are only slighty dependent on the DP. The ratio between the intrinsic viscosity of the cyclic oligomers and that of the linear forms decreases when DP increases from 0.788 for α, to 0.773 for β and 0.733 for γ.

Some results exist in the literature and as an example, the GPC and HPLC of the three cyclodextrins were examined (5-7). GPC was then performed on Sephadex G-15 and G-25. In fact, due to the low resolution obtained in GPC it is sometimes difficult to determine the purity of the cyclodextrins. A chromatogram obtained on Biogel P2 for the mixture of the linear and cyclic oligomers is

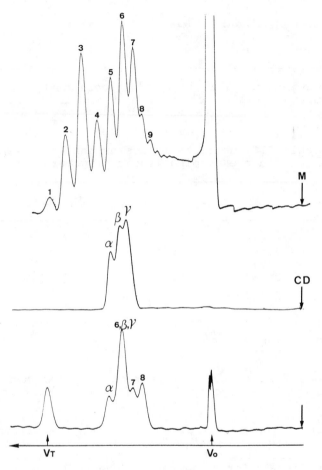

Figure 4. Comparison of linear (M) and cyclic (CD) oligomers separations on a Biogel P_2 column.

given in Figure 4. From these results, it seems that DP 6 linear and β cyclodextrin may not be separated on the basis of [η] M ; the K_d of the cyclodextrins are also reported on the universal calibration (Figure 3).

From the results obtained for the three series, it is concluded that :

- isomaltodextrins are eluted more rapidly than the isomeric maltodextrins (same DP) due to an increased contribution to the hydrodynamic volume from the α 1 → 6 linkage ;

- the cyclodextrins are eluted later than the linear oligomers (α 1 → 4) due to a reduced axial ratio. Nevertheless, the resolution in GPC is often not good enough and HPLC gives a better separation (8).

HPLC. Due to the lack of resolution in SEC, HPLC was used to improve resolution. Furthermore a detector based on optical rotation was adapted to the HPLC equipment and connected in series with the differential refractometer. It is known (9) that molar optical rotation includes contributions from the chemical structure of the molecules [m], its conformation [Λ] and the relative amounts of anomers according to the relation :

$$[\phi] = (n-1) \ ([m] + [\Lambda]) + S(\alpha,\beta) \quad [1]$$

This relation is valid for a homologous series of oligomers with n monomeric units. Here, [m] is the contribution of the monomeric unit ; [Λ] that of the glycosidic linkage ; S(α,β) is the contribution of the reducing end group due to the equilibrium of the α/β forms. For large n : (n-1 ≅ n) the specific optical rotation [α] is given by :

$$[\alpha] = \frac{[\phi]100}{nm_o} = 100 \ \frac{([m] + [\Lambda]) \ (n-1)}{nm_o} + 100 \ \frac{S(\alpha/\beta)}{nm_o} \simeq \frac{([m]+[\Lambda])}{m_o} \quad [2]$$

with m_o the molecular weight of the repeating unit. In the case of large n, [α] is then independent of the degree of polymerization and characterizes polymers.

If an optical rotary power detector is combined with a refractometric detector, then independent of the purety of the samples, the specific optical rotation $[\alpha]_D$ is obtained from the ratio of the two signals. Furthermore, the value [α] is a characteristic of each oligomer and permits its identification, the determination of D or L series, and also the chemical composition for heterogeneous oligomers or polymers (AxBy). It has been demonstrated that the ratio of the two signals gives the ratio M/G in alginates (10) and M/G in galactomannans (11).

A commercial detector is now available and was described by Goodall *et al.* under the name ACS

Chiramonitor (12). It is also convenient to use a flow
cell in a spectropolarimeter Perkin Elmer model 241.
    A first application of double detection was for the
identification of the α/β equilibrium in HPLC
chromatography on a C18 column (13). Each oligomer of the
maltodextrin series gives two peaks in refractive index
over a degree of polymerization of 3 (Figure 5). This is
one of the disadvantage of HPLC compared with GPC. The
oligomers are also well separated for analytical purposes
under their reduced form (3).
    The use of the optical rotation detector in HPLC
also allows one to determine the molar optical rotation
ϕ(DP-1) for the two series (relation 1 ; Figure 6) ; for
these conditions, the term ([m] + [Λ]) is obtained from
the slopes; $[\alpha]_D$ = 192 and 185 is calculated as specific
optical rotation of the corresponding polymers amylose
and dextran, respectively. These values are in good
agreement with the values given in the literature.
    The optical rotation of panose may be considered as
that of a copolymer given by the following additivity :
$[\phi]$ = 162 $[\alpha]_{1 \rightarrow 4}$ + 162 $[\alpha]_{1 \rightarrow 6}$ + 180 $[\alpha]_{glucose}$
taking the contribution determined previously.
    The agreement between the calculated value (705.60)
and the experimental one (732.80) is satisfactory. From
Figure 6, it seems that the difference between the
contribution of $\alpha(1 \rightarrow 4)$ linkage compared with that of
$\alpha(1 \rightarrow 6)$ is too small, to separate the contributions of
amylose and amylopectin in gel permeation chromatography
on the basis of the optical rotation.
    The separation of oligomers in HPLC on a C-18
column is controlled by their interactions with the
hydrophobic matrix. The structure of the oligomer and
especially its solubility is based on -OH groups. Due to
the low solubility of cyclodextrins, the HPLC was
performed in the presence of controlled amounts of
methanol in the eluent. Elution volumes of cyclodextrins
decrease when the content of methanol increases. The same
is observed with the linear oligosaccharides (Figure 7).
    Figure 8 gives the separation of the three
cyclodextrins in 15% MeOH. Elution volume increased in
the order γ < α < β. This seems to indicate different
solubilities of these cyclic oligomers, especially for γ
cyclodextrin. Under these conditions (C-18 columns ; 15%
methanol) the linear oligomers are eluted in the void
volume followed by the cyclodextrins. Optical rotation
should be also useful to confirm the identity of these
oligomers. The ratios between the two signals give the
following numbers : α(0.625), β(0.68), γ(0.803). Thus [α]
allows a good means for identifying cyclodextrins. From
these data, it is clear that HPLC based on specific
interactions allows good separations especially when the
role of the eluent is also considered. For heterogeneous
oligomers, it is nevertheless useful to perform first SEC
fractionation and take each peak for HPLC separation.

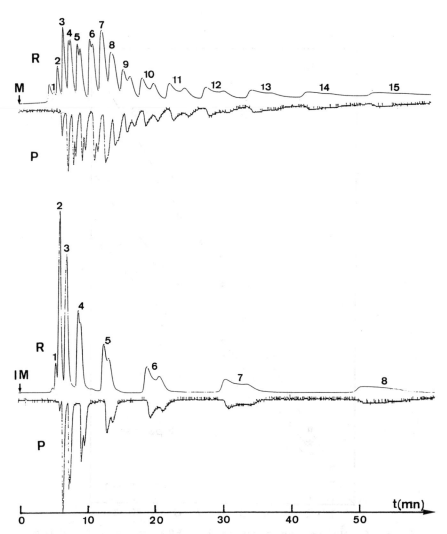

Figure 5. Maltodextrins (M) and isomaltodextrins (IM) separation by HPLC on a C-18 column eluted with H_2O. R = refractometric and P = polarimetric signals.

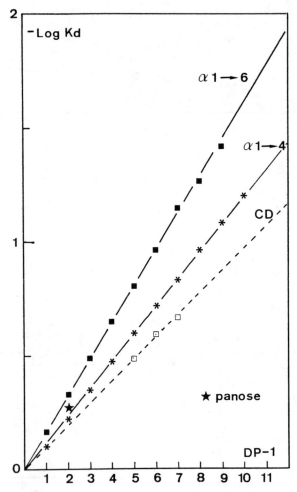

Figure 6. Molar rotation at λ = 589 nm as a function of (DP-1) for $\alpha(1 \to 4)$ and $\alpha(1 \to 6)$ oligomers.

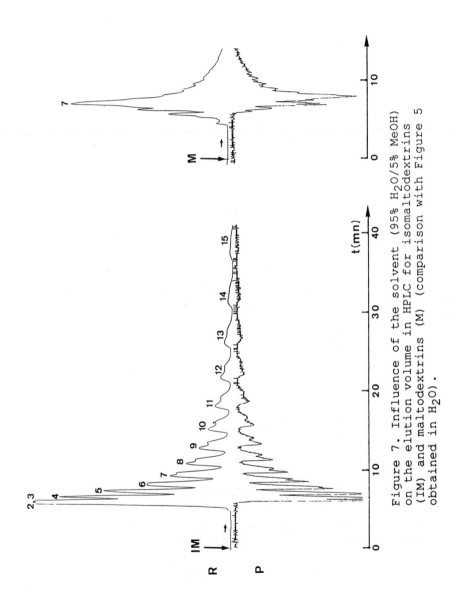

Figure 7. Influence of the solvent (95% H$_2$O/5% MeOH) on the elution volume in HPLC for isomaltodextrins (IM) and maltodextrins (M) (comparison with Figure 5 obtained in H$_2$O).

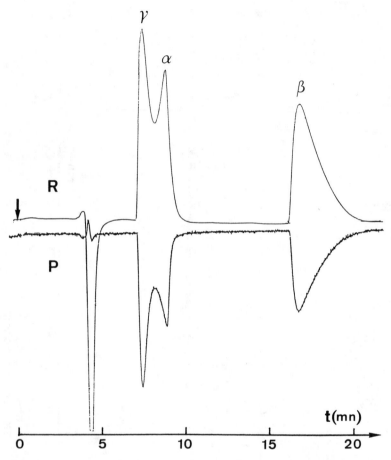

Figure 8. Separation of cyclodextrins by HPLC on a C-18 column in presence of 15% MeOH.

Characterization of Starch

Starch consists of amylopectin (highly branched D-glucose polymer with $\alpha(1 \rightarrow 6)$ linkages) and amylose (linear polymer with $\alpha(1 \rightarrow 4)$ linkages) and a fraction of intermediate material corresponding to amylose with some degree of branching reducing the degree of β-amylolysis. The GPC is well adapted to characterize a polymer by its molecular weight distribution. With starch the molecular weight distribution is superimposed upon a heterogeneous distribution in chemical structure and especially in the degree of branching.

At least two problems must be solved to interpret GPC chromatograms of starch :

- First, starch must be well dissolved with no adsorption on the support. Therefore, starch was dissolved in DMSO, a good solvent of the constitutive polymers. In bad solvents aggregation occurs and loose adsorption greatly disturb the determinations (*14,15*).

- Secondly, a relation between elution volume and the molecular weight must be established which is independent of the polymer morphology. This was obtained with a light scattering detector giving M directly without the use of a calibration curve.

Yamada and Taki (*16*) obtained molecular weight distributions of different starches on Sepharose 2B in the presence of 0.04% $HClO_4$. The iodine index was determined on the chromatogram and seemed to indicate a good separation between high molecular weight amylopectin and amylose. No values of molecular weight were given.

Van Dijk *et al.* (*17*) performed GPC of amylose dissolved in DMSO on a Porasil column (Waters ; silica gel deactivated). By using universal calibration after determination of the viscosity on each fraction, they obtained good agreement between Mw and Mn values obtained from GPC and direct measurements. Following these works, DMSO has been adopted as a good solvent for defatted starch (*14*). GPC experiments were performed on silica gels grafted with diol (Merck, Lichrospher-diol) at 60°C. The eluent was DMSO/MeOH 85/15 (v/v) containing 0.5 M ammonium acetate to reduce interaction (H bonds). The presence of MeOH decreased the viscosity of the solvent. Under these experimental conditions, universal calibration was shown to be valid indicating that there was no adsorption. The larger pores of these supports are even too small to obtain a good filtration of amylopectin which is usually excluded and separated from amylose. All the chromatograms of starch containing amylopectin show a peak with a very large molecular weight. This peak would contribute greatly to the overall weight average molecular weight of the sample.

Some of the values obtained are recalled in the Table IV.

Table IV. Characteristics of some starchs

		Mn	Mw	amylose %	CLI
Amylose AVEBE from potatoes	78% amy-losis	162000	278000	95	19.5
SNOW from waxy maize	>99% amy-lopectin	12×10^6	14.8×10^6	0	0.5
AMILO from Brasil tapioca		7.31×10^6	13.6×10^6	27	5.9

CLI = capacity of iodine fixation expressed in grm of I_2 for 100 grm of starch, taking as reference the first fraction of Avebe cristallized with butanol (ie CLI = 20.5 ; 100% amylose) and waxy maize (CLI = 0.5, 100% amylopectin) in solution I_2 in 0.05 M IK at 20°C.

From these values, it must be pointed out that the fraction of amylose expressed from I_2 fixation is different (generally larger) from the fraction of material crystallized in presence of butanol. In addition, the fraction crystallized from a starch sample usually has a CLI lower than that of pure amylose (18). The importance of the direct molecular weight measurement will be understood if one considers the Mark Houwink relation :

Amylose   DMSO/25°C (19) $[\eta] = 1.25 \times 10^{-3} M^{0.87}$
Amylopectin DMSO/MeOH 85/15 $[\eta] = 2.16 \times 10^{-3} M^{0.67}$
            0.5M AcNH$_4$ or $[\eta] = 1.84 \times 10^{-1} M^{0.36}$
               60°C (18)

The relations given for amylopectin depend on the degree of hydrolysis. The first one is for native starch down to 6.5×10^6. Then one gets a lower dependency with M corresponding to a more compact structure while molecular weight decreases.

Following these values, it is clear that even universal calibration, $[\eta] M(V_e)$, cannot be used directly for analysis of a starch sample. In addition, the K and a values are not well established in the literature due to difficulties in working with these systems. Multidetection GPC allowing the measurements on line of M, $[\eta]$ and concentration should be the best way to follow the characterization of starches from different origins (20). The use of a viscometer as detector should allow determinination of the factor g which is related to the degree of branching .

Experimental

The pure oligosaccharides were obtained by preparative liquid chromatography from partially hydrolyzed polymers. Mixture of maltodextrins is a commercial sample obtained by partial enzymic and acid hydrolysis of starch (Roquette, France). Mixture of isomaltodextrins is prepared by partial acid hydrolysis of dextran. Cyclodextrins are commercial samples from Aldrich.

Gel permeation chromatography was performed on Biogel P_2 eluted by pure water at 65°C as described previously (2, 3).

The HPLC was realized with a Waters solvent delivery system. Separations of oligomers were obtained with a Nucleosil 5 μm C-18 reverse phase column from S.F.C.C. (France) (8). The eluent was distilled water filtered through a 0.45 μm Millipore membrane. The starch was dissolved in DMSO and the eluent was DMSO/MeOH (85/15 v/v) in 0.5 M ammonium acetate. The column set was diol silica gel from Merck : 2 x Si 1000 Diol, 1 x Si 500 Diol, 1 x Si 100 Diol thermostated at 60°C. The detector was either a differential refractometer from Waters (R 401) or an IOTA (Jobin Yvon). The second on line detector was a light scattering detector (Chromatix CMX 100) or a spectropolarimeter (Perkin Elmer model 241 working at 365 nm with a flow cell of path length 10 cm and a 30 μl volume). The value of $[\alpha]_D$ are expressed from $[\alpha]_{365}$ data using a corrective factor. The partition coefficient Kd is expressed as :

$$Kd = \frac{V_e - V_o}{V_{glucose} - V_o}$$

with V_o the void volume and V_e the elution volume. V_o was obtained from the elution volume of a T70 dextran sample (Pharmacia).

Conclusion

In this paper the behavior of oligosaccharides corresponding to models for starch hydrolysis or amylolysis products was discussed. The role of the chemical structure on the elution volume was pointed out in relation to the nature of the glycosidic linkage. Combination with optical rotation improves the quality of analysis. The behavior of cyclic and branched oligomers was also discussed.

In the second part, the main results obtained in the domain of starch characterization were recalled showing the necessity to adopt a multidetection system for the analysis of starch molecular weight distribution. Nevertheless, the major problem remains the preparation of a good solution avoiding aggregates and with no adsorption on the gel matrix.

Literature cited
1.   Heyraud, A., ; Rinaudo, M. *J. Liquid Chromatogr.*, *suppl. 2*, 1981, **4**, 175-293.
2   Heyraud, A. ; Rinaudo, M. *J. Chromatogr.* 1978, **166**, 149-58.
3.   Heyraud, A. Thesis, Grenoble, France, 1978 and 1981.
4.   John, M. ; Trenel, G. ; Dellweg, H. *J. Chromatogr.* 1969, **42**, 476-84.
5.   Hokse, H. *J. Chromatogr.* 1980, **189**, 98-100.
6   Koizumi, K. ; Kubota, Y. ; Okada, Y. ; Utamura, T. ; Hizukuri, S. ; Abe, J. *J. Chromatogr.* 1988, **437**, 47-57.
7   Mattsson, P. ; Mäkelä, M. ; Korpela, T. *J. Chromatoqr.* 1988, **447**, 398-403.
8.   Heyraud, A. ; Rinaudo, M. *J. Liquid Chromatogr.* 1980, **3**, 721-39.
9.   Lawson, C.J.; Rees, D.A. *Nature* 1970, **227**, 390-92
10.   Bouffar Roupe, C. Thesis, Grenoble, France, 1989.
11.   Noble, O. ; Pérez, S. ; Rochas C. ; Taravel, F. *Polymer Bull.* 1986, **16**, 175-80.
12.   Lloyd, D.K. ; Goodall, D.M. ; Scrivener, H. *Analyt. Chem.* 1989, **61**, 1238-43.
13.   Heyraud A. ; Salemis P. *Carbohydr. Res.* 1982, **107**, 123-29.
14.   Salemis, Ph. ; Rinaudo, M. *Polymer Bull.* 1984, **11**, 397-400.
15.   Salemis, Ph. ; Rinaudo, M. *Polymer Bull.* 1984, **12**, 283-85.
16.   Yamada, T. ; Taki, M. *Die Stärke* 1976, **28**, 374-77.
17.   Van Dijk, J.A.P.P. ; Henkens, W.C.M. ; Smit, J.A. *J. Polym. Sci.* 1976, **14**, 1485-93.
18.   Salemis, P. Thesis, Grenoble, 1984.
19.   Cowie, J.M.G. *Makromol. Chem.* 1961, **42**, 230-47.
20.   Tinland, B. ; Mazet, J. ; Rinaudo, M. *Makromol. Chem. Rapid Commun.* 1988, **9**, 69-73.

RECEIVED November 20, 1990

Chapter 13

Phosphorolytic Synthesis of Low-Molecular-Weight Amyloses with Modified Terminal Groups

Comparison of Potato Phosphorylase and Muscle Phosphorylase B

C. Niemann[1,4], W. Saenger[1], B. Pfannemüller[2], W. D. Eigner[3], and A. Huber[3]

[1]Institut für Kristallographie der Freien Universität Berlin, Takustrasse 6, D–1000 Berlin 33, Germany
[2]Institut für Makromolekulare Chemie der Albert-Ludwigs-Universität Freiburg, Stefan-Meier-Strasse 31, D–7800 Freiburg, Germany
[3]Institut für Physikalische Chemie, Karl-Franzens-Universität Graz, Heinrichstrasse 28, A–8010 Graz, Austria

The preparation of low molecular weight amyloses with modified terminal groups in a DP range of 4-25 was carried out by enzymatic synthesis using either potato phosphorylase or phosphorylase b from rabbit muscle. p-Nitrophenyl-α-D-maltooligomers with a minimum chain length of five glucosyl residues served as primers; glucose-1-phosphate was the monomer. The investigation of the products by size exclusion chromatography/low angle laser light scattering and HPLC showed that the behaviour of the enzymes is different in the view of the distribution of oligomers formed under the same reaction conditions. Whereas the synthesis by muscle phosphorylase leads to an expected Poisson distribution the reaction products from potato phosphorylase are altered by a simultaneous pH dependent disproportionation. With both enzymes, significant amounts of the desired p-nitrophenyl-α-D-maltooligosaccharides in the DP range 10-20 can be obtained. These oligomers are of special interest for X-ray single crystal diffraction analyses.

[4]Current address: Whistler Center for Carbohydrate Research, Smith Hall, Purdue University, West Lafayette, IN 47907

Although the formation of supermolecular arrangements
by amylose and the branched amylopectin have been
extensively discussed, the crystallinity of starch
granules and details of retrogradation still remain
somewhat obscure. In order to make progress in this
field, low molecular weight amyloses (LMWA) with
definite chain lengths can serve as model substances
to answer open questions. Compounds with a degree of
polymerization (DP) in the range of 10-20 glucose
units per molecule appear to be particularly useful,
because they represent an intermediate stage between
the "low molecular" maltooligosaccharides and the
"high molecular" amylose and amylopectin. This DP
range corresponds to the length of the outer chains of
the amylopectin molecule, which are probably responsible for starch crystallinity (1). On the other hand,
in preliminary studies a sudden change of the X-ray
powder diffraction pattern of retrograded microcrystalline LMWA from the B-type to the A-type was observed
between DP 13 and 12 (2).
     The best method to elucidate molecular structures
in the solid state is the single crystal X-ray diffraction analysis. However, the production of larger
amounts of LMWA in a purity and quantity required for
crystallization is still a problem. Recently we
succeeded in the crystallization of a maltohexaose (3)
with an α-p-nitrophenyl group at the reducing end as
its polyiodide complexes. We assume that such a derivatization completed with complex formation facilitates
the growth of single crystals at least in this lower
DP range. To study the structure of the amyloses
further, we decided to extend these studies to modified LMWA in the range of DP 10-20.
     The usual methods to obtain LMWA, e.g. acid
hydrolysis of starch (4), amylolytic degradation of
amylose (5) or debranching of amylopectin (6) and
glycogen (7), can not be the method of choice because
a subsequent chemical modification to block the
reducing end in one configuration would be difficult
and ineffective. An alternative way is the chain
elongation of a suitably modified acceptor with the
help of cyclodextrin glucosyl transferase and α-cyclodextrin or by phosphorolytic synthesis with glucose-1-
phosphate as the monomer donor.
     The preparation of p-nitrophenyl-α-D-maltooligo-
saccharides by the former method has been reported by
Wallenfels et al. (8) leading to only low yields of
oligomers in the range beyond DP 7, but with considerable amounts of non-substituted maltooligosaccharides and ß- and γ-cyclodextrin. Meanwhile the p-nitro-
phenylated maltooligomers (up to DP 8) are commer-

cially available and can be used as suitable primers
for chain elongation reactions. Results of our investigations to produce LMWA with modified terminal groups
using cyclodextrin glucosyl transferase will be published elsewhere.

In phosphorolytic synthesis (eq. 1), products of
a rather narrow chain length distribution can be
expected, and unsubstituted oligomers should not be
formed. Using this method, high molecular weight
amyloses (HMWA) carrying different terminal groups
were obtained (9), and in preliminary studies the
reaction conditions to prepare larger quantities of
LMWA were optimised (10).

The aim of this work is the comparison of two of
the most commonly employed phosphorylases, potato
phosphorylase and rabbit muscle phosphorylase b in the
large scale production of LMWA. The decisive criterion
should be the chain length distribution of the oligomers formed.

Results and Discussion

Phosphorolytic synthesis.   In order to synthesize LMWA
with a p-nitrophenyl group in α-position at the
reducing end, p-nitrophenyl-α-D-maltotetraoside was
used as a primer. In these experiments we confirmed
previous results of Emmerling et al. (11) who observed
that an α-substituted maltotetraoside behaves like an
unsubstituted maltotriose which is not a good primer
for potato phosphorylase. The distribution pattern of
the products showed large amounts of non-converted
primer and low yields of higher oligomers (12). Subsequent studies showed that a modified primer suitable
for the reaction has to contain at least five glucosyl
units per molecule (10).

$$\text{pNP } G_n + m \text{ G-1-P} \xrightarrow{\text{phosphorylases}} \text{pNP } G_{n+m} + m P_i \qquad (1)$$

pNP = p-nitrophenyl, G-1-P = glucose-1-phosphate, P_i =
inorganic phosphate, $n \geq 5$, $m = 11$

We used a ratio of primer to monomer of 1:11 because
the conversion of G-1-P at pH 6.0, the pH optimum for
synthesis, should be at about 90 % for potato phosphorylase (13). The same conditions were used in the
experiments with muscle phosphorylase.

Potato phosphorylase (I) was isolated from potato
tubers by precipitation with ammonium sulphate according to the method of Ziegast et al. (9). Further
purification was carried out by hydrophobic interaction

chromatography and subsequent gel filtration (II)
(14). For comparison with these preparations a still
higher purified potato phosphorylase (III) was used
(15). Potato phosphorylase II and III appeared as a
single band in SDS gel electrophoresis. An improved
stability observed for III may result from its preparation in 0.1 M phosphate buffer pH 7.5, whereas I and
II were prepared without any addition of inorganic
phosphate. Muscle phosphorylase b (Sigma), containing
a small amount of inorganic phosphate and 5'-AMP was
used without further purification. 5'-AMP and mercapto-
ethanol were added to transform the inactive b-form
into the active form (16).
   All phosphorylases were used in a ratio of 0.1 U
per μmol primer to reach equilibrium within a relatively short time (10). The conversion of G-1-P, measured
by colorimetric determination of liberated inorganic
phosphate (17), indicates the rate of reaction. Whereas
potato phosphorylase shows a higher initial rate of
reaction, the muscle phosphorylase gives a somewhat
higher conversion at the end (Table I).

Table I. Conversion of glucose-1-phosphate in phospho-
          rolytic synthesis (see text for explanation)

reaction time min	% conversion of total glucose-1-phosphate by	
	potato phosphorylase	muscle phosphorylase
5	14.1	7.5
10	24.4	15.0
20	38.6	26.4
40	58.0	43.8
60	67.2	57.5
120	77.7	78.8
180	83.3	88.8
240	85.3	91.3

The difference in initial rate of reaction could be
due to the higher affinity of potato phosphorylase for
short chain maltooligosaccharides in comparison to
muscle phosphorylase which, according to Fukui et al.
(18), is associated with different amino acid sequences
at the substrate binding sites. The effect obviously
disappears during further progress of reaction.
   Characterization of the products of phosphorolytic synthesis was carried out by size exclusion
chromatography combined with low angle laser light
scattering according to a technique usually applied to
higher molecular weight compounds (19,20). The simultaneous detection of laser scattering intensity (LALLS
detector) and concentration (DRI-detector) provides

information about molecular weight averages Mn, Mw, Mz, polydispersities Mw/Mn, Mz/Mn and molecular weight distributions (MWD) without any external calibration required.

The DRI-signals were of excellent quality and reproducibility. The LALLS-signals were not as good as expected, despite the averaging procedures used. This could be associated with the inherent limits of the technique for the low molecular weight fraction of our oligomer mixtures below DP 12. It could also explain the discrepancy of the DP maxima detected in HPLC and the observed molecular weight averages by SEC/LALLS (Table II).

Table II.   DP Maxima (HPLC); expected and observed molecular weight averages and polydispersities (SEC/LALLS) of products and phosphorolytic synthesis

phosphorylase source	DP max. (HPLC)	M_{av}exp. g/mol	SEC/LALLS data				
			M_w g/mol	M_n g/mol	M_z g/mol	M_w/M_n	M_z/M_w
potato (II)	10	1800	2400	2100	2700	1.14	1.13
potato (III)	11	1950	2800	2500	3000	1.11	1.07
muscle	14	2400	2600	2400	2700	1.08	1.04

The resulting MWD for the LMWA obtained with potato and with muscle phosphorylase are given in Figure 1. The products obtained by the muscle enzyme exhibit a higher uniformity than those prepared by the potato enzyme, independent of the degree of purity of the used potato phosphorylases. Similar results were observed with unmodified LMWA and will be published elsewhere.

Figure 2 shows the molecular weight distribution calculated from HPLC (21) in comparison to a corresponding theoretical Poisson distribution (22). For muscle phosphorylase there is good agreement. The serious deviation shown by potato phosphorylase prompted us to investigate the development of chain length distribution with both enzymes as a function of reaction time. Aliquots were removed from the reaction mixtures at fixed times and the reaction was terminated by treatment with a mixed bed ion exchanger to remove salts and enzymes. The products were analyzed by HPLC.

Figure 3a-h and Tables III and IV illustrate the time dependent differences in the molar composition of the oligomer mixtures. It is obvious that the reaction of both phosphorylases proceeds differently. Whereas the muscle phosphorylase reacts by a straight chain elongation of the primer, the reaction with potato phosphorylase is more complicated. Already after five

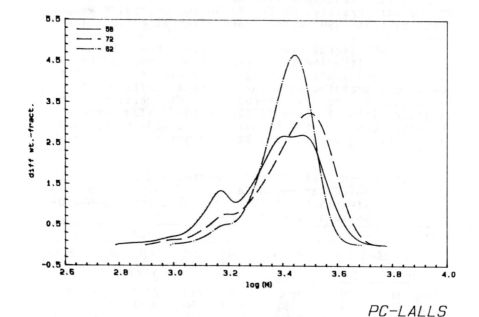

PC-LALLS

Figure 1   Molecular weight distribution (SEC/LALLS)
            of LMWA with modified terminal groups
            obtained by phosphorolytic synthesis
            either by potato phosphorylase (II) (72)
            and (III) (58) or muscle phosphorylase b
            (62).

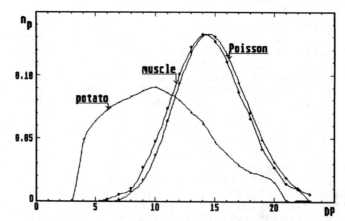

Figure 2   Molar ratios np of LMWA (calculated from
            HPLC) versus DP in comparison to a theore-
            tical Poisson distribution.

Table III. Development of chain length distribution of LMWA with modified terminal groups during phosphorolytic synthesis. Enzyme: potato phosphorlyase. Concentration in mol % calculated from HPLC. DP Maxima written in italic

DP	0	5	10	20	40	60	120	180	240min
3								0.04	0.04
4		2.75	2.00	2.05	2.60	2.33	3.70	4.59	4.12
5	100	21.78	11.40	7.18	5.35	4.75	5.47	5.97	5.71
6		*39.56*	*36.72*	*18.79*	10.70	8.05	6.74	6.79	6.31
7		21.00	27.14	17.90	*11.00*	*8.10*	6.92	7.38	7.08
8		6.48	10.13	8.62	6.71	5.70	7.06	7.81	7.63
9		2.04	4.35	5.65	5.73	5.70	7.34	8.25	*7.70*
10		*2.66*	*5.30*	*8.02*	8.03	7.43	8.39	*8.49*	7.55
11		1.52	3.29	6.90	8.16	7.99	8.60	8.08	7.14
12		0.07	2.58	6.93	*8.69*	*8.54*	*8.70*	7.51	6.85
13			1.62	4.90	7.47	7.88	7.91	6.61	6.54
14			0.06	3.09	5.80	6.75	6.69	5.79	5.65
15				2.10	4.28	5.50	5.50	4.44	4.98
16				1.55	3.53	4.74	4.41	3.54	4.60
17				0.09	2.65	3.83	3.37	2.72	3.54
18					2.00	2.88	2.70	2.16	2.72
19					1.19	2.06	2.38	1.92	2.30
20					0.09	1.32	1.60	1.42	1.85
21					0.05	0.08	1.18	0.09	1.30
22						0.07	1.07	0.08	0.09
23						0.04	0.09	0.07	0.08

Table IV. See Table III. Enzyme: muscle phosphorylase

DP	0	5	10	20	40	60	120	180	240min
5	100	*50.59*	27.10	5.69	0.18	0.04	0	0	0
6		*28.27*	24.92	8.39	0.72	0.08	0	0.13	0.22
7		14.59	*25.54*	19.19	3.81	1.09	0.50	0.95	0.62
8		4.39	17.84	*37.20*	18.36	5.89	1.37	0.95	1.30
9		0.30	3.08	18.98	*22.65*	11.26	2.93	2.59	2.74
10			0.36	7.23	20.83	15.48	5.04	4.61	4.52
11				2.09	16.18	*18.92*	8.08	7.13	6.69
12				0.60	9.84	18.25	11.13	9.75	8.88
13					4.35	13.59	12.95	11.81	10.72
14					1.63	8.43	*13.38*	*12.82*	*11.72*
15					0.40	4.57	1163	12.25	11.64
16						2.05	9.57	10.70	10.67
17						0.89	6.55	8.39	8.80
18						0.23	4.39	6.34	6.63
19							2.59	4.00	4.49
20							1.57	2.59	2.84
21							1.10	1.34	1.57
22							0.82	0.90	1.04
23									0.74
24									0.48

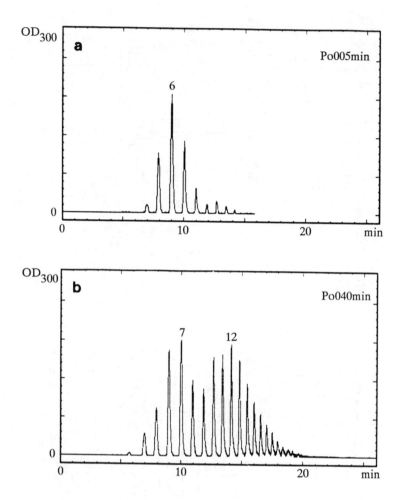

Figure 3. Chain length distribution of LMWA during phosphorolytic synthesis after a, 5 and b, 40 min by potato phosphorylase. *Continued on next page.*

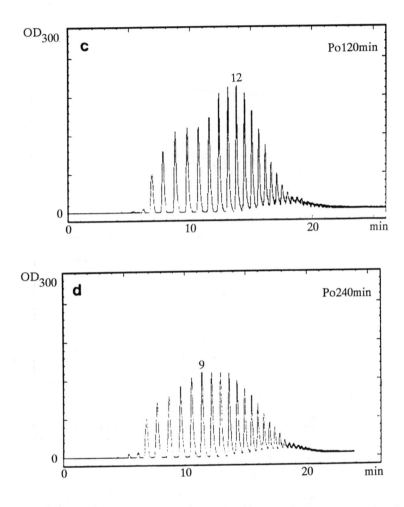

Figure 3. *Continued.* Chain length distribution of LMWA during phosphorolytic synthesis after c, 120 and d, 240 min by potato phosphorylase. *Continued on next page.*

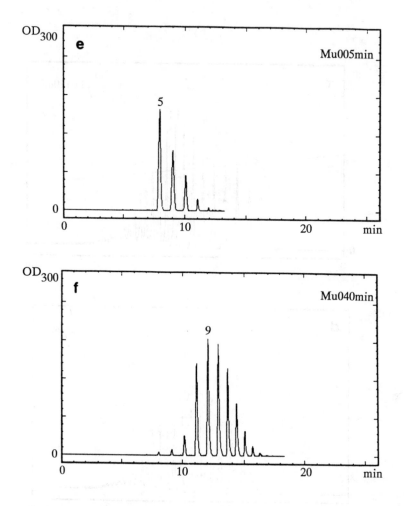

Figure 3. *Continued.* Chain length distribution of LMWA during phosphorolytic synthesis after e, 5 and f, 40 min by muscle phosphorylase. *Continued on next page.*

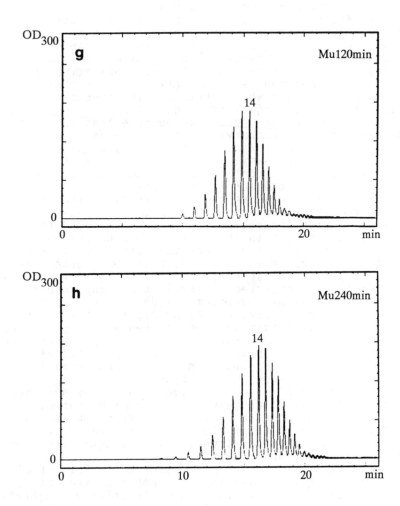

Figure 3. *Continued.* Chain length distribution of LMWA during phosphorolytic synthesis after g, 120 and h, 240 min by muscle phosphorylase.

minutes a remarkable amount of the p-nitrophenyl-α-D-maltotetraoside is formed. Two distinct maxima develop within approximately 40 minutes. This can be correlated with two simultaneous, independent enzymatic activities. After two hours reaction time, the distribution shows a single maximum in the region of DP 12. Even later, the maximum is shifted to DP 10 and small traces of the maltotrioside are detected. The flattening of the distribution continues, indicating that a disproportionation reaction influences the chain length distribution while synthesis is reduced or even discontinued under equilibrium conditions.

These phenomena appear to indicate that the less purified potato phosphorylase (I) is contaminated by a disproportionating enzyme, most probably D-enzyme, or a special kind of α-amylase (16,23). The use of purified enzyme did not confirm this assumption. On the contrary, the effect was rather pronounced.

In order to examine the disproportionation activity separately we incubated the p-nitrophenyl-α-D-maltopentaoside with the enzymes under the same conditions as in synthesis but in the absence of glucose-1-phosphate. The results of HPLC analysis are shown in Fig. 4. After four hours at pH 6.0 with the phosphate free prepared phosphorylase preparation (II), a slow disproportionation is observed. After 24 hours the amounts of p-nitrophenyl-α-D-maltotetraoside and -hexaoside are considerably increased. At pH 7.0 (0.1 M Tris/HCL buffer) the disproportionation is even more significant after four hours. In contrast, the muscle phosphorylase does not attack the primer at pH 6.0 even after 24 hours. At pH 7.0 disproportionation is considerably slower than with the potato enzyme.

Incubation of the pentaoside with highly purified phosphorylase (III) containing 0.03 µmol Pi per ml reaction mixture showed a more rapid reaction even at pH 5.5 and pH 6.0. After two hours more than 50 % of the primer was disproportionated to the tetraoside and higher oligomers up to DP 10.

A similar effect was also observed by Palm et al. (24) with E. coli phosphorylase. They reported that after prolonged incubation of a radioactively labelled maltotetraose at pH 6.5 a small fraction of radioactivity was also found by TLC in the position of maltotriose, maltose and added limit dextrin. Our investigations with 30fold ratio of potato phosphorylase (I) to modified primers at pH 5.8 revealed after 24h as a main product of disproportionation the maltotrioside. The maltoside and glucoside as well as the whole series up to DP 16 were formed (10).

It should be discussed whether inorganic phosphate is actually involved in the glycosyl transfer reaction. Klein et al. (25) postulated a catalytic mechanism of

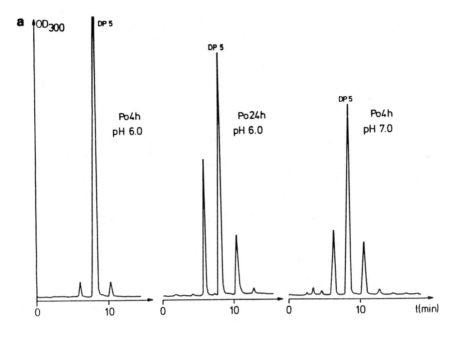

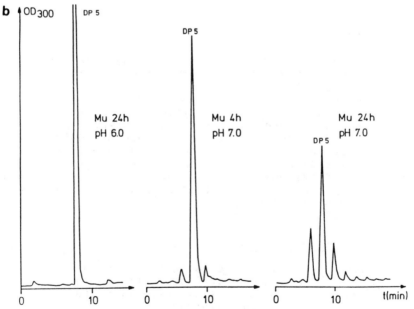

Figure 4   Disproportionation of p-nitrophenyl-α-D-maltopentaoside (DP 5) (a) potato phosphorylase (II) and (b) muscle phosphorylase b at pH 6.0 and 7.0 after 4 h and 24 h.

phosphorylase action in which a "mobile" phosphate
anion plays a versatile role. It could serve as a
proton carrier for substrate activation, stabilizing
the intermediate and acting as a nucleophile which can
accept a glycosyl residue reversibly. However, muscle
phosphorylase was not able to disproportionate the
pentaoside at pH 6.0 even in the presence of 5.5 µmol
Pi. Obviously the high amounts of liberated phosphate
in phosphorolytic synthesis (90 µmol/ml at equilibrium)
did not impair the distribution pattern of the products. The lower rate of disproportionation by phos-
phorylase⁻(II) and the considerable increase of reaction rate at extremely low concentrations of Pi suggest
that inorganic phosphate could have a still unknown
function in enzyme activation, especially in non-regu-
lated plant systems.

The glycosyl transfer activity seems to be more
sensitive with regard to the pH dependence. The pH-
optimum for both enzymes is reported in the range
between pH 4.9 and 8.7 (17,22,26). Particular specifi-
cations of optimum pH values for synthesis and for
degradation differ. In general, phosphorolytic degradation should not occur at pH 5.5 or 6.0. But it has to
be considered that the disproportionation reaction is
due to a synthesis/degradation equilibrium especially
at pH 7.0.

Conclusions

A significant difference was observed between the
chain length distribution of LMWA with modified terminal groups obtained by phosphorolytic synthesis
either with potato phosphorylase or muscle phosphorylase b. Under the given conditions the synthesis by
muscle phosphorylase delivers a much narrower chain
length distribution within the DP range of 10-20. The
products of the synthesis by potato phosphorylase are
altered by a disproportionation of the primer and
formed oligomers which impairs the distribution pattern
to such a degree that the yield of LMWA with modified
terminal groups decreases.

Legend of Symbols

SEC   = size exclusion chromatography
LALLS   = low angle lase light scattering
DRI   = differential refractive index
M_n = number average
M_w = weight average
M_z = z-average

Acknowledgments

We thank Dr. Manfred Buehner for highly purified potato phosphorylase and helpful discussions. We also thank Dr. Dieter Palm for valuable suggestions. This work was supported by the Deutsche Forschungsgemeinschaft and the Fonds der Chemischen Industrie.

Literature Cited

1. French, D. Starch, Chemistry and Technology; Whistler, R.L.; BeMiller, J.N.; Paschall, E.F.; Academic: Orlando, 1984, pp 184-247
2. Pfannemüller, B. Int. J. Biol. Macromol. 1987, 9, 105-108.
3. Hinrichs, W.; Büttner, G.; Steifa, M.; Betzel, Ch.; Zabel, V.; Pfannemüller, B.; Saenger, W. Science 1987, 238, 205-208.
4. Kikomoto, S.; French, D. J. Jpn. Starch. Sci. 1983, 30, 69-75.
5. Emmerling, W.; Pfannemüller, B. Carbohydr. Res. 1980, 86, 321-324.
6. Hizukuri, S. Carbohydr. Res. 1985, 141, 295-306.
7. Gidley, M.J.; Bulpin, P.V. Carbohydr. Res. 1987, 161, 105-108.
8. Wallenfels, K.; Földi, D.; Niermann, H.; Bender, H.; Lindner, D. Carbohydr. Res. 1978, 61, 359-368.
9. Ziegast, G.; Pfannemüller, B. Carbohydr. Res. 1987, 160, 185-204.
10. Niemann, C.; Nuck, R.; Pfannemüller, B.; Saenger, W. Carbohydr. Res. 1989, 197, 187.
11. Emmerling, W. In Mechanism of Saccharide Polymerisation and Depolymerisation; Pfannemüller, B.; Marshall, J.J., Ed.: Academic: New York, 1980, pp 413-420.
12. Pfannemüller, B.; Burchard, W. Makromol. Chem. 1969, 121, 1.
13. Holló, J.; László, E.; Hoschke, A. Stärke/Starch, 1965, 17, 377-381.
14. Niemann, C.; Goyal, B.; Beck, R.H.F.; in preparation
15. Bühner, M. Universität Würzburg, private communication
16. Pfannemüller, B.; Potratz, Ch. Stärke/Starch 1977, 29, 73-80.
17. Husemann, E.; Fritz, B.; Lippert, R.; Pfannemüller, B.; Schupp, E. Makromol. Chem. 1958, 26, 181-198.
18. Fukui, T.; Shimomura, S.; Nakano, K. Mol. Cell. Biochem. 1982, 42, 129-144.

19.   Yu, L.P.; Rollings, J.E. J. Appl. Polymer Sci.
      1987, 33, 1909-1921.
20.   Eigner, W.-D.; Abuja, P.; Beck, R.H.F.; Praznik,
      W. Carbohydr. Res. 1988, 180, 87-95.
21.   Niemann, C.; Nuck, R.; Pfannemüller, B.; Saenger,
      W. Carbohydr. Res. in preparation.
22.   Flory, P.J. Principles of Polymer Chemistry;
      Cornell University Press: Ithaka, 1953, p 337.
23.   Linder, D. Ph.D. Thesis, Universität Freiburg,
      Freiburg 1978/79.
24.   Palm, D.; Blumenauer, G.; Klein, H.W.; Blanc-
      Muesser, M. Biochem. Biophys. Res. Comm. 1983,
      111, 530-536.
25.   Klein, H.W.; Im, H.J.; Palm, D. Eur. J. Biochem.
      1986, 157, 107-114.
26.   Lee, Y.P. Biochem. Biophys. Acta 1960, 43, 18-24.

RECEIVED September 9, 1990

Chapter 14

Maize Starch Sample Preparation for Aqueous Size Exclusion Chromatography Using Microwave Energy

Gregory A. Delgado, David J. Gottneid, and Robert N. Ammeraal

American Maize-Products Company, 1100 Indianapolis Boulevard, Hammond, IN 46320

A convenient method for solubilizing maize starch is needed for characterizing and comparing starches from different genetic varieties of maize as well as their enzyme hydrolysis products. A water-based system for solubilization of starch was developed for size exclusion chromatography (SEC). Samples from four different commercial types of maize starch were prepared for SEC using microwave energy. A microwave method had the advantage of speed and convenience and it did not require addition of any reagent other than water while the chromatography yielded results comparable to other methods of preparation.

A convenient method for solubilizing starch is needed for characterizing and comparing starches from different genetic varieties of maize as well as their enzyme hydrolysis products. This method is needed especially for performing size exclusion chromatography (SEC). SEC allows one to compare the relative distribution of intact polymeric components of starches. A major concern about any solubilization method is that disruption of the granular structure should proceed while minimizing degradation of the starch polymeric structure. Current methods of solubilization include utilizing solvents, such as dimethyl sulfoxide (DMSO) (1), or sodium hydroxide or autoclaving. The autoclave step is usually followed by sonication in either neutral or alkaline aqueous systems (2). A new method for solubilization of starch has been developed and is reported here. This method utilizes a neutral aqueous system and microwave energy. This microwave method has been used to compare starches from seven different genetic varieties of maize.

Materials and Methods

Materials. Seven starches from different genetic varieties of maize

were obtained from American Maize-Products Company. They included
four commercial starches: waxy, common, Amylomaize V and Amylomaize
VII. These starches have approximate apparent amylose contents of
0, 25, 50 and 70%, respectively. Starches from three other genetic
varieties of corn were also used. They were dull waxy (duwx), dull
horny (duh) and amylose extender dull (aedu). The approximate
apparent amylose contents of these starches are 0, 30 and 50%,
respectively.

Chromatographic System. SEC was performed on a Hewlett-Packard 1090
Chromatograph. Two Waters linear Ultra hydrogel columns (crosslinked methacrylate, 7.8 mm diameter x 300 mm length) were connected
in series preceded by a guard column. The columns were maintained at
45°C. The mobile phase was 0.1M sodium nitrate in water pumped at a
flow rate of 0.8 milliters per minute. Temperature of the
refractive index detector was maintained at 40°C. The injection
volume was 100 microliters. Commercial standards were used. Low
molecular weight materials were obtained from Sigma. Higher
molecular weight standards were obtained from Polymer Laboratories.
They are derived from pullulan, and, as such, were used only as
general indicators of molecular weight range rather than as precise
standards. Standard solutions were prepared (0.1 wt. per volume) in
0.1M sodium nitrate aqueous mobile phase with 0.2% sodium azide
added as a preservative. All solutions were passed through a 0.45
micron filter before injection.
     Starch solubility was determined by measuring total refractive
index response from SEC-HPLC chromatography. Detector response was
compared to that of a standard soluble 5 DE waxy maize hydrolysate
(Lo-Dex 5) of known concentration. Insoluble material was removed
by filtration. The mass of total starting material was measured
gravimetrically. Solubility is reported as the percent of soluble
starch to the total starting dry solids.

Autoclave/Sonication Method for Solubilization. Solubilization of
maize starches by autoclaving followed by sonication in both an
aqueous and an alkali system was performed according to the method
of Jackson et al. (2).

Solvent (DMSO) Method for Solubilization. A modified method for
solubilizing starch in solvent was used (1). Maize starch samples
were prepared in a test tube by dissolving 16 mg. of starch in 4 ml.
of 90% dimethyl sulfoxide solution. A suspension was formed by
mixing with a vortex mixer for 5 sec. It was then allowed to stand
at room temperature for 30 min. It was again vortex-mixed for 5
sec. Then it was placed in a boiling water bath for 5 min. and
allowed to cool to room temperature. It was again vortex-mixed for
5 sec. and centrifuged for 5 min. at 3000 x g. Aliquots of 100 ul.
were then carefully removed from the top of the supernatant liquid
and chromatographed.

Microwave Energy Method For Solubilization. A microwave oven was
preheated (3). A 0.1% starch sample was prepared by suspending 5.5
mg. of starch (dry basis) in 5.0 ml. of deionized water. The sample
was prepared in a teflon microwave digestion bomb (model no. 4782,

Parr Instrument Company). The bomb was placed in a microwave oven
and subjected to the desired radiation power level for the appropriate time interval. The bomb was then cooled in a cold water bath.
After the solution had reached room temperature it was filtered thru
a 0.45 micron filter. The solution was then ion exchanged by adding
ion exchange resin and filtered again thru a 0.45 micron filter. A
sample of filtrate (100 ul.) was then chromatographed by SEC.

Results and Discussion

Our initial objective was to evaluate the amylose component of
different genetic varieties of starch. The exclusion volume of the
particular column system was selected to end at 13 to 14 minutes;
the early peak in the chromatogram is actually excluded material.
This peak represents higher molecular weight materials and probably
contains amylopectin. Molecular weight approximations were limited
to the highest pullulan molecular weight of 853,000 at a retention
time of about 18 minutes.

Sample Preparation Observations. The observations outlined below
were made:
A. During the autoclave/sonication method:
   1) a gel was observed at the bottom of the test tube after
      boiling in the aqueous system and it persisted even
      after autoclaving,
   2) sonication for longer time intervals showed increased
      dispersion of the gel,
   3) microscopy revealed both heat swollen and unswollen
      intact granules in the sediment after centrifugation,
   4) gelatinized starch was present after boiling in the
      alkali system; it dispersed after autoclaving.
B. When the solvent (DMSO) method for solubilization was used,
   after centrifugation:
   1) a gel-like sediment formed at the bottom of the test
      tube and
   2) the supernatant was clear.
C. Increased clarity of solution and increased ease of
   filtration were observed as the heating time and microwave
   power output were increased during the solubilization of
   starch using microwave energy.

Autoclave Method. Amylomaize V was solubilized in an autoclaved
aqueous system (a 10 ml. aliquot of a 0.5% by weight starch
solution) and was sonicated at different time intervals. As
sonication time increased, the relative ratio of high molecular
weight material to low molecular weight material in the soluble
phase increased. This suggests that precipitated higher molecular
weight material was being solubilized by sonication. Table I shows
that solubilization in an aqueous system increased with increasing
sonication time. Solubilization is also enhanced as the apparent
amylose content of the starch decreases.

Table I. Effects of Sonication Time: % Solubilized

Sonication Time	Starch Type		
	Waxy	Common	Amy V
30 seconds	56.5%	47.4%	39.5%
80 seconds	79.0%	75.3%	46.5%

When amylomaize 5 is autoclaved and sonicated in alkali, the concentration of solubilized components increased. As the alkali concentration increased, a shift in the distribution of components occurred with the appearance of more low molecular weight material. This may be due to degradation of the starch polymer.

Solvent Method. Unlike the autoclaved aqueous system, solubility of the components in 90% DMSO/water increased as the apparent amylose content increased. The lower molecular weight fraction is more easily solubilized in DMSO than is the higher molecular weight fraction. Overall, this method yielded low percentages of solubility. Only 32.5% (by weight) was solubilized for the most soluble starch, Amylomaize VII.

Microwave Method. Common maize starch was heated by microwave for 6 minutes at different power levels. As the wattage increased, the concentration of solubilized polymers also increased. At 1200 watts, 96% of the starch was solubilized. However, the peak corresponding to the higher molecular weight fraction was reduced, suggesting that degradation occurred. All of the following microwave method results were obtained by irradiating starch samples with 720 watts of microwave energy. As the microwave heating interval increased, the amount of starch solubilized increased. This was generally true for all the starches used in this study. In the case of starch from corn with the dull horny mutation, as the microwave heating interval increased there was a significant shift toward a higher concentration of lower molecular weight materials. Chromatographic analysis of common and dull horny starches yielded similar component distribution patterns at approximately 70% solubility (Figure 1).

Waxy starch, which is essentially all amylopectin, had an increase in solubilization as the duration of heating with microwave energy was increased. The longest exposure time, however, revealed a reduction in the higher molecular weight peak suggesting degradation. Starch obtained from dull waxy corn is also comprised of amylopectin. The physical properties of this starch are significantly different, however, from those of conventional waxy starch. Chromatograms of dull waxy starch also show a shift toward lower molecular weight materials with longer exposure to microwave energy, similar to waxy starch. Waxy and dull waxy starches yielded similar molecular weight distribution at similar levels of solubilization.

The distribution patterns of components for all of the starches studied contained an early amylopectin peak (Figure 1) or shoulder (Figure 2). At a retention time of 19-20 minutes a peak is evident which includes the higher molecular weight amyloses and intermediate

WEIGHT % DISTRIBUTION

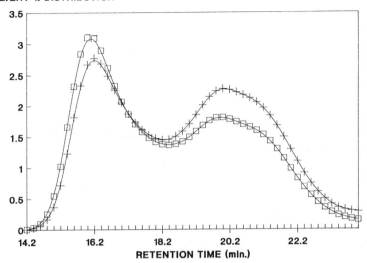

Figure 1. High-performance size exclusion chromatographic profiles of common and dull horny maize starch solubilized by microwave energy; ▣ common, ✚ dull horny.

WEIGHT % DISTRIBUTION

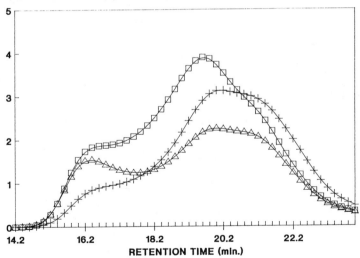

Figure 2. High-performance size exclusion chromatographic profiles at 62.7% solubilized amylomaize V, 74.0% solubilized amylomaize VII and 92.0% solubilized amylose extender dull (aedu) maize starch solubilized by microwave energy; ▲ amy V, ✚ amy VII, ▣ aedu.

Table II. Microwave Energy Method for Solubilization

Advantages	Disadvantages
Short Analysis Time	Indirect Temperature Control
Low Equipment Cost	Indirect Pressure Control
Low Energy Cost	
High Solubilization	
No Special Reagents	

Table III. Comparison of Solubilization Methods

	Method			
Features	AC(Water)	AC(Akali)	DMSO	Microwave
Preparation Time	135 min.	135 min.	105 min.	90 min.
Equipment Needed	Hot Plate	Hot Plate	Hot Plate	Microwave Oven
	Autoclave	Autoclave	Vortex Mixer	Digestion Bomb
	Sonicator	Sonicator	Centrifuge	
	Centrifuge	Centrifuge		
	Waterbath	Waterbath		
Steps	8	9	10	4
Degradation	No	Yes	No	No

molecular weight branched molecules. This is followed by a shoulder or broad distribution of the lower molecular weight amyloses and smaller branched molecules. While there was shift toward production of lower molecular weight fractions with overheating, the heating time at which this degradation begins is different for each of the starches studied.

Starch obtained from corn with the amylose extender dull (aedu) mutation also contains approximately 50% apparent amylose. The aedu starch did not show a shift in molecular weight distribution to a higher concentration of low molecular weight material as the duration of heating with microwave energy increased. This is evident by a high concentration of material about the middle of the distribution (Figure 2). These features are unique when compared to Amylomaize V and Amylomaize VII molecular weight distributions, (Figure 2). In general, as the apparent amylose content of starches increased, solubilization with microwave energy increased. This is evident with 60% of the Amylomaize V starch solubilized after only 4 minutes of heating time.

The microwave energy method for starch solubilization for size exclusion chromatography had many advantages (Table II). A minimum amount of time is needed for sample preparation and equipment required is relatively inexpensive. A high degree of solubility can be achieved but no special reagents or solvents are needed. The main disadvantages are that the temperature and pressures within the digestion bomb are controlled only indirectly. Excessive heating appears to damage the starch polymers. The autoclave/sonication method yielded a desired high solubility but its disadvantages outweighed this advantage. It required a long analysis time and it had a high equipment cost. High energy input and a special reagent were needed. The solvent (DMSO) method had a moderate analysis time and energy need. But, again, there was a high equipment cost and a special solvent was needed. This method's main disadvantage was that it could not produce a high solubility in a short analysis time. Table III compares features of the four solubilization methods reviewed. A new microwave energy method was presented that is convenient, quick and effective for the solubilization of starch. Current methods for solubilization were reviewed and the microwave method was compared to them. The microwave method was also used for comparing solubilization of starches from different genetic varieties of maize. It should be noted that individual methodologies must be varied depending on the type of starch and the apparent amylose and amylopectin content.

Literature Cited

1. Kobayashi, S.; Schwartz, S. J.; Lineback, D. R.; Cereal Chem. 1986, Vol. 63, No. 2, 71-74.
2. Jackson, D. S.; Choto-Owen, C., Waniska, R. D.; Rooney, L. W. Cereal Chem. 1988, Vol. 65, No. 6, 493-496.
3. Schiffman, R. F. Microwave World 1987, Vol. 8, No. 1, 7-9, 14-15.

RECEIVED October 3, 1990

Chapter 15

Distribution of the Binding of A Chains to a B Chain in Amylopectins

Susumu Hizukuri and Yuji Maehara

Department of Agricultural Chemistry, Faculty of Agriculture, Kagoshima University, Korimoto–1, Kagoshima-shi 890, Japan

B chains of amylopectin were classified into Ba and Bb chains by whether they carry A chains (Ba) or not (Bb). A Ba chain of wheat amylopectin bound 2.1 A chains on an average but this number was variable in the range of 1-4. The longer Ba chain carried more A chains. A similar distribution was also found in waxy rice amylopectin. This was analyzed by determining the ratios of the non-reducing to reducing residues of the maltosyl-Ba chain fragments, which were obtained by the stepwise hydrolysis of amylopectin with β-amylase, isoamylase, and β-amylase. The A to B chain and the Ba to Bb chain ratios in wheat amylopectin were found to be 1.26 and 1.7, respectively.

Starch is comprised of two major components, amylose and amylopectin. The former is composed of linear and branched molecules with very limited branches, and the latter is composed of highly branched molecules with hundreds or thousands of branches. The branched component of amylose may be called an intermediate molecule or the third component (1). These molecules are built up of only glucose but their molecular structures are not well elucidated yet, in spite of the many efforts by various investigators. Here, we present our recent findings on the molecular structure of amylopectin.

Several models have been proposed for amylopectin structure (2), but only the cluster models (3,4) have been accepted widely. The cluster model shown in Figure 1 has been presented based on the chain-length distributions of some amylopectins (5). The chains are named as A, B, and C by Peat et al. (6): that is, A chains link to

0097–6156/91/0458–0212$06.00/0

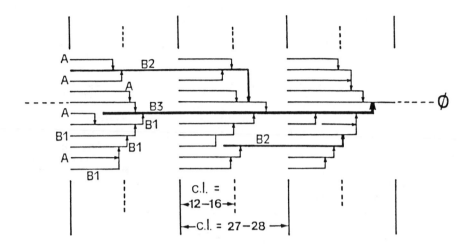

Figure 1. Cluster model for amylopectin. (Reproduced with permission from Ref. 5. Copyright 1986 Elsevier.)

other chains only by their reducing residues; B chains link to another B or C chains at their reducing residues, and in addition they carry one or more A and/or B chains as branches; C chains carry the reducing residues in their molecules. B_1, B_2, and B_3 chains mean that the B chains comprise of one, two, and three clusters, respectively (5). These chains make clusters by linking in some complicated manners through α-(1→6)-glucosidic linkages but details remain to be solved. The chain-length distribution as revealed by size exclusion chromatography supports the cluster structure (5) but gives little information on the inter-chain linking. In this study, the binding mode of the A and B chains was examined by a new enzymic analysis.

Theory and Experimental Procedures
The amounts of A chains, which link to a Ba chain, were analyzed by the following procedures. Here, Ba chains are defined as B chains which carry at least one A chain and the rest of the B chains are defined as Bb chains. Amylopectin is first processed into β-limit dextrin (β-LD) with sweet potato β-amylase, then with Pseudomonas isoamylase, and finally again with β-amylase (2nd β-amylolysis), successively. A chains are trimmed into maltosyl or maltotriosyl residues depending on the even or odd numbers of the chain length

on the 1st β-amylolysis (7). The following isoamylolysis splits all the α-(1→6)-glucosidic linkages except maltosyl-Ba chain linkages which are highly resistant to the enzyme. The inter-chain linkages between Bb chains result in linear chains. Maltotriose and the linear chains are hydrolyzed into maltose and glucose by the 2nd β-amylolysis. However, the majority of maltosyl-Ba chain fragments remain intact after the 2nd β-amylolysis and they are easily separable from maltose and glucose by conventional gel filtration. Therefore, nearly half of the connecting segments of A and Ba chains are recovered as maltosyl-Ba chain fragments. The number (N) of A chains binding to a Ba chain in amylopectin is determined from the number (n) of the binding maltosyl residues to Ba chain fragments or the ratio (R) of non-reducing to reducing residues of the maltosyl-Ba chain fragments. The n value is determined by the increase of reducing power after hydrolysis of the maltosyl-Ba chain fragments with pullulanase. Non-reducing residues are determined by enzymic assay of glycerol after Smith degradation (8). Reducing residues are determined colorimetrically by the method of Somogyi and Nelson with extended heating to 30 min (9) or by modifications of the Park and Johnson ferricyanide method (10). R value is equal to (n + 1). The n is not equal to N because half of the linkages between A and B chains involving maltotriose are liberated with isoamylolysis. Mostly, R value was assayed instead of n value. The relationship between N and R is expressed by "Equation 1" where C means combinations, and is shown in Table I and Scheme 1.

$$R = \frac{{}_NC_1(1+1) + \cdots + {}_NC_N(1+N)}{{}_NC_i + \cdots + {}_NC_N} = \frac{\sum_{i=1}^{N} {}_NC_i(1+i)}{2^N - 1} \qquad (1)$$

Table I. Relationship between R and N

R	2.00	2.33	2.71	3.13	3.58
N	1	2	3	4	5

Results and Discussion

Wheat amylopectin was processed by a series of enzyme actions, β-amylase, isoamylase and β-amylase again. The resulting maltosyl-Ba chain fragments were fractionated into four fractions of nearly equal amounts, by weight, on Bio-Gel P-2, and the R value of each fraction was determined. The R values were in the range of 2.1-2.8, higher with increase in molecular size, which correspond to the N

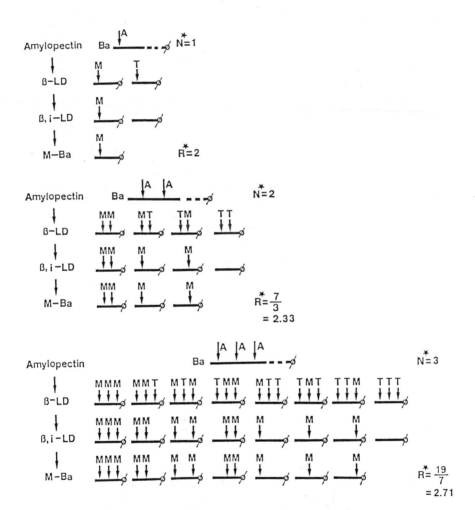

Scheme 1. Structural changes of the binding segments of the A and B chains by a series of hydrolysis. M and T denote maltosyl and maltotriosyl stubs, respectively. * See the text.

values of 1.2 - 3.2. This implies a Ba chain in amylopectin carries
1-4 A chains, and longer Ba chains carry more A chains (Hizukuri,
S., Maehara, Y., Carbohydr. Res., in press). The R value of the
unfractionated maltosyl-Ba chain fragments was found to be 2.4,
which means a Ba chain in amylopectin carries 2.1 A chains on an
average. The value agrees well with that of the revised Meyer model
by Gunja-Smith et al. (11), but the distribution of the R value is
not in accord with the model and favors the cluster structure with
random branching.

The A to B chain ratio of wheat amylopectin was found to be
1.26:1 from the amounts of maltose and maltotriose yielded from its
β-LD by debranching with isoamylase and pullulanase. From the A to
B chain and the A to Ba chain ratios, the Ba to Bb chain ratio was
calculated to be 1.7 : 1.

The maltosyl-Ba chain fragments of waxy rice amylopectin were
fractionated into two fractions, A and B. They were separated from
glucose and maltose (C, Figure 2) by Bio-Gel P-2 filtration after
the 2nd β-amylolysis and the properties of these fractions are
summarized in Table II. The Ba chain of the amylopectin had 2 A
chains on an average, similar to wheat amylopectin, but the number
is variable in the range of 1-3. This would be probably in a more
broad range as in wheat amylopectin, if the maltosyl-Ba chains were
fractionated into more fractions. These results support the random
branching of cluster structure.

Table II. Properties of Maltosyl-Ba Chain Fragments of Waxy Rice
            Amylopectin

Fraction	A	B	Whole
D.p.	33.2	13.0	20.9
C.l.	12.7	6.1	9.0
R*	2.6	2.1	2.3
N*	2.7	1.2	2.0
Amount			
Mole(%)	39.2	60.8	(100)
Weight(%)	61.5	38.5	(100)

*See text

The maltosyl-Ba chain fragments of fraction B were debranched
with pullulanase and the chain-length distribution of the resultant
Ba chain fragments were examined by HPLC on NH_2-bonded silica. The
most frequent chain length was 8, and 60% of the fragments had
chain lengths between 6-11 (Figure 3). This implies that the inner
chain length of the majority of Ba chains ranged between 3.5-8.5,
because most of the Ba chains of the fraction carry a single A
chain and one or two glucosyl residue(s) remain(s) on the outer
chains of B chains after β-amylolysis. These short inner chains
are considered to form non-crystalline domains of clusters.

A problem of this enzymic analysis is that maltosyl-Ba chain
fragments are slightly susceptible to the isoamylase. The
hydrolysis of maltosyl-Ba chain fragments affect only slightly or

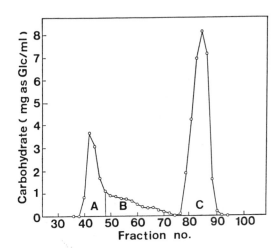

Figure 2. Fractionation of Maltosyl-Ba chain fragments of waxy rice amylopectin by a Bio-Gel P-2 column.

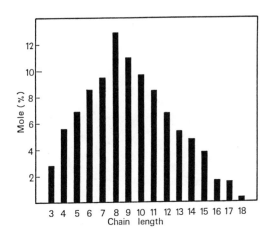

Figure 3. Chain-length distribution of the Ba chain fragments of fraction B in waxy rice amylopectin.

negligibly the final result when the hydrolysis is below 20% and the analysis is done within the limit. However, accurate determination of R value is important because the error of the value is amplified 2-3 fold in N value.

This enzymic analysis for the numbers of A chain binding to a Ba chain provides basic information for elucidation and characterization of the intricate branching of amylopectin.

Acknowledgment
This work was supported in part by Asahi Breweries Foundation and a Grant-in-Aid for Co-operative Research from the Ministry of Education, Science and Culture, Japan.

Literature Cited
1. Hizukuri, S.; Takeda, Y.; Maruta, N.; Juliano, B. O. Carbohydr. Res. 1989, 227-235.
2. Manners, D. J. Carbohydr. Polym. 1989, 11, 87-112.
3. Nikuni, Z. Chori Kagaku 1969, 2, 6-14.
4. French, D. Denpun kagaku 1972, 19, 8-25.
5. Hizukuri, S. Carbohydr. Res. 1986, 147, 342-347.
6. Peat, S.; Whelan, W. J.; Thomas, G. J. J. Chem. Soc. 1956, 3025-3030.
7. Summer, R.; French, D. J. Biol. Chem. 1956, 222, 469-477.
8. Hizukuri, S.; Osaki, S. Carbohydr. Res. 1978, 63, 261-264.
9. Hizukuri, S.; Tabata, S.; Nikuni, Z. Starch/Stärke 1970, 22, 338-343
10. Hizukuri, S.; Takeda, Y.; Yasuda, M.; Suzuki, A. Carbohydr. Res. 1984, 94, 205-213.
11. Gunja-Smith, Z.; Marshall J. J. Mercier, C.; Smith, E. E.; Whelan, W. J. FEBS Letters 1970, 12, 101-104.

RECEIVED November 9, 1990

Chapter 16

Polymer Physicochemical Characterization of Oligosaccharides

Harry Levine and Louise Slade

Fundamental Science Group, Nabisco Brands, Inc., P.O. Box 1944, East Hanover, NJ 07936–1944

This paper reviews the development and technological applications of a polymer characterization method, based on low
temperature differential scanning calorimetry (DSC), to
analyze the structure-physicochemical property relationships of linear, branched, and cyclic oligosaccharides. The
DSC method, based on analog derivative thermograms, is used
to measure Tg', the characteristic subzero glass transition
temperature of a maximally freeze-concentrated aqueous solution. Solute-specific Tg' values have been measured in
order to evaluate the behavior of 91 commercial starch hydrolysis products (SHPs) (i.e. dextrins, maltodextrins,
corn syrup solids, and corn syrups, of dextrose equivalent
(DE) 0.3-100 and polydisperse molecular weight (MW)) and 84
other polyhydroxy compounds (PHCs) (i.e. sugars, polyhydric
alcohols, and derivatized glycosides, of monodisperse MW
62-1153) (1,2). For the commercial SHPs, an inverse linear
correlation exists between Tg' and DE. A plot of Tg' vs.
number-average MW ($\overline{Mn}$) demonstrated the classical behavior
of polydisperse SHPs as a homologous family of amorphous
glucose oligomers and polymers, and revealed an "entanglement coupling" capability for polymeric SHPs of Tg' $\geq$ −8°C
and DE $\leq$ 6 (i.e. $\overline{Mn}$ $\geq$ 3000 and number-average degree of
polymerization, $\overline{DPn}$ $\geq$ 18). In contrast to the higher $\overline{Mn}$
SHPs, the quasi-homologous series of lower MW, monodis-
perse, monomeric and oligomeric PHCs, which included a homologous family of linear malto-oligosaccharides of DP 1-7
and MW 180-1153, exhibited an inverse linear correlation
between Tg' and 1/MW, and thus evidenced no capability for
intermolecular entanglement in the maximally freeze-concentrated aqueous glass at Tg'. The structure-property relationship between intermolecular entanglement (leading to
three-dimensional network formation) and the functional
behavior of polymeric SHPs as food ingredients in applications involving gelation, encapsulation, frozen-storage
stabilization, thermomechanical stabilization, or facilita-

0097–6156/91/0458–0219$11.50/0

tion of drying processes has been demonstrated and is discussed. This underlying physicochemical basis for the em-
pirically demonstrated technological utility of low DE SHPs
in inhibiting various "collapse" phenomena, which affect
the processing/storage stability of many foods, is described and explained, as is the contrasting role typically
played by small sugars and polyols in promoting these usually
detrimental phenomena.

In the decade of the 1980s, the value of a polymer science approach
to the study of structure-property relationships in food materials,
products, and processes has been increasingly recognized by a growing
number of food scientists (3-7 and refs. therein). In this respect,
food science has followed the compelling lead of the synthetic polymers field. As reviewed recently in detail elsewhere (3-7), the
emerging research discipline of "food polymer science" emphasizes the
fundamental and generic similarities between synthetic polymers and
food molecules, and provides a new theoretical and experimental framework for the study of food systems which are kinetically constrained.
On a theoretical basis of established structure-property relationships from the field of synthetic polymer science (8-13), this inno-
vative discipline has developed to unify structural aspects of foods,
conceptualized as kinetically-metastable, completely amorphous or
partially crystalline, homologous polymer systems, with functional
aspects, dependent upon mobility and conceptualized in terms of
"water dynamics" and "glass dynamics" (3-7). These unified concepts
have been used to explain and predict the functional properties of
food materials during processing and product storage (1-7,14-29).
Key elements of this theoretical approach to investigations of food
systems, with relevance to moisture management and water relationships, include recognition of:
1) the behavior of foods and food materials as classical polymer
   systems, and that the behavior is governed by dynamics rather
   than energetics;
2) the importance of the characteristic temperature Tg, at which
   the glass-rubber transition occurs, as a physicochemical parameter which can determine processibility, product properties, quality, stability, and safety of food systems;
3) the central role of water as a ubiquitous plasticizer of natural
   and fabricated amorphous food ingredients and products;
4) the effect of water as a plasticizer on Tg and the resulting
   non-Arrhenius, diffusion-limited behavior of amorphous polymeric, oligomeric, and monomeric food materials in the rubbery liquid state at T > Tg;
5) the significance of non-equilibrium glassy solid and rubbery
   liquid states (as opposed to equilibrium thermodynamic phases)
   in all "real world" food products and processes, and their effects on time-dependent structural and mechanical properties
   related to quality and storage stability.
   In previous reports and reviews (1-7,14-29 and refs. therein),
we have described how the recognition of these key elements of the
food polymer science approach and their relevance to the behavior of
a broad range of different types of foods (e.g. intermediate-moisture
foods (IMFs), low-moisture foods, frozen foods, starch-based foods,

gelatin-, gluten-, and other protein-based foods) and corresponding
aqueous model systems has increased markedly during this decade. We
have illustrated the perspective afforded by using this conceptual
framework and demonstrated the technological utility of this new approach to understand and explain complex behavior, design processes,
and predict product quality, safety, and storage stability, based on
fundamental structure-property relationships of food systems viewed
as homologous families (i.e. monomers, oligomers, and high polymers)
of partially crystalline glassy polymer systems plasticized by water.
Referring to the food polymer science approach, John Blanshard (University of Nottingham, personal communication, 1987.) has stated that
"it is not often that a new concept casts fresh light across a whole
area of research, but there is little doubt that the recognition of
the importance of the transition from the glassy to the crystalline
or rubbery state in foodstuffs, though well known in synthetic polymers, has opened up new and potentially very significant ways of
thinking about food properties and stability." In a recent lecture on
historical developments in industrial polysaccharides, James BeMiller
(Whistler Center for Carbohydrate Research, Purdue University) has
echoed Blanshard's words by remarking that a key point regarding the
future of polysaccharide research and technology is "the potential,
already partly realized, in applying ideas developed for synthetic
polymers to polysaccharides; for example, the importance of the
glassy state in many polysaccharide applications" (30).
     In this article, we illustrate the theory and practice of food
polymer science by highlighting the development and technological
applications of a polymer characterization method, based on low temperature DSC, to analyze the structure-physicochemical property relationships of linear, branched, and cyclic mono-, oligo-, and polysaccharides. These studies have demonstrated the major opportunity offered by this food polymer science approach to expand not only our
quantitative knowledge but also, of broader practical value, our
qualitative understanding of the structure-function relationships of
such carbohydrates in a wide variety of food products and processes.

Theoretical Basis of the Characterization Method - Tg'; Glass Dynamics; the State Diagram as a Dynamics Map Used to Trace "Collapse"
Phenomena in Foods

Our method of characterizing the structure-function relationships of
mono-, oligo-, and polysaccharides derives from the fact that their
functional properties in foods depend on mobility and can be understood in terms of the integrated concepts of "water dynamics" and
"glass dynamics" (1-7). Through this integration, the appropriate
kinetic description of the non-equilibrium thermomechanical behavior
of such food materials has been illustrated in the context of a "dynamics map" (23), an example of which is shown in Figure 1 (7). This
map was derived from a generic solute-solvent state diagram (31,32),
in turn based originally on a more familiar equilibrium phase diagram
of temperature vs. composition (23). The dynamics map has been used
to describe mobility transformations (23) in water-compatible food
polymer systems that exist in kinetically-metastable glassy and rubbery states always subject to conditionally beneficial or detrimental
plasticization by water (3-7).
     Glass dynamics deals with the time- and temperature-dependence

of relationships among composition, structure, thermomechanical properties, and functional behavior. As its name implies, glass dynamics
focuses on 1) the glass-forming solids in an aqueous food system, 2)
the Tg of the resulting aqueous glass that can be produced by cooling
to T < Tg, and 3) the effect of the glass transition and its Tg on
food processing and process control, via the relationships between Tg
and the temperatures of individual processing steps (which may be
deliberately chosen to be first above and then below Tg) (6,7). This
concept emphasizes the operationally immobile, stable, and unreactive
situation (actually one of kinetic metastability) that can obtain
during product storage (of a practical duration) at temperatures below Tg and moisture contents below Wg (the amount of water in the
glass at its Tg) (3,6). It has been used to describe a unifying concept for interpreting "collapse" phenomena, which govern, e.g., the
time-dependent caking of amorphous food powders during storage (1,2).
Collapse phenomena in completely amorphous or partially crystalline
food systems (33-37,115), examples of which are listed in Table I (1-
6), are diffusion-limited consequences of a material-specific structural and/or mechanical relaxation process (1). The microscopic and
macroscopic manifestations of these consequences occur in real time
at a temperature about 20°C above that of an underlying molecular
state transformation (23). This transformation from kinetically-metastable amorphous solid to unstable amorphous liquid occurs at Tg (1).
The critical effect of plasticization (leading to increased free volume (10) and mobility in the dynamically constrained glass) by water
on Tg is a key aspect of collapse and its mechanism (2).

A general physicochemical mechanism for collapse has been described (1), based on occurrence of the material-specific structural
transition at Tg, followed by viscous flow in the rubbery liquid
state (36). The mechanism was derived from Williams-Landel-Ferry
(WLF) free volume theory for (synthetic) amorphous polymers (10). It
has been concluded that Tg is identical to the phenomenological transition temperatures observed for structural collapse (Tc) and recrystallization (Tr). The non-Arrhenius kinetics of collapse and/or recrystallization in the high viscosity (η) rubbery state are governed
by the mobility of the water-plasticized polymer matrix. These kinetics depend on the magnitude of $\Delta T = T - Tg$ (1), as defined by a
temperature-dependent exponential relationship derived from WLF theory. Glass dynamics has proved a useful concept for elucidating the
physicochemical mechanisms of structural/mechanical changes involved
in various melting and (re)crystallization processes (3). Such phenomena are observed in many partially crystalline food polymers and
processing/storage situations (4,15).

One particular location among the continuum of Tg values along
the reference glass curve in Figure 1 results from the behavior of
water as a crystallizing plasticizer and corresponds to an operationally invariant point (called Tg') on a state diagram for any particular solute (1,18-20,22,23,31,38-42). Tg' represents the solute-specific subzero Tg of the maximally freeze-concentrated, amorphous sol-
ute/unfrozen water (UFW) matrix surrounding the ice crystals in a
frozen solution. As illustrated in the idealized state diagram in
Figure 1, the Tg' point corresponds to, and is determined by, the
point of intersection of the kinetically-determined glass curve for
homogeneous solute-water mixtures and the non-equilibrium extension
of the equilibrium liquidus curve for the Tm of ice (1,18-20,22,23).

Table I. Collapse Processes Governed by Tg and Dependent on Plasticization by Water

---

A. Processing and/or storage at T < 0°C

---

1. Ice recrystallization ("grain growth") ≥ Tr (*)
2. Lactose crystallization ("sandiness") in dairy products ≥ Tr (*)
3. Enzymatic activity ≥ Tg' (*)
4. Structural collapse, shrinkage, or puffing (of amorphous matrix surrounding ice crystals) during freeze-drying (sublimation stage) ≡ "melt-back" ≥ Tc (*)
5. Structural collapse or shrinkage due to loss of entrapped gases during frozen storage ≥ Tg' (*)
6. Solute recrystallization during freeze-drying (sublimation stage) ≥ Td (*)
7. Loss of encapsulated volatiles during freeze-drying (sublimation stage) ≥ Tc (*)
8. Reduced survival of cryopreserved embryos, due to cellular damage caused by diffusion of ionic components ≥ Tg'
9. Reduced viability of cryoprotected, frozen concentrated cheese-starter cultures ≥ Tg'
10. Reduced viability of cryoprotected, vitrified mammalian organs due to lethal effects of ice crystallization ≥ Td
11. Staling due to starch retrogradation via recrystallization in breads and other high-moisture, lean baked products during freez-er storage ≥ Tg' (*)

==

B. Processing and/or storage at T ≥ 0°C

---

1. Cohesiveness, sticking, agglomeration, sintering, lumping, caking, and flow of amorphous powders ≥ Tc (*)
2. Plating, coating, spreading, and adsorbing of, e.g., coloring agents or other fine particles on the amorphous surfaces of gran-ular particles ≥ Tg
3. (Re)crystallization in amorphous powders ≥ Tc (*)
4. Recrystallization due to water vapor adsorption during storage of dry-milled sugars (i.e. grinding → amorphous particle surfaces) ≥ Tg
5. Structural collapse in freeze-dried products (after sublimation stage) ≥ Tc (*)
6. Loss of encapsulated volatiles in freeze-dried products (after sublimation stage) ≥ Tc (*)
7. Oxidation of encapsulated lipids in freeze-dried products (after sublimation stage) ≥ Tc (*)
8. Enzymatic activity in amorphous solids ≥ Tg
9. Maillard browning reactions in amorphous powders ≥ Tg (*)
10. Sucrose inversion in acid-containing amorphous powders ≥ Tg
11. Stickiness in spray-drying and drum-drying ≥ T sticky point (*)
12. Graining in boiled sweets ≥ Tg (*)
13. Sugar bloom in chocolate ≥ Tg
14. Color uptake due to dye diffusion through wet fibers ≥ Tg
15. Gelatinization of native granular starches ≥ Tg (*)

Continued on next page

Table I. *Continued*

16. Sugar-snap cookie spreading (so-called "setting") during baking $\geq$ Tg (*)
17. Structural collapse during baking of high-ratio cake batter formulated with unchlorinated wheat flour or with reconstituted flour containing waxy corn starch in place of wheat starch (due to lack of development of leached-amylose network Tg) $\geq$ Tg'
18. Recrystallization of amorphous sugars in ("dual texture") cookies at the end of baking vs. during storage $\geq$ Tg
19. "Melting" (i.e. flow) of bakery icings (mixed sugar glasses) due to moisture uptake during storage $\geq$ Tg (*)
20. Staling due to starch retrogradation via recrystallization in breads and other high-moisture, lean baked products during storage $\geq$ Tg' (*)

(*) = Examples exist in the food science and technology literature of stabilization against collapse through the use of low DE SHPs. See (22,26) for previous versions of this Table that list those references.

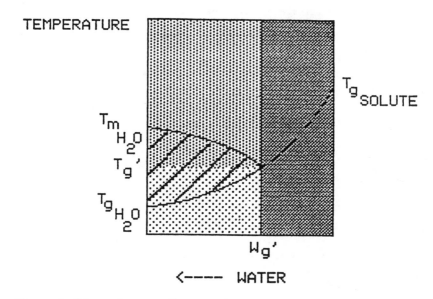

Figure 1. Schematic state diagram of temperature vs. w% water for an aqueous solution of a hypothetical, glass-forming, small carbohydrate (representing a model frozen food system), illustrating how the critical locations of Tg' and Wg' divide the diagram into three distinguishable structure-property domains. (Reproduced with permission from reference 7. Copyright 1991 Plenum.)

This solute-specific location defines the composition of the glass that contains the maximum practical amount of plasticizing water (called Wg', expressed as g UFW/g solute or weight % (w%) water, or alternatively designated in terms of Cg', expressed as w% solute) (1,3) and represents the transition from concentrated fluid to kinetically-metastable, dynamically constrained solid that occurs on cooling to T < Tg' (25). In this homogeneous, freeze-concentrated solute-water glass, the water represented by Wg' is not "bound" energetically but rather rendered unfreezable in a practical time frame due to the immobility imposed by the extremely high local viscosity of about 10^{12} Pa s at Tg' (1-7,18-20,22,23,38-42). Marsh and Blanshard (43) have recently documented the technological importance of freeze-concentration and the practical implication of the description of water as a readily crystallizable plasticizer, characterized by a high value of Tm/Tg ratio ≈ 2 (23,44). A theoretical calculation (43) of the Tg of a typically dilute (i.e. 50%) wheat starch gel fell well below the measured value of about −5 to −7°C for Tg' (15), because the theoretical calculation based on free volume theory did not account for the formation of ice and freeze-concentration that occurs below about −3°C. Recognition of the practical limitation of water as a plasticizer of water-compatible solutes, due to the phase separation of ice, reconciled the difference between theoretical and measured values of Tg (23,43). Moreover, the theoretical calculations supported the measured value of ≈ 27% water for Wg' (15), the maximum practical water content of an aqueous wheat starch glass. The calculated water content of the wheat starch glass with Tg of about −7°C is about 28%.

A critical point implicit in the idealized state diagram in Figure 1 is that the structure-property relationships of water-compatible food polymer systems are dictated by a moisture-temperature-time superposition (1,23,36,45). Visualizing Figure 1 as a dynamics map of mobility, one sees that the Tg curve represents a boundary between non-equilibrium glassy solid (at T < Tg) and rubbery liquid (at T > Tg) physical states (1-3,42). In these non-equilibrium states, various diffusion-limited processes (e.g. collapse phenomena involving mechanical and structural relaxations) either can (at T > Tg and W > Wg', the high moisture portion of the water dynamics domain corresponding to the upper-left part of Figure 1, or T > Tg and W < Wg', the low moisture portion of the water dynamics domain corresponding to the upper-right part of Figure 1) or cannot (at T < Tg, in the domain of glass dynamics corresponding to the bottom part of Figure 1) occur over realistic times (1-3,42). The WLF equation defines the kinetics of molecular-level relaxation processes, which will occur in practical time frames only in the rubbery state above Tg, in terms of an exponential, but non-Arrhenius, function of ΔT above this boundary condition (1,21).

Physicochemical Basis of the Experimental DSC Characterization Method for Determining Tg' Values of Carbohydrates

Figure 2 (1) shows typical low temperature DSC thermograms for 20 w% solutions of a) glucose and b) a 10 DE maltodextrin (Staley Star Dri 10). In each, the heat flow curve begins at the top (endothermic down), and the analog derivative trace (endothermic up and zeroed to the temperature axis) at the bottom. For both thermograms, instru-

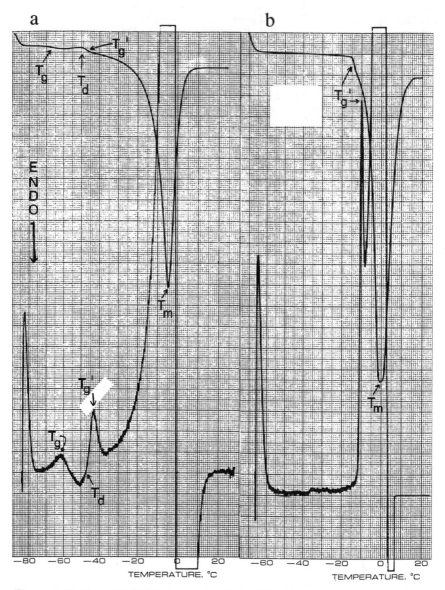

Figure 2. DuPont 990 DSC thermograms for 20 w% solutions of a) glucose, and b) Star Dri 10 (10 DE) maltodextrin. In each, the heat flow curve begins at the top (endothermic down), and the analog derivative trace (endothermic up and zeroed to the temperature axis) at the bottom. (Reproduced with permission from reference 1. Copyright 1986 Elsevier.)

mental amplification and sensitivity settings were identical, and sample weights comparable. [For details of the experimental DSC methodology, see (1,2).] It is evident that the direct analog derivative feature of the DSC (DuPont Model 990) greatly facilitates deconvolution of sequential thermal transitions, assignment of precise transition temperatures (to ± 0.5°C for Tg' values of duplicate samples), and thus overall interpretation of thermal behavior, especially for such frozen aqueous solutions exemplified by Figure 2A. We commented in 1986 (1) on the surprising absence of previous reports of the use of derivative thermograms, in the many earlier DSC studies of such systems with water content > Wg' (see Franks (38) for an extensive bibliography), to sort out the small endothermic and exothermic changes in heat flow that typically occur below 0°C. Most modern commercial DSC instruments provide a derivative feature, but its use for increased interpretative capability still appears to remain much neglected in the thermal analysis of foods in general, and frozen aqueous food systems in particular (22,46).

Despite the handicap of such instrumental limitations in the past, the theoretical basis for the thermal properties of aqueous solutions at subzero temperatures has come to be increasingly understood (31,32,38-40,47-49). As shown in Figure 2A, after rapid cooling (about 50°C/min) of the glucose solution from room temperature to < −80°C, slow heating (5°C/min) reveals a minor Tg at −61.5°C, followed by an exothermic devitrification (a crystallization of some of the previously UFW) at Td = −47.5°C, followed by another (major) Tg, namely Tg', at −43°C, and finally the melting of ice, beginning at T > Tg' and ending at Tm. In Figure 2B, the maltodextrin solution thermogram shows only an obvious Tg' at −10°C, followed by Tm. These assignments of characteristic transitions (i.e. the sequence Tg < Td < Tg' < Tm) and temperatures have been reconciled definitively with actual state diagrams previously reported for various solutes, including small sugars and water-soluble polymers (18,20,38,49). It has been demonstrated (1,20) that the thermogram for the glucose solution in Figure 2A represents a characteristic example, if somewhat trivial case (23), of the unusual phenomenon of multiple values of Tg in glass-forming systems, which is a subject of increasing current interest in the cryotechnology field (1,2,18,19,20,22,50-53). Due to incomplete phase separation (50-53) in an incompletely frozen aqueous solution, two distinguishable dynamically-constrained glasses, with local domains of sufficient dimension (i.e. > 100 Å (9)) and cooperativity to allow ready detection, may coexist (23,53). One is a "bulk" glass with the same spatial homogeneity and solute concentration as the original dilute solution and a corresponding low value of Tg. The other, surrounding the ice crystals, is the freeze-concentrated glass with a higher value of Tg, which is Tg' (1,2,18-23). The lower limiting value of Tg for the dilute bulk glass is Tg of pure amorphous solid water itself (about −135°C (55)), and the upper limiting value of Tg' for the freeze-concentrated glass is Tm of pure crystalline water (20). The observation of such a Tg + Tg' doublet depends on sample moisture content, cooling/heating history, and pressure history (23,53), and represents an example of the difficulty that can be encountered in deconvoluting the non-equilibrium effects of sample history (54,119), and the resulting potential for misinterpretation that can arise when experiments on frozen aqueous systems are not designed from a knowledge of the operative reference state (50-52).

The idealized state diagram shown in Figure 3 ([20]), modified from MacKenzie and Rasmussen ([49]), exemplifies those previously reported and reveals the various distinctive cooling/heating paths that can be followed by solutions of monomeric (glucose) vs. polymeric (maltodextrin) saccharide solutes during typical low temperature DSC experiments. As demonstrated by the DSC thermograms in Figure 2, in either general case, and regardless of initial cooling rate, rewarming from T < Tg' forces the system through a solute-specific glass transition at Tg' ([1]). As illustrated in the state diagram in Figure 3, the Tg'-Cg' point represents a "universal crossroads" on this map, in that all cooling/heating paths eventually lead to this point ([20]). As shown by one of the idealized paths in Figure 3, slow cooling of a stereotypical sugar solution from room temperature (point X) to a temperature corresponding to point Y can follow the path XVSUWY, which passes through the Tg'-Cg' point, W. In the absence of undercooling (e.g. upon deliberate nucleation), freezing (ice formation) begins at point V (on the equilibrium liquidus curve, at a subzero temperature determined by the MW and concentration of the particular solute, via colligative freezing point depression) and ends at point W (on the non-equilibrium extension of the liquidus curve). Due to vitrification of the Tg'-Cg' glass at point W, some of the water in the original solution (i.e. an amount defined as Wg') is left unfrozen in the time frame of the experiment. This UFW is not "bound" to the solute nor "unfreezable" on thermodynamic grounds, but simply experiences retarded mobility in the Tg'-Cg' glass. The extremely high local viscosity of this kinetically-metastable, dynamically constrained glass prevents diffusion of a sufficient number of water molecules to the surface of the ice lattice to allow measurement of its growth in real time ([1,38-42]). As exemplified by the thermogram for the maltodextrin solution in Figure 2B, rewarming from point Y to point X can follow the reversible path YWUSVX, passing back through the Tg'-Cg' point at W ([20]).

In contrast to the slow-cooling path XVSUWY in Figure 3, quench-cooling can follow the direct path from point X to point Z, whereby vitrification can occur at T = Tg, the temperature corresponding to point A, *without* any freezing of ice or consequent change in the initial solution concentration ([49]). However, unlike path XVSUWY, path XZ is not realistically reversible in the context of practical warming rates ([56]). Upon slow, continuous rewarming from point Z to point X, the glass (of composition Cg-Wg rather than Cg'-Wg') softens as the system passes through the Tg at point A, and then devitrifies at the Td at point D ([49]). Devitrification leads to disproportionation, which results in the freezing of pure ice (point E) and revitrification via freeze-concentration of the non-ice matrix to Cg' (point F) during *warming* ([49]). Further warming above Td causes the glass (of composition Cg'-Wg' rather than Cg-Wg) to pass through the Tg'-Cg' point at W (where ice melting begins), after which the solution proceeds along the liquidus curve to point V (where ice melting ends at Tm), and then back to point X. The rewarming path ZADFWUSVX ([20]) is exemplified by the thermogram for the glucose solution in Figure 2A.

The third cooling path illustrated in Figure 3, XQSUWY, is the one most relevant to the practical cooling and warming rates involved in commercial frozen food processes ([20]). Cooling of a solution from point X can proceed beyond point V (on the liquidus curve) to point

Q, because the system can undercool to some significant extent before heterogeneous nucleation occurs and freezing begins (38). Upon freezing at point Q, disproportionation occurs, resulting in the formation of pure ice (point R) and freeze-concentration of the solution to point S (38). The temperature at point S is above that at point Q due to the heat liberated by the freezing of ice (38). The freeze-concentrated matrix at point S concentrates further to point U, because more ice forms as the temperature of the system relaxes to that at point U. Upon further cooling beyond point U, ice formation and freeze-concentration continue as the system proceeds along the liquidus curve to point W. Vitrification of the Tg'-Cg' glass occurs at point W, and further cooling of this glass can continue to point Y without additional ice formation in real time. Rewarming of the kinetically-metastable glass from point Y to point X follows the path YWUSVX, which passes through the Tg'-Cg' point at W. The above descriptions of the various cooling/warming paths illustrated in Figure 3 demonstrate the critical fact that, regardless of cooling/warming rates (within practical limits), every aqueous system of initial concentration ≤ Cg', cooled to T ≤ Tg', must pass through its own characteristic and operationally invariant Tg'-Cg' point (20). If, in commercial practice, a food product is not cooled to T ≤ Tg' after freezing, but rather is maintained within the temperature range between points V and W, that system would track back and forth along the liquidus curve as Tf fluctuates during storage.

The technological significance of Tg' to the storage stability of frozen food systems, implicit in the preceding description of Figure 3, will be discussed later with regard to Cryostabilization technology (1-3,18-22). Suffice it to say for now that Tg' (of the freeze-concentrated solution), rather than Tg (of the original solution), is the *only* glass transition temperature relevant to freezer-storage stability at a given freezer temperature Tf (20), because almost all "frozen" products contain at least some ice. Consistent with the description of the cooling path XVQSUWY, most commercial food-freezing processes, regardless of cooling rate, induce ice formation beginning at point Q (via heterogeneous nucleation after some extent of undercooling). Since the temperature at point Q (generally in the neighborhood of −20°C (42)) is well above that at point A, the lower Tg, that of the glass with the original solute(s) concentration in a typical high moisture product, is never attained and therefore has no practical relevance (20). Once ice formation occurs in a frozen product, the predominant system-specific Tg' becomes the one and only glass transition temperature that controls the product's behavior during freezer storage at any Tf below Tm and either above or below Tg' (20).

The Effect of Molecular Weight on Tg

For pure synthetic polymers, in the absence of diluent, Tg is known to vary with MW in a characteristic and theoretically predicted fashion, which has a significant impact on resulting mechanical and rheological properties (5,10). For a homologous series of amorphous linear polymers, Tg increases with increasing $\overline{M}n$, due to decreasing free volume contributed by chain ends (10), up to a plateau limit for the region of "entanglement coupling" in rubber-like viscoelastic random networks (typically at $\overline{M}n = 1.25 \times 10^3$ to 10^5 daltons) (57), then levels off with further increases in $\overline{Mn}$ (10,12). Below the entanglement $\overline{Mn}$ limit, there is a theoretical linear relationship between increasing Tg and decreasing inverse $\overline{Mn}$ (13). [For polymers with constant values of $\overline{Mn}$, Tg increases with increasing weight-average MW ($\overline{Mw}$), due to increasing local viscosity (23). This contribution of local viscosity is reported to be especially important when comparing different MWs in the range of low MWs (10).] The difference in three-dimensional morphology and resultant mechanical and rheological properties between a collection of non-entangling, low MW polymer chains and a network of entangling, high MW, randomly coiled polymer chains can be imagined as analogous to the difference between masses of elbow macaroni and spaghetti (5). For synthetic polymers, the $\overline{Mn}$ at the boundary of the entanglement plateau often corresponds to about 600 backbone chain atoms (13). Since there are typically about 20-50 backbone chain atoms in each polymer segmental unit involved in the cooperative translational motions at Tg (58), entangling high polymers are those with at least about 12-30 segmental units per chain (5). Figure 4 (13) illustrates the characteristic dependence of Tg on $\overline{Mn}$ (expressed in terms of DP) for several homologous series of synthetic amorphous polymers. In this semi-log plot, the Tg values for each polymer reveal three distinguishable intersecting linear regions: III) a steeply rising region for non-entangling small oligomers; II) an intermediate region for non-entangling low polymers; and I) the horizontal plateau region for entangling high polymers (59). From extensive literature data for a variety of synthetic polymers, it has been concluded that this three-region behavior is a general feature of such Tg vs. log $\overline{Mn}$ plots, and demonstrated that the data in the non-entanglement regions II and III show the theoretically predicted linear relationship between Tg and inverse $\overline{Mn}$ (59).

Within a homologous food polymer family (e.g. from the glucose monomer through maltose, maltotriose, and higher malto-oligosaccharides (e.g. maltodextrins) to the amylose and amylopectin high polymers of starch), Tg' increases in a characteristic fashion with increasing solute MW (1,2,18). This finding has been shown to be in full accord with the established variation of Tg with MW for homologous families of pure synthetic amorphous polymers (10,12,13), described above. The insights resulting from this finding have proved pivotal to the characterization of structure-function relationships in many different types of completely amorphous and partially crystalline food polymer systems (1-7,14-29). The relationship between Tg' and solute MW has been established from DSC measurements of Tg' values for over 150 different food carbohydrates (1,2,18,19). Eighty four of these were small carbohydrates (referred to as PHCs) of known, monodisperse MWs (i.e. $\overline{Mw}/\overline{Mn}$ = 1). These PHCs represented a comprehensive but non-homologous series of mono-, di-, and small oligosaccharides and their derivatives, including many common sugars, polyols, and glycosides, covering a MW range of 62-1153 daltons. Ninety one others were SHPs, including monomeric, oligomeric, and high-polymeric saccharides, representing a homologous family of glucose polymers. These SHPs represented a spectrum of commercial products (including modified starches, dextrins, maltodextrins, corn syrup solids, and corn syrups), with polydisperse MWs (i.e. $\overline{Mw}/\overline{Mn} \gg 1$), covering a very broad range of DE values (where DE = 100/($\overline{Mn}$/180.2)) from 0.3-100. [For tabulated Tg' values for specific SHPs and PHCs, see (1,2,18,19).]

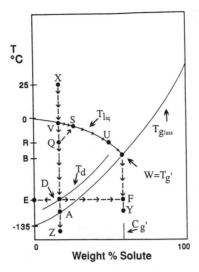

Figure 3. Schematic state diagram of temperature vs. weight percent solute for an aqueous solution of a hypothetical small carbohydrate (representing a model frozen food system), illustrating various cooling/heating paths and associated thermal transitions measurable by low temperature differential scanning calorimetry (e.g. as shown by the thermograms in Figure 2). See the text for explanation of symbols. (Reproduced with permission from reference 20. Copyright 1989 Gordon & Breach.)

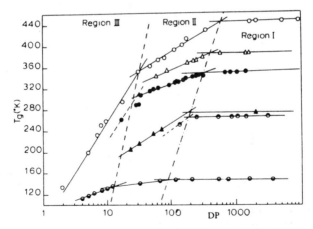

Figure 4. Plot of Tg as a function of log DP (degree of polymerization) [a measure of $\overline{M}n$], for poly(alpha-methyl-styrene) (open circles); poly(methylmethacrylate) (open triangles); poly(vinyl chloride) (solid circles); isotactic polypropylene (solid triangles); atactic polypropylene (circles, top half solid); and poly(dimethylsiloxane) (circles, bottom half solid). (Reproduced with permission from reference 13. Copyright 1986 Wiley-Interscience.)

For the SHPs, a homologous series of glucose oligomers and polymers, Tg' values range from −43°C for glucose (the monomer itself, of
DE = 100) to −4°C for a 0.5 DE maltodextrin. A plot of Tg' vs. DE
(shown in Figure 5 (1)) revealed a linear correlation between increasing Tg' and decreasing DE (r = −0.98) for all SHPs with manufac-
turer-specified DE values (1). Since DE is inversely proportional to
$\overline{DP}n$ and $\overline{M}n$ for SHPs (60), these results demonstrated that Tg' increases with increasing solute $\overline{M}n$ (from $\overline{M}n$ = 180 for glucose to 36000
for 0.5 DE maltodextrin) (1). Such a linear correlation between Tg
and 1/$\overline{M}n$ is the general rule for any homologous family of pure,
glass-forming polymers (12). The equation of the regression line is
DE = −2.2(Tg', °C) − 12.8, and the plot of Tg' vs. DE in Figure 5 has
proved useful as a calibration curve for interpolating DE values of
new or "unknown" SHPs (2).
        Results for polymeric SHPs have demonstrated that Tg' depends
rigorously on linear, weight-average DP ($\overline{DP}w$) for such highly polydisperse solutes, so that linear polymer chains (e.g. amylose) give
rise to a higher Tg' than branched chains (e.g. amylopectin, with
multiple chain ends) of equal MW (1,2). Due to the variable polydispersity and solids composition of commercial SHPs (60,61), the range
of Tg' values for SHPs of the same specified DE can be quite broad.
This behavior was shown by several pairs of SHPs (1,2). For each
pair, of the same DE and manufacturer, the hydrolysate from waxy
starch (all amylopectin) had a lower Tg' than the corresponding one
from normal starch (containing amylose). This behavior was also exemplified by the Tg' data for thirteen 10 DE maltodextrins (1,2), for
which Tg' ranged from −7.5°C for a normal starch product to −15.5°C
for a product derived from waxy starch, a ΔTg' of 8°C. Such a ΔTg'
is greater than that between maltose (DP 2) and maltotriose (DP 3)
(18). Further evidence was gleaned from Tg' data for selected glucose oligomers (2). Comparisons of the significant Tg' differences
among maltose (1→4-linked dimer), gentiobiose (1→6-linked), and isomaltose (1→6-linked), and among maltotriose (1→4-linked trimer), panose (1→4, 1→6-linked), and isomaltotriose (1→6, 1→6-linked), have
suggested that 1→4-linked (linear amylose-like) glucose oligomers
manifest greater "effective" linear chain lengths in solution (and,
consequently, larger hydrodynamic volumes) than oligomers of the same
MW which contain 1→6 (branched amylopectin-like) links (26). These
results have also been used to illustrate the sensitivity of the Tg'
parameter to molecular configuration, in terms of linear chain
length, as influenced by the nature of the glycosidic linkages in
various non-homologous saccharide oligomers (not limited to glucose
units) and the resultant effect on solution conformation (19). Further evidence was seen in the Tg' values for other PHCs, where, for
sugars of equal MW (e.g. 164), ΔTg' is as large as 10°C, a spread
even larger than for the thirteen 10 DE maltodextrins (18). Another
interesting comparison was that between Tg' values for the linear and
cyclic α-(1→4)-linked glucose hexamers, maltohexaose (−14.5°C) and α-
cyclodextrin (−9°C). In this case, the higher Tg' of the cyclic oligomer led to a suggestion (3) that the ring of α-cyclodextrin apparently has a much larger hydrodynamic volume (due to its relative ri-
gidity) than does the linear chain of maltohexaose, which is relatively flexible and apparently can assume a more compact conformation
in aqueous solution. The above comparisons have been discussed in
the past to emphasize the subtleties of structure-property analyses

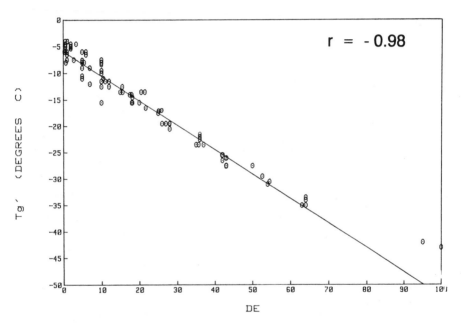

Figure 5. Variation of a glass transition temperature, Tg′, for maximally frozen 20 w% solutions against DE value for an extensive series of commercial SHPs. (Reproduced with permission from reference 1. Copyright 1986 Elsevier.)

of SHPs and PHCs by our DSC characterization method (22). The unavoidable conclusion, concerning the choice of a suitable carbohydrate ingredient for a specific product application, is that one SHP
(or PHC) is not necessarily interchangeable with another of the same
nominal DE (or MW). Characterization of fundamental structure-property relationships, in terms of Tg', has been strongly advised before
selection of such ingredients for fabricated foods (1,2).

The Tg' results for the commercial SHPs demonstrated exactly the
same Tg vs. $\overline{M}n$ behavior as described earlier for synthetic amorphous
polymers. Tg' values for this series of SHPs (of polydisperse MWs in
the range from 180 for glucose to about 60000 for a 360-DP polymer)
thus demonstrated their classical behavior as a homologous family of
amorphous glucose oligomers and polymers (1,2). The plot of Tg' vs.
solute $\overline{M}n$ in Figure 6 (1) clearly exhibits the same three-region behavior as shown in Figure 4: I) the plateau region indicative of the
capability for entanglement coupling by high polymeric SHPs of DE ≤ 6
and Tg' ≥ -8°C; II) the intermediate region of non-entangling, low
polymeric SHPs of 6 < DE < 20; and III) the steeply rising region of
non-entangling, small SHP oligomers of DE > 20 (5). The plot of Tg'
vs. $1/\overline{M}n$ in the inset of Figure 6, with a linear correlation coefficient r = -0.98, demonstrates the theoretically predicted linear relationship for all the SHPs in regions II and III, with DE values >
6. The plateau region evident in Figure 6 identified a lower limit
of $\overline{M}n ≈ 3000$ ($\overline{DP}n ≈ 18$) for entanglement leading to viscoelastic network formation (8,62) by such polymeric SHPs in the freeze-concen-
trated glass formed at Tg' and Cg'. This $\overline{M}n$ is within the typical
range of 1250-19000 for minimum entanglement MWs of many pure synthetic amorphous linear high polymers (57). The corresponding $\overline{DP}n$ of
about 18 is within the range of 12-30 segmental units in an entangling high polymer chain, thus suggesting that the glucose repeat in
the glucan chain (with a total of 23 atoms/hexose ring) may represent
the mobile backbone unit involved in cooperative solute motions at
Tg' (5). The entanglement capability has been suggested to correlate
well with various functional attributes (see the labels on the plateau region in Figure 6) of low DE SHPs, including a predicted (1)
and subsequently demonstrated (2) ability to form thermoreversible,
partially crystalline gels from aqueous solution (63-72,114). It has
been suggested that SHP gelation occurs by a mechanism involving
crystallization-plus-entanglement in concentrated solutions undercooled to T < Tm, as described elsewhere (3).

In contrast to the commercial SHPs, the series of quasi-homologous, monodisperse PHCs, including a homologous set of malto-oligosaccharides from glucose up to DP 7, was found to manifest Tg' values
which fall below the Tg' limit defined by SHPs for entanglement and
the onset of viscoelastic rheological properties and to be incapable
of gelling from solution (2,19). The plot of Tg' vs. MW for all the
PHCs in Figure 7 (2), drawn conventionally as a smooth curve through
all the points (12), can easily be visualized to represent two intersecting linear regions (III for MW < 300 and II for 300 < MW < 1200)
(5). From the fair linearity of the Tg' vs. 1/MW plot for all the
data in the inset of Figure 7, it was concluded that these diverse
low MW sugars, polyols, and glycosides show no evidence of entanglement in the freeze-concentrated glass at Tg'. The corresponding results for glucose and malto-oligosaccharides of DP 2-7 (shown in Figure 8 (3)) demonstrated a better linear correlation, with r = -0.99

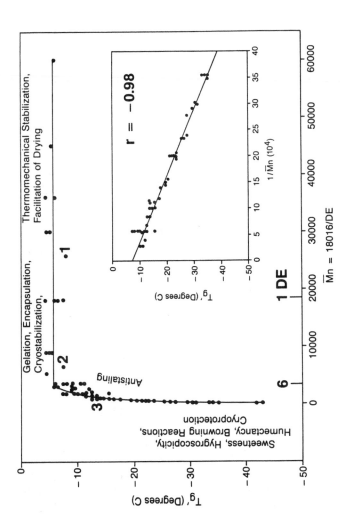

Figure 6. Variation of a glass transition temperature, Tg', for maximally frozen 20 w% solutions against M̄n (expressed as a function of DE) for the commercial SHPs in Figure 5. DE values are indicated by numbers marked above the x-axis. Areas of specific functional attributes, corresponding to three regions of the diagram, are labeled. [Inset: plot of Tg' vs. 1/Mn (X 10 000) for SHPs with M̄n values below entanglement limit, illustrating the theoretically predicted linear dependence.] (Reproduced with permission from ref. 1. Copyright 1986 Elsevier.)

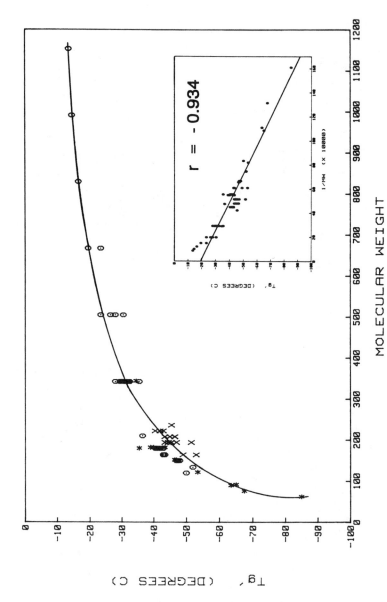

MOLECULAR WEIGHT

Figure 7. Variation of the glass transition temperature, Tg', for maximally frozen 20 w% solutions against MW for an extensive collection of sugars (⊙), glycosides (X), and polyols (*). [Inset: plot of Tg' vs. 1/MW (X 10 000), illustrating the theoretically predicted linear dependence.] (Reproduced with permission from reference 2. Copyright 1988 Butterworths.)

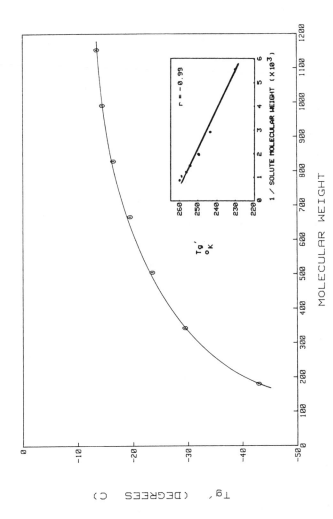

Figure 8. Variation of the glass transition temperature, Tg', for maximally frozen 20 w% solutions against MW for a homologous series of malto-oligosaccharides from glucose through maltoheptaose. [Inset: plot of Tg' vs. 1/MW (X 1000) of solute, illustrating the theoretically predicted linear dependence.] (Reproduced with permission from reference 3. Copyright 1988 Cambridge University Press.)

for a plot of Tg' vs. $1/MW$, shown in the inset of Figure 8. This linear dependence (9) of the Tg' results for the malto-oligosaccharides in aqueous solution exemplified the theoretical glass-forming behavior (i.e. diluent-free Tg vs. $1/MW$) characteristic of a homologous family of non-entangling, linear, monodisperse oligomers (10, 12). For all the PHCs, none larger than a heptamer of MW 1153, the main plots in Figures 7 and 8 show that region I, representing the entanglement plateau where Tg remains constant with increasing MW, has not been reached, in accord with the MW (and corresponding DP) range cited above as the lower limit for polymer entanglement.

Saccharide Polymers: Entanglement and Network Formation - Network Tg

There is a profound technological importance of MWs above the entanglement MW limit, as illustrated earlier for commercial SHPs (as a model for other homologous families of amorphous saccharide oligomers and polymers) by the structure-function relationships defined by the entanglement plateau in Figure 6 (1,2). Ferry (10) has described the generic behavior observed for all polymer systems with respect to the relationships between linear DP of the backbone chain, polymer concentration, and viscosity. MW is a relative measure of linear DP of the primary chain, when the polymer has a uniform structure along its entire length. At any given concentration, there is a minimum DP required for entanglement and network formation. For very dilute solutions (such that the solution viscosity, measured as a relative flow rate, is similar to that of the solvent alone), high MW polymers are necessary to form gels or networks (characterized by very high macroscopic viscosity, measured as a relative firmness). For example, 1.5 w% gelatin solutions in water can form firm gel networks (which exhibit resistance to dehydration, due to mechanical resistance to shrinkage (7)), through entanglement followed by crystallization of junction zones, if the linear DP is ≥ 1000 (MW $\geq 10^5$) (73). Similarly, 1.5 w% amylose solutions in water can form firm gel networks if the linear DP is about 3000 ($\overline{Mw} \approx 5 \times 10^5$) (70,71). At intermediate chain lengths, greater concentrations of chains are required for entanglement and network formation. In the case of SHPs, such as the low DE maltodextrins patented as partially crystalline, fat-mimetic gels (63,64,69), concentrations must be increased to at least about 20-25 w% in water (i.e. typical of Cg' of the freeze-concentrated glass at Tg') as linear DP is decreased to approach 18 glucose units (MW ≈ 3000) (2,3). In contrast, oligomers of hydrolyzed gelatin (peptones) or hydrolyzed starch (corn syrup solids or higher DE maltodextrins with MW ≤ 3000) are incapable of gel network formation via entanglement at any concentration (2,3,24). However, recrystallization of such oligomers can occur due to concentration above the saturation limit or to a change of solvent. For carbohydrate polymers based on primary chains of $\alpha-1,4$ glucans, the critical DP required for network formation via entanglement is ≥ 18 (1).

Network formation, especially in the absence of crystallization, depends on the ability of flexible chains to entangle (10). [The contribution of crystallization to network formation and gelation, described elsewhere (3-7), will be discussed below in the specific context of saccharide polymers, with regard to the question - when is retrogradation synonymous with recrystallization and with gelation?] One convenient diagnostic test for entanglement relies on the fact,

previously illustrated in Figures 4 and 6, that the Tg values of a
homologous family of polymers increase with increasing linear DP up
to the chain length sufficient to allow entanglement. Entanglement
networks consist of internode chains and network junction zones
(nodes) that are transient topological constraints to chain motion
(57,74). The probability of formation of (non-crystalline) junctions
depends on chain length and concentration. The greater the number of
junctions, the shorter the inter-node chain length (for a fixed parent chain length). Thus, there is a limiting length for any chain
that exhibits translational freedom, and a limiting molecular Tg for
that DP (7).

A second important diagnostic test for entanglement (7) is illustrated in Figure 9 (10), a log-log plot of viscosity vs. MW for a
series of synthetic polymers. For undiluted polymers or for polymer
solutions studied at constant total concentration, a critical chain
length can be demonstrated, above or below which the dependence of
viscosity on MW changes dramatically (10). Above the critical chain
length, entanglement results in a drastic sensitivity of viscosity to
chain length. In the absence of entanglement, chains shorter than
the critical length show solution behavior with relative insensitivity of viscosity to chain length. The topological constraints of the
(non-crystalline) entanglement network are not due to any particular
chemical interactions (such as hydrogen bonds or dipolar or charge
interactions), nor to any particular structural features (5). As
demonstrated in Figure 9, entanglement is a generic behavior of polymers of sufficient chain length and can be seen equally in poly-
(ethylene glycol) and in non-polar, structurally featureless polymers
such as poly(iso-butylene). The important lesson to be learned from
Figure 9 has been described as follows (7). The entanglement MW limit, coinciding with the critical linear DP required for intermolecular network formation, corresponds to the point at which the slope
changes abruptly. In the region of MW above the critical DP, the
slope of log viscosity vs. log MW is 3.4. In this region, cutting
molecules (e.g. SHPs) of DP 300 in half, to obtain the same total
concentration of molecules with DP 150, would result in a 10-fold
reduction in viscosity. In contrast, in the absence of entanglement,
the slope of log viscosity vs. log MW in the region below the critical DP is 1. In this region, cutting molecules of DP $\leq$ 18 in half
would result in only a 2-fold reduction in viscosity.

It has been pointed out (23) that the low values of local viscosity typically found in glass-forming polymer systems at T = Tg +
100°C (10) compare to a macroscopic viscosity of about 10^9 Pa s for
an entanglement network, and even higher viscosities if the network
is crosslinked (10). This point has been used to underline the importance of research on small PHC-water systems (75), based on a polymer science approach (23). Synthetic high polymers, as well as many
high-polymeric food materials, often suffer from the handicaps of unknown, polydisperse MW and MW distribution, and MWs above their entanglement limit, in which case local viscosity is *not* equivalent to
macroscopic viscosity. For such cases of MWs above the entanglement
limit, as mentioned above, a halving of MW results in a 10-fold reduction in the macroscopic viscosity of the network (10). In contrast, small PHCs have known, monodisperse values of MW, all below
the entanglement limit, so that local viscosity *is* equivalent to macroscopic viscosity, and a halving of MW results only in a halving of

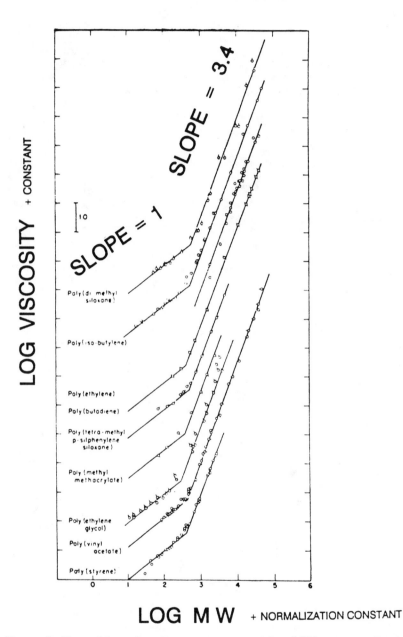

Figure 9. Plot of log viscosity + constant vs. log MW + normalization constant for a series of synthetic polymers, illustrating the generic behavior of polymers with MWs above and below the critical DP required for intermolecular entanglement and network formation. (Reproduced with permission from reference 10. Copyright 1980 John Wiley & Sons.)

local viscosity (as also illustrated in Figure 9 for synthetic polymers with MWs below their entanglement limits) (10). Such small PHCs
offer a great variety and selection of glass-forming materials for
the study of various aspects of the non-equilibrium behavior of foods
(23).

In the context of starch retrogradation as a collapse process
(5), retrogradation of gelatinized starch involves the recrystallization of both amylopectin and amylose (76-80,117,118). It has been
demonstrated for SHPs that the minimum linear chain length required
for intermolecular entanglement upon concentration to Cg' corresponds
to $\overline{DP}n \approx 18$ and $\overline{M}n \approx 3000$ (1). Sufficiently long linear chain length
($\overline{DP}n \gtrsim 15$-20) has also been correlated with intermolecular network
formation and thermoreversible gelation of SHPs (2,3,68,114) and
amylopectin (81), and with starch (re)crystallization (78,79,81-83).
It has been suggested that, in a partially crystalline starch, SHP,
or amylopectin gel network, the existence of random interchain entanglements in amorphous regions and "fringed micelle" (3) or chain-folded microcrystalline junction zones (68) each represents a manifestation of sufficiently long chain length (1). This suggestion was supported by other work (70,71,79,120) which has shown that amylose
gels, which are partially crystalline (83), are formed by cooling
solutions of entangled chains. For aqueous solutions of both high MW
amylose (70,71,84) and amylopectin (81,85), intermolecular entanglement and network formation have been evidenced by log-log plots of
viscosity vs. concentration with a characteristic break in the curve
(analogous to the break in the curves of log viscosity vs. log MW for
the synthetic polymers in Figure 9), such that the slope of the linear portion above the so-called "coil overlap" concentration is steep-
er than the slope of the other linear portion at lower concentrations. From such a plot, Miles et al. (70) have identified a critical minimum concentration ($\gtrsim 1.5$ w% amylose) for entanglement of
high-polymeric amylose ($\overline{M}w \approx 5 \times 10^5$). These workers have stated
that amylose gelation requires network formation, and this network
formation requires entanglement, and they have concluded that "polymer entanglement is important in understanding the gelation of amylose" (70). A more recent study of aqueous amylose gelation by Gidley et al. (84,86,87), using nearly monodisperse amyloses of DP 250-
2800, has identified a somewhat lower critical gelling concentration
of ≈ 1.0 w%. This finding has been corroborated in a subsequent rheological study by Doublier and Choplin (120). Gidley et al., while
accepting the concept of intermolecular entanglement in "semi-dilute"
amylose solutions advanced by Miles et al. (70), have suggested that
the lower gelling concentration of 1.0 w% results from the predominant contribution of crystalline junction zone formation to the gelation mechanism for amylose (84,86,87).

The time-dependent gelation of amylose from dilute aqueous solution is generally agreed to occur in two stages: a relatively fast
but finite stage due to viscoelastic network formation via entanglement (which is reversible by dilution but not thermoreversible); followed closely by a slower, but continually maturing, crystallization
(in a chain-folded or extended-chain morphology) process (which is
thermoreversible above 100°C) (70,71,78,79,81,88-90,120). In contrast, in partially crystalline, thermoreversible (below 100°C), aqueous amylopectin gels, viscoelastic network formation (which is relatively slow and time-dependent) is more closely related to the pres-

ence of microcrystalline junctions than to entanglements, although
entanglement does occur (78,81,82,117,118). Since most normal
starches are 70-80% amylopectin (91), their gelatinization and retrogradation processes are dominated by the non-equilibrium melting and
recrystallization behavior of amylopectin (15,77,117,118), although
contributions due to amylose can be observed (80,92,120). Generally,
the early stages of starch retrogradation are dominated by chain-
folded amylose (of DP from about 15 to about 50 and fold length about
100 Å (78,81,93)); the later stages by extended-chain amylopectin
(43) outer branches (of DP about 12-16 (81,94)) (15).
     Experimental evidence, which supports these conclusions about
the thermoreversible gelation mechanism for partially crystalline
polymeric gels of starch, amylopectin, amylose, and SHPs, has come
from DSC studies (15), the favored technique for evaluating starch
retrogradation (95). Analysis of 25 w% SHP gels, set by overnight
refrigeration, has revealed a small crystalline melting endotherm
·with Tm ≈ 60°C (3), similar to the characteristic melting transition
of retrograded B-type wheat starch gels (15). Similar DSC results
have been reported for 20 w% amylopectin (from waxy maize) gels (82,
96). The small extent of crystallinity in SHP gels can be increased
significantly by an alternative two-step temperature-cycling gelation
protocol (12 hr at 0°C, followed by 12 hr at 40°C) (15), adapted from
the one originally developed by Ferry (73) for gelatin gels, and subsequently applied by Slade et al. (29) to retrograded starch gels.
In many fundamental respects, the thermoreversible gelation of aqueous solutions of polymeric SHPs, amylopectin, amylose, and gelatinized starch is analogous to the gelation-via-crystallization of synthetic homopolymer and copolymer-organic diluent systems, described
elsewhere (3,15). For the latter partially crystalline gels, the
possibly simultaneous presence of random interchain entanglements in
amorphous regions (97) and microcrystalline junction zones (98) has
been reported. However, controversy exists (97,98) (as it also does
in the case of amylose (70,84,86,87,120)) over which of the two conditions (if either alone) might be the necessary and sufficient one
primarily responsible for the structure-viscoelastic property relationships of such polymeric systems. Part of this controversy could
be resolved by a simple dilution test (99), which could also be applied to polysaccharide gels (e.g. amylose); i.e. entanglement gels
can be dispersed by dilution at room temperature, while microcrystalline gels cannot be when room temperature is < Tm (3). In the context of SHPs as inhibitors of collapse processes (1), it is worth
mentioning that the literature on SHPs as anti-staling ingredients
for starch-based foods (reviewed elsewhere (15)) includes a report by
Krusi and Neukom (100) that (non-entangling) SHP oligomers of $\overline{DP}n$ 3-8
(i.e. within the intermediate region II of Figure 6) are effective in
inhibiting, and *not* participating in, starch recrystallization.
     An important consequence of entanglement and network formation
is the effect on the Tg that determines all diffusion-limited structural and mechanical relaxation processes of the system. As shown
schematically in Figure 10 (7), while the molecular or segmental Tg
remains constant above the entanglement MW limit, the *network* Tg,
i.e. the macroscopic, controlling Tg of the supramolecular network
(that would affect Instron measurements of the modulus, for instance
(5)), continues to increase with increasing MW above the entanglement
MW, because of the increased probability of crosslinks (97,101).

This fact has major structural and textural implications for food
polymer systems, because such systems with MWs above the entanglement
limit are capable of forming fibers, networks, and gels: i.e. macroscopic structures that can reinforce and support their own weight
against gravity (5). A noteworthy example is the gelatinized wheat
starch-gluten matrix of baked bread. The effective network Tg responsible for mechanical firmness of freshly baked bread would be near
room temperature for low extents of network formation (i.e. the contribution due to low extents of starch retrogradation), well above
room temperature for mature networks (i.e. greater extents of retrogradation), and equivalent to Tgel (gel melting temperature) near
60°C for staled bread (i.e. Tm of fully retrograded B-type wheat
starch), even though the underlying Tg for segmental motion (of either starch or gluten at plasticizing moisture contents $\geq$ Wg'), responsible for the predominant second-order thermal transition, remains below 0°C at Tg' (4,5).

Saccharide Oligomers and Polymers as Moisture Management Agents. The
relationship between the solute concentration and linear DP requirements for entanglement and network formation, and its resultant effect on Tg (molecular vs. network), also has important implications
for moisture management by saccharide oligomers and polymers (6,7).
As exemplified for the homologous family of amorphous glucose oligomers and polymers represented by the commercial SHPs in Figure 6,
there are three distinguishable regimes of moisture management, which
are analogous to the three regions of Tg vs. MW behavior shown in
Figure 4 and mirrored in Figure 6. Aspects of moisture management
relevant to these three regimes of functional behavior include, e.g.,
hydration, freezing, drying, moisture migration, and biological stability (6). The first regime includes non-entangling solutes of linear DP < 3. For such small sugars at a given water concentration,
this moisture management regime is characterized by low apparent
(non-equilibrium) relative vapor pressure (RVP) (3), very large osmotic driving force to take up water, low local viscosity, and only a
small barrier to local translational and rotational diffusion. The
second regime, for non-entangling solutes of DP ≈ 3 to 17, is characterized by high apparent RVP, small osmotic driving force to take up
water, high local viscosity, and a large barrier to local translational and rotational diffusion. The third regime, for entangling
solutes of DP >> 17, is characterized by very high apparent RVP, very
large *local* osmotic driving force to swell, very low local viscosity,
and essentially no barrier to local translational and rotational diffusion (7). In this context, it is interesting to note that our finding of $\overline{DP}n$ ≈ 18 for the minimum DP for entanglement and network formation by commercial SHPs, a result identified from the polymer characterization analysis represented by Figure 6, has been confirmed by
a revealing finding recently reported by Radosta et al. (114). From
their study of the water sorption behavior of maltodextrins, they
concluded that "the transition between "polymer" and "oligomer" behavior under sorption conditions is located in the region of $\overline{DP}n$ values between 60 and 16. In this $\overline{DP}n$ region, the shifting from restricted swelling [by what Radosta et al. refer to as a "sorption
gel"] to solution under sorption conditions takes place" (114).
    A conceptual representation of the three regimes of moisture
management is shown in Figure 11 (7). For a given temperature (e.g.

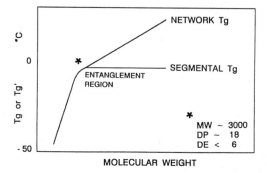

Figure 10. Schematic plot of Tg (or Tg') vs. molecular weight (modeled after the data plot for SHPs in Figure 6), which illustrates that, while the segmental Tg remains constant with increasing MW for MWs above the entanglement limit, the network Tg continues to increase monotonically with increasing MW above the entanglement MW limit. (Reproduced with permission from reference 7. Copyright 1991 Plenum.)

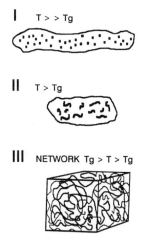

Figure 11. Conceptual representaion of the three regimes of moisture management by a homologous family of saccharide polymers (e.g. the SHPs in Figure 6):I. at T >> Tg of a concentrated solution of non-entangling saccharide oligomers of linear DP < 3; II. at T > Tg of a less concentrated solution of non-entangling saccharide polymers of linear DP = 3 to 17.; III. at network Tg > T > Tg of a dilute solution (but above the critical concentration for entanglement) of entangling saccharide high polymers of linear DP >> 17. (Reproduced with permission from reference 7. Copyright 1991 Plenum.)

room temperature) and time (e.g. a practical experimental time frame), T is well above Tg for concentrated solutions of regime I solutes, and such a system would be subject to viscous liquid flow due to gravity. In regime II at T > Tg, apparent RVP is not depressed as much as in regime I, because there are fewer solute molecules in solution. However, their higher linear DP results in higher local viscosity, which in turn results in a larger barrier to local diffusion and reduced viscous liquid flow due to gravity. In regime III (for a solute concentration high enough to allow entanglement), at network Tg > T > molecular Tg, there is elastic resistance to flow, and the gel is able to support its own weight against the force of gravity (7). In this regime, there is essentially no barrier to local diffusion, so a small molecule such as water or a dye molecule can diffuse freely in the gel network (102). Hence, regime III manifests very high apparent RVP. Despite this, however, there is a very large local osmotic driving force to take up water, not only via hygroscopicity but via swelling, the latter due to the mechanical resistance of the entangled network to shrinkage (7). This mechanical resistance to shrinkage, which is analogous to hydraulic resistance to water removal, has an effect on the local chemical potential of the solvent, analogous to an addition to the osmotic pressure (103). Thus, while there is a very large driving force to take up water via swelling in regime III, and a normal, classical osmotic driving force to take up water via hygroscopicity in regime I, there is a much lower driving force to take up water via classical osmotic pressure effects in regime II, because of the absence of a swelling force due to entanglement (7).

To illustrate the consequences of the three regimes of functional behavior of saccharide moisture management agents, one could use results from, e.g., drying or freezing experiments. Drying and freezing are equivalent diffusion-limited processes in the sense that both involve removal of water via phase separation; in drying by increasing the temperature to produce water vapor, and in freezing by decreasing the temperature to produce ice (6,7). Muhr and Blanshard (104) have measured the relative rates of linear ice front advancement at subzero temperatures in aqueous solutions of 35 w% sucrose with and without added polysaccharide "stabilizers". Their results showed conclusively that the rates depend critically on the presence or absence of a gel network, even for exactly the same formulation. For the solution of sucrose alone at T >> Tg' [regime I], the relative rate of ice front advancement was 6.0. [It would have been essentially zero at T < Tg'.] At the same temperature, the rate was 4.1 in a sucrose solution containing 0.75 w% non-entangling (i.e. non-gelling) Na alginate [regime II], but only 1.0 in a sucrose solution containing 0.75 w% entangling (i.e. gelling) Ca alginate [regime III]. Thus, when there was a hydraulic resistance to water removal, due to the resistance of the Ca alginate gel network to shrinkage, the rate of ice front advancement was dramatically reduced (7).

By analogy to other moisture management problems involving diffusion-limited processes (e.g. "water activity" control, textural stabilization), entangling, network-forming saccharide high polymers from regime III can be used as functional additives to, e.g., retard moisture migration in baked goods, retain crispness of breakfast cereals, and reduce sogginess of pastries and pie crusts (6,7).

Cryostabilization Technology - Collapse Phenomena - Functionality of
Saccharide Polymers as Collapse Inhibitors

"Cryostabilization technology" (1,2,18-22) represents a new conceptual approach to a practical industrial technology for the stabilization during processing and storage of frozen, freezer-stored, and
freeze-dried foods. This technology emerged from our food polymer
science research approach and developed from a fundamental understanding of the critical physicochemical and thermomechanical structure-
property relationships that underlie the behavior of water in all
non-equilibrium food systems at subzero temperatures (38,40). Cryostabilization provides a means of protecting products, stored for
long periods at typical freezer temperatures (e.g. Tf = −18°C), from
deleterious changes in texture (e.g. "grain growth" of ice, solute
crystallization), structure (e.g. collapse, shrinkage), and chemical
composition (e.g. enzymatic activity, oxidative reactions such as fat
rancidity, flavor/color degradation). Such changes are exacerbated
in many typical fabricated foods whose formulas are dominated by low
MW saccharides. The key to this protection, and resulting improvement in product quality and storage stability, lies in controlling
the structural state, by controlling the physicochemical and thermomechanical properties, of the freeze-concentrated amorphous matrix
surrounding the ice crystals in a frozen system. As alluded to earlier, the importance of the glassy state of this maximally freeze-
concentrated solute-UFW matrix and the special technological significance of its particular Tg, i.e. Tg', relative to Tf, have been described and illustrated by solute-water state diagrams such as the
idealized one in Figure 12 (18). Upon a foundation of pioneering
studies of the low temperature thermal properties of frozen aqueous
model systems by Luyet, MacKenzie, Rasmussen (32,34,47-49,105), and
Franks (31,38-41), an extensive cryostabilization technology data
base of DSC results for carbohydrate and protein food ingredients has
been built (1-5,18-23,26). As reviewed earlier, DSC results for the
characteristic Tg' values of individual saccharide solutes have demonstrated that Tg' is a function of MW for both homologous and quasi-
homologous families of water-compatible monomers, oligomers, and high
polymers. Examples of how the selection and use of appropriate ingredients (e.g. low DE SHPs) in a fabricated product have allowed the
food technologist to manipulate the composite Tg', and thus deliberately formulate to elevate Tg' relative to Tf and so enhance product
stability, have been described (1-5,18-23,26-28), as reviewed below.
    Much of our understanding of the thermal and thermomechanical
properties of concentrated aqueous solutions has been derived from
extensive studies of small glass-forming saccharides at subzero temperatures. These studies, which began over 50 years ago with the
seminal work of Luyet (105), have established that Tr, the microscop-
ically observed temperature of irruptive ice recrystallization in
such glass-forming systems of low MW sugars and polyols (31,32,38,47-
49), coincides with the solute-specific Tg' measured by thermal or
thermomechanical analysis (1,2,19,22,40,42,106,107). It has also
been recognized that ice recrystallization is but one of many possible manifestations (referred to as collapse phenomena) of the dynamically controlled behavior of aqueous glasses and rubbers, which exist
at subzero temperatures in kinetically constrained, metastable states
rather than equilibrium thermodynamic phases (1-5,18-23). Generic

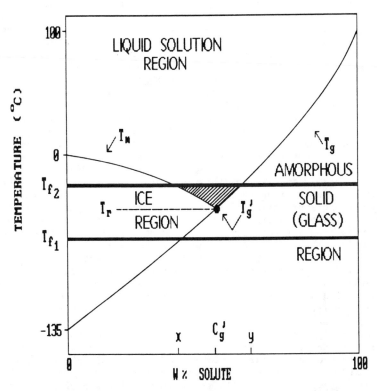

Figure 12. Schematic state diagram of temperature vs. w% solute for an aqueous solution of a hypothetical small carbohydrate (representing a model frozen food system), illustrating the critical relationship between Tg′ and freezer temperature (Tf), and the resulting impact on the physical state of the freeze-concentrated amorphous matrix. (Reproduced with permission from reference 18. Copyright 1988 Cambridge.)

use of the term "rubber" in this context describes all glass-forming liquids at Tg < T < Tm, including both molecular rubbers (viscous liquids) of low MW monomers and oligomers and viscoelastic network-forming rubbers of entangling high polymers.

A comparison of literature values of subzero collapse transition temperatures (Tc) for an extensive list of water-compatible monomers and polymers has established the fundamental identity of Tg' with the minimum onset temperatures observed for various structural collapse and recrystallization processes (1,2,115), for both model solutions and real systems of foods, as well as pharmaceuticals and biologicals (38,42,106,108,109). The Tc and Tr values of frozen or vitrified samples, tabulated elsewhere (1,2,22), are typically determined by cryomicroscopy, thermal analysis, or electrical resistence measurements (22 and refs. therein), on an experimental time scale similar to that for Tg' by our DSC method (106,109). A comprehensive list of collapse processes (Table I), all of which are governed by Tg' of frozen systems (or a higher Tg pertaining to low moisture systems processed or stored at T > 0°C) and involve potentially detrimental plasticization by water, has been identified and elucidated (1-5,18-22,26). The two parts of Table I, taken together, emphasize how the Tg values relevant to both low and high temperature collapse processes are systematically related through the corresponding water contents, thus illustrating how this interpretation of collapse phenomena has been generalized to include both high temperature/low moisture and low temperature/high moisture food products and processes (1,2,22,26). Previous versions of Table I (22,26) contained extensive references from the food science and technology literature to specific examples of both low temperature/high moisture and high temperature/low moisture collapse processes. In all cases, a partially or completely amorphous system in the mechanical solid state at T < Tg and local viscosity > ηg at Tg would be stable against collapse, within the period of experimental measurements of Tg, Tc, and/or Tr. Increased moisture content (and concomitant plasticization) would lead to decreased stability and shelf-life, at any particular storage temperature (37,115). The various phenomenological threshold temperatures for the diverse collapse processes listed in Table I all correspond to the particular Tg' or other Tg relevant to the solute(s) system and its content of plasticizing water.

Our interpretation of the physicochemical basis of collapse has also provided insights to the empirical countermeasures traditionally employed to inhibit collapse processes (2). In practice, collapse in all its different manifestations can be prevented, and food product quality, safety, and stability maintained, by the following measures (1): 1) storage at a temperature below or sufficiently near Tg (110); 2) deliberate formulation to increase Tc (i.e. Tg) to a temperature above or sufficiently near the processing or storage temperature, by increasing the composite M̄w of the water-compatible solids in a product mixture, often accomplished by adding "polymeric (cryo)stabilizers" (1,2,18) such as low DE SHPs or other high MW carbohydrates, proteins, or cellulose and polysaccharide gums to formulations dominated by low MW solutes such as sugars and/or polyols (35-37,110); and 3) in hygroscopic glassy solids and other low moisture amorphous food systems especially prone to the detrimental effects of plasticization by water (including various forms of "candy" glasses) (110), a) reduction of the residual moisture content to ≲ 3% during process-

ing, b) packaging in superior moisture-barrier film or foil to prevent moisture uptake during storage, and c) avoidance of excessive
temperature and humidity ($\geq$ 20% R.H.) conditions during storage (36,
110). The successful practice of the principles of cryostabilization
technology has often been shown to rely on the critical role of high-
polymeric saccharides and proteins in preventing collapse (by raising
the composite $\overline{M}w$ and resulting Tg' of a frozen product relative to
Tf) and to apply equally well to low moisture foods, such as amorphous, freeze-dried powders (1-5,18-22,26-28,42).

Collapse processes during freezer storage are promoted by the
presence of high contents of low MW saccharides of characteristically
low Tg' and high Wg' in the composition of many frozen foods (e.g.
desserts) (1,2,18-22). The fundamental physicochemical basis of the
cryostabilization of such products has been illustrated by the idealized state diagram (modeled after one for fructose-water (3)) shown
in Figure 12, which has also been used to explain why Tg' is the key-
stone of the conceptual framework of this technology (18-22). As
shown in Figure 12, the matrix surrounding the ice crystals in a maximally frozen solution is a supersaturated solution of all the solute
in the fraction of water remaining unfrozen. This matrix exists as a
glass of constant composition at any temperature below Tg', but as a
rubbery fluid of lower concentration at higher temperatures between
Tg' and the Tm of ice. If this amorphous matrix is maintained as a
mechanical solid, as at Tfl < Tg' and local viscosity > η at Tg, then
diffusion-limited processes that typically result in reduced quality
and stability can be virtually prevented or, at least, greatly inhibited. This physical situation has been illustrated by scanning electron microscopy photographs (18) of frozen model solutions, which
show small, discrete ice crystals embedded and immobilized in a continuous amorphous matrix of freeze-concentrated solute-UFW which exists as a glassy solid at T < Tg'. The situation has been described
by analogy to an unyielding block of window glass with captured air
bubbles (18). In contrast, storage stability is reduced if a natural
material is improperly stored at too high a temperature, or a fabricated product is improperly formulated, so that the matrix is allowed
to exist as a rubbery fluid at Tf2 > Tg' (see Figure 12), in and
through which diffusion is free to occur. Thus, the Tg' glass has
been recognized as the manifestation of a kinetic barrier to any diffusion-limited process (1,40), including further ice formation (within the experimental time frame), despite the continued presence of
UFW at all temperatures below Tg'. The delusive "high activation
energy" of this kinetic barrier to relaxation processes has been identified as the extreme temperature dependence that governs changes in
local viscosity and free volume just above Tg (22). This perspective
on the glass at Tg'-Cg' as a mechanical barrier has provided a long-
sought theoretical explanation of how undercooled water can persist
(over a realistic time period) in a solution in the presence of ice
crystals (18,20,22). Recognizing these facts, and relating them to
the conceptual framework described by Figure 12, one can appreciate
why the temperature of this glass transition is so important to aspects of frozen food technology involving freezer storage stability,
freeze-concentration, and freeze-drying (38,40,42,107,115,116), which
are all subject to various recrystallization and collapse phenomena
at T > Tg' (1,2).

The optimum Tf for a natural material or optimum formula for a

fabricated product is dictated by the Tg' characteristic of a particular combination of solutes and UFW in the matrix composition of the
glass at Tg'-Cg' (1,2,18-22). Tg' is governed in turn by the $\overline{M}w$ of
this particular matrix combination in a complex food system (1,23,
40). Moreover, the dynamic behavior of rubbery frozen food products
during storage above Tg' is dramatically temperature-dependent, and
the rates of diffusion-limited deterioration processes are quantitatively determined by the temperature difference ΔT = Tf − Tg' (in °C)
(3,18). These rates have been shown to increase exponentially with
increasing ΔT, in agreement with WLF, rather than Arrhenius, kinetics
(18-22). Results of a cryomicroscopy experiment (18), in which the
increase in ice crystal diameter was measured as a function of Tg'
for a series of model sugar/maltodextrin solutions (Tg' range −9.5 to
−31°C) after 4 weeks of storage in a −18°C home freezer, illustrated
this dynamic behavior and the cryostabilizing (i.e. Tg'-elevating)
effect of a high MW SHP (1,2). When Tf was below Tg' (−18 < −9.5°C),
the ice crystal size remained nearly the same as for the initially
frozen samples. When Tf was above Tg', the increase in ice crystal
size (i.e. decreasing stability) demonstrated a striking correlation
with decreasing Tg' (thus, increasing ΔT), and the temperature dependence was clearly greater than that expected for Arrhenius kinetics.

For the cryostabilization of real frozen food products such as
ice cream (with desirable smooth, creamy texture) against ice crystal
growth over storage time (at Tf > Tg'), inclusion of polymeric cryostabilizers such as low DE maltodextrins elevates the composite Tg'
of a mix of soluble solids that is typically dominated by low MW sugars (1,18). In practice, a retarded rate of migratory ice recrystallization ("grain growth" of pre-existing ice crystals) (47) at Tf and
an increase in observed Tr result. Such behavior has been documented
in several "soft-serve" ice cream patents (27,28,111). In such products (and in a variety of other frozen foods and aqueous model systems), ice recrystallization has been shown to involve a diffusion-
limited maturation process with a mechanism analogous to "Ostwald
ripening", whereby larger crystals grow with time at the expense of
smaller ones which eventually disappear (18,20,22 and refs. therein).
The ripening rate at Tf is reduced (due to reduced ΔT above Tg') by
formulation with low DE maltodextrins of high Tg' (1,18). Low DE
maltodextrins have also been used to stabilize frozen dairy products
against lactose crystallization during storage (112), another collapse process (see Table I) which can occur in a practical time frame
at Tf > Tg'. In fact, as indicated by the footnote to Table I (22,
26), high-polymeric SHPs have been used as effective stabilizers
against collapse in the majority of the listed collapse processes.
As noted earlier, whether for cryostabilization, encapsulation, thermomechanical stabilization, or facilitation of drying processes, the
utility of such SHPs derives from the structure-function relationship
defined by the entanglement plateau region in Figure 6.

The cryostabilizing functionality of saccharide polymers has
also been demonstrated by DSC measurements of Tg' values for various
diagnostic three-component aqueous solutions as model systems for
frozen foods (18,20,22). These results illustrated very clearly the
dependence of Tg' on the weight-average composition of a compatible,
multi-solute mixture (1). In this context, an interesting illustra-
tion of what has been gleaned from an analysis of glass curves for
such complex aqueous mixtures is shown in Figure 13 (3). In this

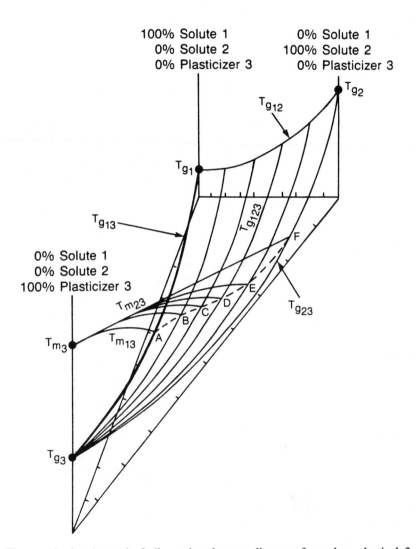

Figure 13. A schematic 3-dimensional state diagram for a hypothetical 3-component aqueous system. The two solutes (e.g. polymer + monomer) are both non-crystallizing, interacting, and plasticized by water, which is the crystallizing solvent. The diagram illustrates the postulated origin of a sigmoidal curve of Tg' vs. w% solute composition. (Reproduced with permission from reference 3. Copyright 1988 Cambridge University Press.)

artist's rendering of a three-dimensional state diagram for a hypothetical three-component system, both solutes (e.g. a polymer, 2, and
its monomer, 1) are non-crystallizing, interacting (i.e. compatible),
and plasticized by water, which is the crystallizing solvent, 3. The
diagram revealed the postulated origin of a sigmoidal Tg'(c) curve,
i.e. the Tg' glass curve ABCDEF (3). In fact, similar sigmoidal
curves of Tc vs. w% concentration, for collapse during freeze-drying
of analogous three-component aqueous systems, had previously been reported (Figure 5 in MacKenzie (34)), but "the basis for [their non-
linearity] ha[d] not been determined" (34). Four series of model
solutions at 10 w% total solids composed of mixtures with varying
ratios of maltodextrins (0.5-15 DE) to fructose demonstrated an experimental verification of the postulated sigmoidal shape of Tg'(c)
curves, as shown by the plots of Tg' vs. w% maltodextrin in Figure 14
(18). It has been found that such sigmoidal glass curves represent
the general behavior of compatible mixtures of both homologous and
non-homologous polymeric (including Na caseinate protein) and low MW
(including various sugars and acids) solutes (18). In all such
curves, the low- and high-Tg' tie points were determined by the Tg'
values of the individual low and high MW solutes, respectively. Figure 14 illustrated that the composite Tg' value characteristic of a
given amorphous solute(s)-UFW composition is governed by the $\overline{M}w$ of
the particular combination of compatible water-soluble solids in a
complex frozen system. It has also been used to illustrate the principle of polymeric cryostabilization: the stabilizing influence on
the structural state of a complex amorphous matrix derives from the
high MW of polymeric cryostabilizers and the resulting elevating effect on Tg' of a food product (1,18).

Saccharide Polymers as Collapse Inhibitors in Low Moisture Systems.
As explained elsewhere (23), if the relative shapes of the polymer-
diluent glass curves are similar within a homologous polymer series,
increases in MW (of the diluent-free polymer) lead to proportional
increases in both Tg and Tg'. This fact has been recently demonstrated by the aqueous glass curves for maltose, maltotriose, and maltohexaose published by Orford et al. (113), coupled with Tg'-Wg' values
for these oligosaccharides (7). Prior to this confirmation, it had
been assumed that a plot of Tg vs. MW for dry PHCs or SHPs would reflect the same fundamental behavior as that of Tg' vs. solute MW
shown in Figures 6, 7, and 8 (26). Earlier evidence supporting this
assumption had been provided by To and Flink (35), who reported a
plot of Tc vs. DP for a series of low moisture, fractionated SHP oligomers of 2 ≤ DP ≤ 16 (i.e. non-entangling (1)), similar in shape to
the plot of Tg' vs. MW for the non-entangling PHCs in Figure 7. It
had been pointed out that Tc for low moisture SHPs, which increases
monotonically with increasing DP, represents a good quantitative ap-
proximation of dry Tg (1). The basic assumption was verified for the
homologous series of glucose and its pure malto-oligomers of DP 2-7
(26). A plot of Tg vs. MW showed that dry Tg increases monotonically
with increasing MW of the monodisperse sugar, from Tg = 31°C for glucose (in good agreement with several other published values (7 and
refs. therein)) to Tg = 138.5°C for maltoheptaose (26). The plot
showed the same qualitative curvature (and absence of an entanglement
plateau) as the corresponding Tg' plot in Figure 8, and the plot of
dry Tg vs. 1/MW showed the same linearity and r value as the corre-

sponding Tg' plot in the inset of Figure 8 (26). [These results were subsequently corroborated by Orford et al. (113), who recently reported a similar curve of dry Tg vs. DP for glucose and its malto-oligomers of DP 2-6.] Further verification of the assumption was demonstrated by a plot (shown in Figure 15 (26)) of Tg vs. w% composition for a series of spray-dried, low moisture powders (about 2 w% water) prepared from solution blends of commercial SHPs, Lodex 10 and Maltrin M365. This plot showed that Tg increases from 58°C for Maltrin M365 (36 DE, Tg' = -22.5°C) to 121°C for Lodex 10 (11 DE, Tg' = -11.5°C) for these SHPs at about 2 w% moisture. Here again, the characteristic monotonic increase of Tg with $\overline{M}w$ ($\equiv$ increasing composition as w% Lodex 10) and curvature expected and theoretically predicted for homologous (mixtures of) oligomers with $\overline{M}w$ values below the entanglement plateau limit were evident.

As mentioned earlier, the principles of cryostabilization technology have also been applied successfully to the prevention of collapse processes in low moisture, glass-forming food systems, through the use of saccharide polymers as stabilizing ingredients to raise the composite $\overline{M}w$ and resulting Tg of a mixture of water-compatible solids, otherwise typically dominated by low MW, low Tg components such as sugars and acids (26). As illustrated in Figure 16 (data from Fouad Saleeb, General Foods Corp., personal communication, 1984.), the thermomechanical stabilization of amorphous food powders by low DE SHPs has been demonstrated in model system studies of spray-dried, three-component mixtures. Figure 16 shows that typical low MW glass-formers (e.g. common food ingredients such as maltose monohydrate, mannose, and citric acid), at low moisture contents, often have low Tg values, around and even below room temperature. For this reason, such amorphous food materials are particularly sensitive to moisture uptake and prone to collapse (e.g. caking) during ambient shelf storage. However, as shown in Figure 16, in mixed glasses containing a polymeric stabilizer such as 10 DE maltodextrin, the composite Tg increases dramatically with increasing w% stabilizer, due to increasing $\overline{M}w$. Thus, in general, 10 DE maltodextrin is an excellent thermomechanical stabilizer, which is capable of elevating the Tg of an amorphous mixture to a "safe" temperature well above room temperature, and thereby inhibiting collapse processes during ambient storage and increasing product shelf-life.

Conclusion

In this chapter, we have reviewed the development and highlighted some technological applications of a polymer characterization method, based on low temperature DSC and derived from a food polymer science research approach, to analyze the structure-physicochemical property relationships of linear, branched, and cyclic mono-, oligo-, and polysaccharides. Through the use of this DSC characterization method, important subtleties of the structure-property relationships of SHPs and PHCs have been revealed. Our DSC results have emphasized the unavoidable conclusion, concerning the choice of a suitable saccharide ingredient for a specific food application, that one SHP (or PHC) is not necessarily interchangeable with another of the same nominal DE (or MW). We have shown that characterization of fundamental structure-property relationships, in terms of Tg', is highly advisable before selection of such ingredients for use in fabricated food prod-

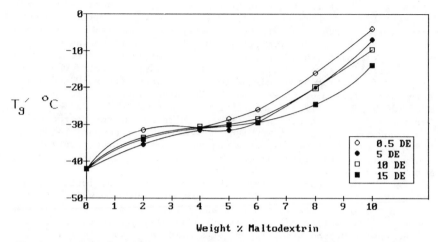

Figure 14. Variation of the glass transition temperature, Tg', for maximally frozen solutions against weight % maltodextrin in 10 w% total solids solutions of maltodextrin + fructose, for four different low DE maltodextrins. (Reproduced with permission from reference 18. Copyright 1988 Cambridge.)

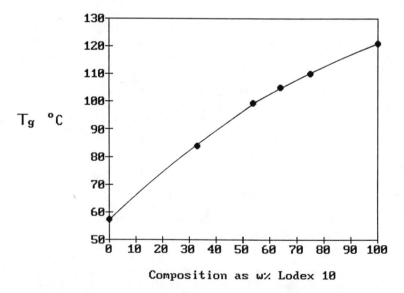

Figure 15. Variation of the glass transition temperature, Tg, against weight % composition for spray-dried, low moisture powders prepared from aqueous solution blends of Lodex 10 (10 DE maltodextrin) and Maltrin M365 (36 DE corn syrup solids) SHPs. (Reproduced with permission from reference 26. Copyright 1989 Elsevier.)

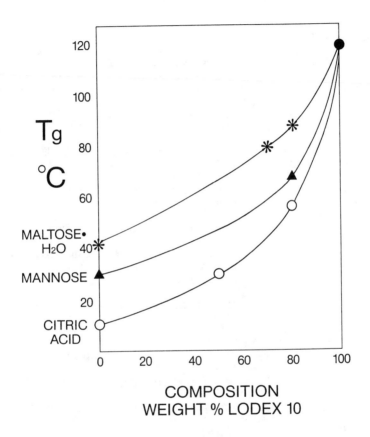

Figure 16. Variation of the glass transition temperature, Tg, against weight % composition for spray-dried, low moisture powders prepared from aqueous solution blends of Lodex 10 and a low MW, glass-forming food ingredient, such as citric acid, mannose, or maltose monohydrate.

ucts. The studies reviewed in this chapter have also served to demonstrate the major opportunity offered by the food polymer science approach to expand not only our quantitative knowledge but also, of broader practical value, our qualitative understanding of, and ability to correctly predict, the structure-function relationships of linear, branched, and cyclic mono-, oligo-, and polysaccharides in a wide variety of food products and processes.

Literature Cited

1. Levine, H.; Slade, L. Carbohydr. Polym. 1986, 6, 213-44.
2. Levine, H.; Slade, L. In Food Structure - Its Creation and Evaluation; Mitchell, J.R.; Blanshard, J.M.V., Eds.; Butterworths: London, 1988; pp 149-80.
3. Levine, H.; Slade, L. In Water Science Reviews; Franks, F., Ed.; Cambridge University Press: Cambridge, 1988; Vol. 3, pp 79-185.
4. Slade, L.; Levine, H.; Finley, J.W.; In Protein Quality and the Effects of Processing; Phillips, D.; Finley, J.W., Eds.; Marcel Dekker: New York, 1989; pp 9-124.
5. Levine, H.; Slade, L. In Dough Rheology and Baked Product Texture; Faridi, H.; Faubion, J.M., Eds.; Van Nostrand Reinhold/ AVI: New York, 1989; pp 157-330.
6. Slade, L.; Levine, H. CRC Crit. Revs. Food Sci. Nutr. 1990, in press.
7. Slade, L.; Levine, H.; In Water Relationships In Foods; Levine, H.; Slade, L., Eds.; Plenum: New York, 1991; pp 29-101.
8. Flory, P.J. Principles of Polymer Chemistry; Cornell University Press: Ithaca, 1953.
9. Wunderlich, B. In Thermal Characterization of Polymeric Materials; Turi, E.A., Ed.; Academic Press: Orlando, 1981; pp 91-234.
10. Ferry, J.D. Viscoelastic Properties of Polymers, 3rd edn.; John Wiley & Sons: New York, 1980.
11. Sears, J.K.; Darby, J.R. The Technology of Plasticizers; Wiley-Interscience: New York, 1982.
12. Billmeyer, F.W. Textbook of Polymer Science, 3rd edn.; Wiley-Interscience: New York, 1984.
13. Sperling, L.H. Introduction to Physical Polymer Science; Wiley-Interscience: New York, 1986.
14. Slade, L.; Levine, H. Carbohydr. Polym. 1988, 8, 183-208.
15. Slade, L.; Levine, H. In Industrial Polysaccharides - The Impact of Biotechnology and Advanced Methodologies; Stivala, S.S.; Crescenzi, V.; Dea, I.C.M., Eds.; Gordon and Breach Science: New York, 1987; pp 387-430.
16. Slade, L.; Levine, H. In CRA Scientific Conference; Corn Refiners Assoc.: Washington, DC, 1988; pp 169-244.
17. Slade, L.; Levine, H. In Frontiers in Carbohydrate Research-1: Food Applications; Millane, R.P.; BeMiller, J.N.; Chandrasekaran, R., Eds.; Elsevier Applied Science: London, 1989; pp 215-70.
18. Levine, H.; Slade, L. Cryo-Lett. 1988, 9, 21-63.
19. Levine, H.; Slade, L. J. Chem. Soc., Faraday Trans. I 1988, 84, 2619-33.
20. Levine, H.; Slade, L. Comments Agric. Food Chem. 1989,1,315-96.

21. Levine, H.; Slade, L. Cryo-Lett. 1989, 10, 347-70.
22. Levine, H.; Slade, L. In Thermal Analysis of Foods; Ma, C.-Y.; Harwalkar, V.R., Eds.; Elsevier Applied Science: London, 1990; chapter 9.
23. Slade, L.; Levine, H. Pure Appl. Chem. 1988, 60, 1841-64.
24. Slade, L.; Levine, H. In Advances in Meat Research, Vol. 4 - Collagen as a Food; Pearson, A.M.; Dutson, T.R.; Bailey, A., Eds.; AVI: Westport, 1987; pp 251-66.
25. Slade, L.; Levine, H. In Food Structure - Its Creation and Evaluation; Mitchell, J.R.; Blanshard, J.M.V., Eds.; Butterworths: London, 1988; pp 115-47.
26. Levine, H.; Slade, L. In Water and Food Quality; Hardman, T.M., Ed.; Elsevier: London, 1989; pp 71-134.
27. Cole, B.A.; Levine, H.I.; McGuire, M.T.; Nelson, K.J.; Slade, L. U.S. Patent 4 374 154, 1983.
28. Cole, B.A.; Levine, H.I.; McGuire, M.T.; Nelson, K.J.; Slade, L. U.S. Patent 4 452 824, 1984.
29. Slade, L.; Altomare, R.; Oltzik, R.; Medcalf, D.G. U.S. Patent 4 657 770, 1987.
30. BeMiller, J.N. (as reported by Hill, M.A.) Carbohydr. Polym. 1989, 10, 64.
31. Franks, F.; Asquith, M.H.; Hammond, C.C.; Skaer, H.B.; Echlin, P. J. Microsc. 1977, 110, 223-38.
32. MacKenzie, A.P. Phil. Trans. Royal Soc. London B. 1977, 278, 167-89.
33. Karel, M. In Properties of Water in Foods; Simatos, D.; Multon, J.L., Eds.; Martinus Nijhoff: Dordrecht, 1985; pp 153-69.
34. MacKenzie, A.P. In Freeze Drying and Advanced Food Technology; Goldlith, S.A.; Rey, L.; Rothmayr, W.W., Eds.; Academic Press: New York, 1975; pp 277-307.
35. To, E.C.; Flink, J.M. J. Food Technol. 1978, 13, 551-94.
36. Flink, J.M. In Physical Properties of Foods; Peleg, M.; Bagley, E.B., Eds.; AVI: Westport, 1983; pp 473-521.
37. Karel, M.; Flink, J.M. In Advances in Drying; Mujumdar, A.S., Ed.; Hemisphere: Washington, 1983; Vol. 2, pp 103-53.
38. Franks, F. In Water: A Comprehensive Treatise; Franks, F., Ed.; Plenum Press: New York, 1982; Vol. 7, pp 215-338.
39. Franks, F. Biophysics and Biochemistry at Low Temperatures; Cambridge University Press: Cambridge, 1985.
40. Franks, F. In Properties of Water in Foods; Simatos, D.; Multon, J.L., Eds.; Martinus Nijhoff: Dordrecht, 1985; pp 497-509.
41. Franks, F. J. Microsc. 1986, 141, 243-49.
42. Franks, F. Process Biochem. 1989, 24(1), R3-R7.
43. Marsh, R.D.L.; Blanshard, J.M.V. Carbohydr. Polym. 1988, 9, 301-17.
44. Soesanto, T.; Williams, M.C. J. Phys. Chem. 1981, 85, 3338-41.
45. Starkweather, H.W. In Water in Polymers; Rowland, S.P., Ed.; ACS Symp. Ser. 127; American Chemical Society, Washington, DC, 1980; pp 433-40.
46. Hofer, K.; Hallbrucker, A.; Mayer, E.; Johari, G.P. J. Phys. Chem. 1989, 93, 4674-77.
47. Luyet, B. Ann. NY Acad. Sci. 1960, 85, 549-69.
48. Rasmussen, D.; Luyet, B. Biodynamica 1969, 10, 319-31.
49. MacKenzie, A.P.; Rasmussen, D.H. In Water Structure at the Wa-

ter-Polymer Interface; Jellinek, H.H.G., Ed.; Plenum Press: New York, 1972; pp 146-71.

50. Vassoille, R.; El Hachadi, A.; Vigier, G. Cryo-Lett. 1986, 7, 305-10.

51. MacFarlane, D.R. Cryo-Lett. 1985, 6, 313-18.

52. Boutron, P.; Kaufmann, A. Cryobiol. 1979, 16, 557-68.

53. Bohon, R.L.; Conway, W.T. Thermochim. Acta 1972, 4, 321-41.

54. Vrentas, J.S.; Hou, A.C. J. Appl. Polym. Sci. 1988, 36, 1933-34.

55. Mayer, E. Cryo-Lett. 1988, 9, 66-77.

56. Fahy, G.M.;   Levy, D.I.; Ali, S.E. Cryobiol. 1987, 24, 196-213.

57. Graessley, W.W. In Physical Properties of Polymers; Mark, J.E.; Eisenberg, A.; Graessley, W.W.; Mandelkern, L.; Koenig, J.L., Eds.; American Chemical Society, Washington, DC, 1984; pp 97-153.

58. Brydson, J.A. In Polymer Science; Jenkins, A.D., Ed.; North Holland, Amsterdam, 1972; pp 194-249.

59. Shalaby, S.W. In Thermal Characterization of Polymeric Materials; Turi, E.A., Ed.; Academic Press, Orlando, 1981; pp 235-364.

60. Dziedzic, S.Z.; Kearsley, M.W. In Glucose Syrups: Science and Technology; Dziedzic, S.Z.; Kearsley, M.W., Eds.; Elsevier Applied Science, London, 1984, pp 137-68.

61. Brooks, J.R.; Griffin, V.K. Cereal Chem. 1987, 64, 253-55.

62. Flory, P.J. Faraday Disc. Chem. Soc. 1974, 57, 7-18.

63. Richter, M.; Schierbaum, F.; Augustat, S.; Knoch, K.D. U.S. Patent 3 962 465, 1976.

64. Richter, M.; Schierbaum, F.; Augustat, S.; Knoch, K.D. U.S. Patent 3 986 890, 1976.

65. Braudo, E.E.; Belavtseva, E.M.; Titova, E.F.; Plashchina, I.G.; Krylov, V.L.; Tolstoguzov, V.B.; Schierbaum, F.R.; Richter, M. Starke 1979, 31, 188-94.

66. Braudo, E.E.; Plashchina, I.G.; Tolstoguzov, V.B. Carbohydr. Polym. 1984, 4, 23-48.

67. Bulpin, P.V.; Cutler, A.N.; Dea, I.C.M. In Gums and Stabilizers for the Food Industry 2; Phillips, G.O.; Wedlock, D.J.; Williams, P.A., Eds., Pergamon Press, Oxford, 1984; pp 475-84.

68. Reuther, F.; Damaschun, G.; Gernat, C.; Schierbaum, F.; Kettlitz, B.; Radosta, S.; Nothnagel, A. Coll. Polym. Sci. 1984, 262, 643-47.

69. Lenchin, J.M.; Trubiano, P.C.; Hoffman, S. U.S. Patent 4 510 166, 1985.

70. Miles, M.J.; Morris, V.J.; Ring, S.G. Carbohydr. Res. 1985, 135, 257-69.

71. Ellis, H.S.; Ring, S.G. Carbohydr. Polym. 1985, 5, 201-13.

72. German, M.L.; Blumenfeld, A.L.; Yuryev, V.P.; Tolstoguzov, V.B. Carbohydr. Polym. 1989, 11, 139-46.

73. Ferry, J.D. J. Amer. Chem. Soc. 1948, 70, 2244-49.

74. Mitchell, J.R. J. Text. Stud. 1980, 11, 315-37.

75. Franks, F. Pure Appl. Chem. 1987, 59, 1189-202.

76. Zobel, H.F. Starke 1988, 40, 1-7.

77. Russell, P.L. J. Cereal Sci. 1987, 6, 147-58.

78. Mestres, C.; Colonna, P.; Buleon, A. J. Cereal Sci. 1988, 7, 123-34.

79. Miles, M.J.; Morris, V.J.; Orford, P.D.; Ring, S.G. Carbohydr. Res. 1985, 135, 271-81.
80. Matsukura, U.; Matsunaga, A.; Kainuma, K. J. Jpn. Soc. Starch Sci. 1983, 30, 106-13.
81. Ring, S.G.; Colonna, P.; l'Anson, K.J.; Kalichevsky, M.T.; Miles, M.J.; Morris, V.J.; Orford, P.D. Carbohydr. Res. 1987, 162, 277-93.
82. Ring, S.G.; Orford, P.D. In Gums and Stabilizers for the Food Industry 3; Phillips, G.O.; Wedlock, D.J.; Williams, P.A., Eds., Elsevier Applied Science, London, 1986; pp 159-65.
83. Welsh, E.J.; Bailey, J.; Chandarana, R.; Norris, W.E. Prog. Fd. Nutr. Sci. 1982, 6, 45-53.
84. Gidley, M.J. Macromolecules 1989, 22, 351-58.
85. Steeneken, P.A.M. Carbohydr. Polym. 1989, 11, 23-42.
86. Gidley, M.J.; Bulpin, P.V. Macromolecules 1989, 22, 341-46.
87. Clark, A.H.; Gidley, M.J.; Richardson, R.K.; Ross-Murphy, S.B. Macromolecules 1989, 22, 346-51.
88. Ring, S.G. Starke 1985, 37, 80-83.
89. l'Anson, K.J.; Miles, M.J.; Morris, V.J.; Ring, S.G.; Nave, C. Carbohydr. Polym. 1988, 8, 45-53.
90. Gidley, M.J.; Bulpin, P.V.; Kay, S. In Gums and Stabilizers for the Food Industry 3; Phillips, G.O.; Wedlock, D.J.; Williams, P.A., Eds., Elsevier Applied Science, London, 1986; pp 167-76.
91. Whistler, R.L.; Daniel, J.R. In Starch: Chemistry and Technology; Whistler, R.L.; BeMiller, J.N.; Paschall, E.F., Eds.; Academic Press, Orlando, 1984; 2nd edn., pp 153-82.
92. Jankowski, T.; Rha, C.K. Starke 1986, 38, 6-9.
93. Buleon, A.; Duprat, F.; Booy, F.P.; Chanzy, H. Carbohydr. Polym. 1984, 4, 161-73.
94. Hizukuri, S. Carbohydr. Res. 1986, 147, 342-47.
95. Atwell, W.A.; Hood, L.F.; Lineback, D.R.; Varriano-Marston, E.; Zobel, H.F. Cereal Foods World 1988, 33, 306-11.
96. Ring, S.G. Int. J. Biol. Macromol. 1985, 7, 253-54.
97. Boyer, R.F.; Baer, E.; Hiltner, A. Macromolecules 1985, 18, 427-34.
98. Domszy, R.C.; Alamo, R.; Edwards, C.O.; Mandelkern, L. Macromolecules 1986, 19, 310-25.
99. Burchard, W. Progr. Colloid Polym. Sci. 1988, 78, 63-67.
100. Krusi, H.; Neukom, H. Starke 1984, 36, 300-05.
101. Keinath, S.E.; Boyer, R.F. J. Appl. Polym. Sci. 1981, 26, 2077-85.
102. Wesson, J.A.; Takezoe, H.; Yu, H.; Chen, S.P. J. Appl. Phys. 1982, 53, 6513-19.
103. Durning, C.J.; Tabor, M. Macromolecules 1986, 19, 2220-32.
104. Muhr, A.H.; Blanshard, J.M.V. J. Food Technol. 1986, 21, 683-710.
105. Luyet, B.J. J. Phys. Chem. 1939, 43, 881-85.
106. Reid, D.S. Cryo-Lett. 1985, 6, 181-88.
107. Blanshard, J.M.V.; Franks, F. In Food Structure and Behaviour; Blanshard, J.M.V.; Lillford, P., Eds.; Academic Press, London, 1987; pp 51-65.
108. Pikal, M.J.; Shah, S.; Senior, D.; Lang, J.E. J. Pharmaceut. Sci. 1983, 72, 635-50.
109. Pikal, M.J.; Shah, S. Int. J. Pharm. 1990, in press.
110. White, G.W.; Cakebread, S.H. J. Food Technol. 1966, 1, 73-82.

111. Holbrook, J.L.; Hanover, L.M. U.S. Patent 4 376 791, 1983.
112. Kahn, M.L.; Lynch, R.J. U.S. Patent 4 552 773, 1985.
113. Orford, P.D.; Parker, R.; Ring, S.G.; Smith, A.C. Int. J. Biol. Macromol. 1989, 11, 91-96.
114. Radosta, S.; Schierbaum, F.; Reuther, F.; Anger, H. Starke 1989, 41, 395-401.
115. Karel, M. In Food Properties and Computer-Aided Engineering of Food Processing Systems; Singh, R.P.; Medina, A.G., Eds.; Kluwer, Dordrecht, 1989; pp 135-155.
116. Franks, F. Int. Indust. Biotechnol. 1990, in press.
117. Russell, P.L.; Oliver, G. J. Cereal Sci. 1989, 10, 123-138.
118. Lund, D.B. In Food Properties and Computer-Aided Engineering of Food Processing Systems; Singh, R.P.; Medina, A.G., Eds.; Kluwer, Dordrecht, 1989; pp 299-311.
119. Aras, L.; Richardson, M.J. Polymer 1989, 30, 2246-2252.
120. Doublier, J.L.; Choplin, L. Carbohydr. Res. 1989, 193, 215-226.

RECEIVED November 2, 1990

Chapter 17

Solution Properties and Composition of Dextrins

Gordon G. Birch, M. Nasir Azudin[1], and John M. Grigor

Department of Food Science and Technology, University of Reading, Whiteknights, P.O. Box 226, Reading, Berkshire RG6 2AP, United Kingdom

Solution properties of dextrins, such as NMR, apparent specific volume and intrinsic viscosity, are similar to those of their components, glucose maltose and higher saccharides. However, fine differences between the individual components of glucose syrups can be precisely monitored by modern solution chemistry techniques and usefully employed to elucidate their mode of interaction with water structure and to predict their behaviour in food systems. Examples are the determination of DE by high resolution NMR and "equivalent DE" (in hydrogenated glucose syrups) by combined measurement of refractometric solids and osmotic pressure, without recourse to volumetric chemical methods. Solution measurements of glucose syrups or dextrins derived from glucose syrups are based on average molecular weight. In a DE17 glucose syrup, for example, the average apparent specific volume of the dextrin increases from 0.62-0.63 cm³/g as the concentration increases from 5-50% w/w. These figures are high compared to glucose (0.615cm³/g) and maltose (0.612cm³/g) but low compared to β-cyclodextrin (0.668cm³/g). [1]H-pmr pulse relaxation analysis of the 17 DE dextrin in concentrated solution (65% w/w) allows ring protons, hydroxyl protons and water to be distinguished, and in this regard the dextrin behaves similarly to sucrose. Thus solution properties provide a useful insight of solute-water interaction.

[1]Current address: Faculty of Food Science and Biotechnology, University Pertanian Malaysia, Serdang 43400 UPM, Selangor, Malaysia

0097–6156/91/0458–0261$06.00/0

Some solution properties of carbohydrates in homologous series (eg. densities) appear to be similar among the members of that series, whereas others (eg. specific rotations) vary markedly and systematically with degree of polymerisation. It is recognised (1,2) that water stabilises the pyranose ring which, along with the helical order of the amylodextin, contributes to the value of the observed specific rotation $[\alpha]_D$. Thus, a plot of $[\alpha]_D$ against DP, for the lowest members of the linear maltodextrin series (Table 1), gives a curve which approaches a limiting value of $[\alpha]_D$ of about +200.

Table 1

Specific Rotations and DP
for lowest members of the linear maltodextrin series (3)

DP	$[\alpha]_D^{15}(H_2O)$
1	+52.6
2	+136.0
3	+160.0
4	+177.0
5	+180.3
6	+184.7
7	+186.4

Computations from typical glucose syrup compositions would predict similar results for dextrose equivalent (DE) against $[\alpha]_D$, as actually observed. Such findings illustrate the insignificant contributions to $[\alpha]_D$ of successive coils in the amylose helix; polarimetric measurements are always made at low concentration of solute to avoid distortions due to solute-solute interaction. Polarimetry has been traditionally used by carbohydrate chemists for identification and determination of sugar molecules and is a sensitive tool for following the kinetic course of hydrolysis (4). It has the advantage of providing a property of a small amount of solute in a large amount of solvent (water), but the disadvantage of not identifying the role of water itself makes polarimetry less attractive to the chemist than spectrometry in studying solution properties of dextrins and related substances.

Determination of DE by High Resolution Nuclear Magnetic Resonance (NMR)

High resolution NMR allows individual protons of the glucose moiety in glucose syrups to be distinguished. If the NMR spectrum is obtained first in D_2O and second in D_2O/D_2SO_4 solution the α- and β-anomeric proton doublets (from the free reducing groups) can be easily distinguished from the multiplets of remaining protons. The ratio of the areas of the doublets to the multiplets thus gives a measure of the degree of hydrolysis and DE (5).

Table 11 summarises some typical successes with this type of approach.

Table 11 - DE of Glucose Syrups by NMR (5)

DE by Lane & Eynon Titration	DE by NMR
21.0	21.5
31.0	29.0
40.0	38.5
50.0	50.0
65.0	66.5

Obviously the time-saving of an instrumental technique over a wet chemical method is of interest for industrial control but D_2O solution is necessary to avoid overlapping problems from OH signals.

Determination of DE by Refractive Index/Osmotic Pressure

An obvious method for determining DE rapidly is to utilize a colligative property such as osmotic pressure. As DE of a glucose syrup increases so does the total number of molecules. Cryoscopic determination of osmotic pressure, combined with total solids determination with a refractometer, suffice to determine DE (6) within a total time of about 3 minutes.

Nowadays hydrogenated glucose syrups constitute important dextrin derivatives used for manufacturing products with specialised technological and biological properties. Since hydrogenated glucose syrups by definition possess no DE, it is important to be able to assess their "equivalent DE", or in other words the type of parent syrup from which they were derived. The above method, utilising only refractometry and cryoscopy provides an excellent combination of solution properties to achieve that end.

Precision Densitometry and Apparent Specific Volume

Although dextrins resemble all carbohydrates in their similar solution densities, there is good reason to suppose that differences in structure should cause fine differences in solution packing characteristics of solutes and hence density differences.

Modern precision densitometry allows measurements of density yielding six significant figures and thus reveals fine differences between sugars possessing different arrangements of axial and equatorial hydroxyl groups. Equatorial hydroxyl groups are more easily hydrated than axial and the resulting hydrated structure is better able to 'fit' with surrounding bulk water structure.

The 'fit' of a solute with surrounding water structure is probably a prime determinant of apparent molar volume (V) which is calculated from solution density using the expression:

$$\text{apparent molar volume } V = \text{mol. wt} \left\{ \dfrac{\dfrac{1}{s} - W_1}{W_2} \right\}$$

where s = density of sample
where   = density of water at 20° C
where W_1 = mass fraction of water
where W_2 = mass fraction of solute

Apparent Molar volume (V) increases accordingly with Mwt so a more useful measure of degree of 'fit' with water structure is Apparent Specific Volume (V/Mwt) which allows direct comparison of different molecular architectures on a weight basis. The apparent specific volumes of all simple sugars lie within the range 0.58-0.62 $cm^3 g^{-1}$ and remarkably good agreement occurs between analogues with corresponding axial and equatorial arrangements of hydroxyl groups. Among the aldopyranose pairs D- glucose:D-xylose, D-galactose:L- arabinose and D-fructose:D-arabinose, for example, the ratios of apparent specific volumes of hexose to pentose is always 1.2 in accordance with molecular weight (7).

Apparent molar volumes of solutes result from the positive displacement of water molecules (depending on molecular size and shape) and the negative electrostriction effect caused by the hydration of suitably disposed hydroxyl groups (8). For this reason heavily hydrated structures, such as sugars and oligosaccharides, have much smaller apparent specific volumes than poorly hydrated structures, such as benzene or tetrahydropyran.

Moreover, the hydroxyl substituents at different positions around the sugar ring contribute differently to the overall apparent specific volume (7). In glucopyranose types of structure, as found in the dextrins, it is the 3,4 α-glycol system which fulfills this role, indicating that this is the mosty heavily hydrated region of the molecule. It is also significant that the 3,4 α-glycol system appears to be responsible for sweetness (9). This is why the sweetness of malto-oligosaccharides decreases with increasing DP (10). The 3,4 α-glycol group of the terminal non-reducing residue of the malto-oligosaccharide appears to stimulate the sweet receptor and the remainder of the molecule is neither involved nor sterically excluded from the receptor. Table III lists some values of sweetness of malto-oligosaccharides according to DP. All of the oligosaccharides appear to be sweeter than D-glucose on a molar basis but maltotriose appears to be anomalously less sweet than the other members of the series.

Table III Sweetness equivalents of Malto-oligosaccharides
in 0.33 M solution

DP	Sweetness Equivalent Sucrose concentration (M)
1	0.090
2	0.126
3	0.115
4	0.136
5	0.133
6	0.133

Solute-Solute Interaction as revealed by solution studies

As the concentration of a sugar increases in solution, solute-solute interaction gradually replaces solute-solvent interaction and apparent specific volumes increase accordingly. Although the apparent specific volumes of simple sugars are generally below 0.62 cm^3 g^{-1} when measured at low concentrations, they may rise above this figure for solutions of 50% (w/v) or more. The solute-solute interaction deprives the sugar molecule of water of hydration and a correspondingly greater disturbance of water structure occurs.

Table IV lists some apparent specific volumes (V/MWT) for glucose, maltose, maltodextrin (DE=17) and β-cyclodextrin at different

Table IV   Apparent Specfic Volumes and Concentrations
of Sugars and dextrins *

Solute	V/MWT cm^3 g^{-1}				
	0.2% w/w	0.4% w/w	0.6% w/w	5.0% w/w	50.0% w/w
D-glucose	–	–	–	0.615	0.634
Maltose	–	–	–	0.612	0.646
Maltodextrin (DE=17)	–	–	–	0.620	0.630
β-Cyclodextrin	0.651	0.651	0.668	–	–
Sorbitol	–	–	–	0.655	0.667

* Measured with an Anton-Parr Density Meter as in earlier publications (11)

concentrations but comparisons are impossible with β-Cyclodextrin at high concentration due to its low solubility.

The markedly high apparent specific volume of β-Cyclodextrin reflects poor compatibility with water structure. Apparent specific volumes of sugars in solution reflect the packing characteristics of sugar molecules among water molecules mediated by hydrogen-bonding networks. Ideal measurements (free of solute-solute interactions) are made at low concentrations but high concentration measurements provide important technological information as well as data for biological applications. In the microenvironment of a taste receptor, for example, water activity may be minimal and solute-solute interaction may play a large part in the receptor response.

Finally, Table IV shows a high apparent specific volume for sorbitol. This must be due to lack of cyclic structure rather than lack of an anomeric centre in the molecule because the apparent specific volume of myoinositol is smaller than that of the sugars, i.e. 0.560 cm^3 g^{-1} (11).

Low resolution ^{1}H NMR Pulse Relaxation Studies of Dextrins and Related sugars

In exploring the hydration characteristics of solutes, NMR pulse relaxation studies provide information about loss of energy of protons depending on their immediate environment. Low resolution NMR spectrometers, such as the Bruker Minispec, give average spin-spin relaxation times (T_2 values) and spin-lattice relaxation times (T_1 values). At low concentrations, these mainly reflect the average state of water protons under the influence of the solute. In Fig.1 the T_1 values of a 17 DE maltodextrin (M) are shown, together with enzymic digests of the dextrin, A (after 1 hour) and E (after 5 hours), at increasing concentrations. Obviously the T_1 values drop as the solute protons gradually constitute a greater proportion of total protons. This is likely because the hydrated solute hydroxyl groups exchange protons less readily than do bulk water molecules and the O-H protons of the solute do not exchange at all.

In Fig.2 the same trend is observed for T_2 values, but there are no clear differences between the values of the dextrin and its digests. This is because fine differences between oligosaccharides are obscured by their average distributions. Fig.3 shows the T_1 and T_2 values of individual oligosaccharides (at 3% w/v) and illustrates an interesting peak for maltotriose. This indicates that maltotriose is disturbing water structure more than the other members of the series and parallels its anomalous sweetness shown in Table III. It may be significant that the 3,4 - glycol group of these oligosaccharides is responsible for their sweetness and that it also constitutes the most heavily hydrated region of the molecule. For some reason in maltotriose the compatibility with water structure is lower than the other members of the series.

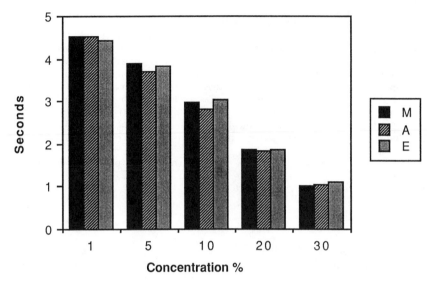

Fig 1 T_1 Values of dextrin (M = DE 17) and its digests (A = 1h digestion; E = 5h digestion)

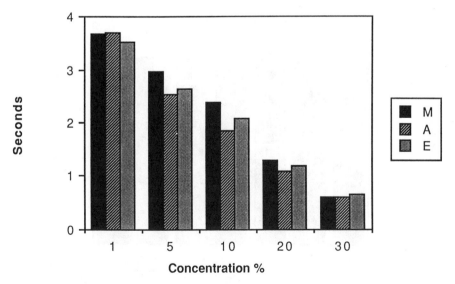

Fig 2 T_2 Values of dextrin (M) and digests (as in Fig 1)

Fig. 4 shows that α, β and γ -cyclodextrins vary neither in T_1 values in the range 1-3% w/v nor among themselves. Also only in the case of cyclodextrin (Fig.5) does the T_2 value turn out to be significantly lower than the others. It should be noted that the T_1 and T_2 values shown in Figs. 3,4 and 5 relate to quite dilute solutions and the proton signals measured are therefore mainly due to water protons. The Figures thus reflect changes in water structure due to the effects of the solute. When NMR pulse relaxation studies are carried out in more concentrated solutions, it can be observed (12) that the log spin echo amplitude is not monophasic over the course of time. Fig.6 shows one such curve for the DE17 maltodextrin and Fig.7 illustrates its separate components. These equate clearly to the different types of proton (exchange-rates) of the solute as well as the water protons.

Fastest decay occurs for the ring C-H protons of each oligosaccharide. Then come the hydroxyl protons, with the signal for water protons decaying most slowly.

Conclusion

Characteristics of dextrins and glucose syrups can be explored by combinations of different solution properties. Although the solution properties of each depict different aspects of dextrin behaviour they are probably all mediated by hydrogen-bonding and are of direct relevance to the applications of these molecules in food and biotechnology.

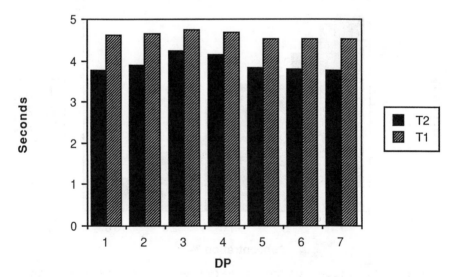

Fig 3 T_1 and T_2 Values of Oligosaccharides

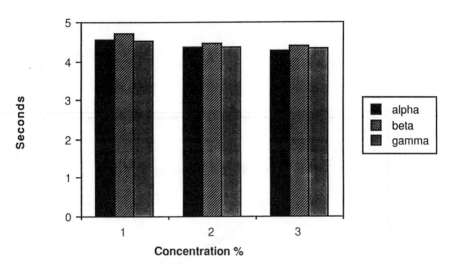

Fig 4 T_1 Values of Cyclodextrins (g solute/100g water)

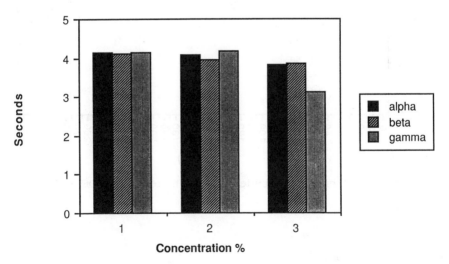

Fig 5 T_2 Values of Cyclodextrins (g solute/100g water)

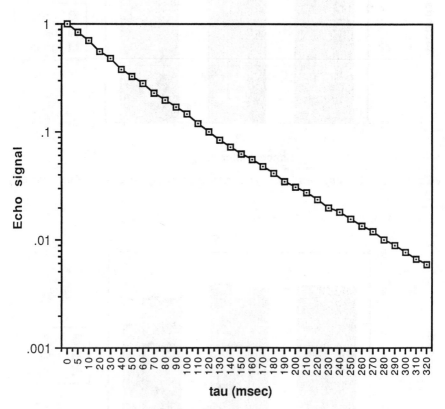

Fig 6   Echo Amplitudes of Maltodextrin (DE = 17)

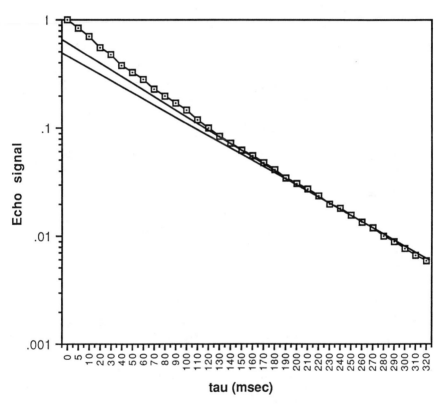

Fig 7   Echo Amplitudes of Maltodextrin (DE = 17)

Acknowledgments

The authors thank the Royal Society, The Association of Commonwealth Universities and The American Chemical Society for financial assistance.

Literature Cited

1.   Franks, F. Pure Appl. Chem. 1987. 59, 1189-1202.

2.   Shallenberger, R.S. 1982, Advanced Sugar Chemistry, AVI Connecticut.

3.   Whelan, W.J.; Bailey, J.M.; Roberts, P.J.P. J. Chem. Soc. 1953,   1293-1297.

4.   Wienen, W.: Shallenberger, R.S.; Food Chem. 1988 29, 51-56.

5.   Birch, G.G.; Kheiri , M.S.A. Carbohydrate Res. 1971, 16, 215-21.

6.   Kearsley, M.W. J. Assoc. Publ Analysts 1978, 16 85-88.

7.   Birch, G.G.; Shamil, S. J. Chem. Soc. (Farad Trans). 1988, 84, 2635- 2640.

8.   Shahidi, F.; Farrell, P.G.; Edward, J.T.; J. Solution Chem 1976, 5, 807-816.

9.   Birch, G.G.; Lee, C.K. J. Food Sci 1974, 39, 947-949.

10. O'Donnell, K. Ph.D. Thesis, Reading University, UK, 1983.

11. Shamil, S.; Birch, G.G.; Mathlouthi, M.; Clifford, M.N. Chem. Senses 1987, 12, 397-409.

12. Birch, G.G.; Grigor, J.M., Derbyshire, W. J. Solution Chem 1989, 18, 795-801.

RECEIVED October 3, 1990

Chapter 18

Linear Dextrins

Solid Forms and Aqueous Solution Behavior

S. G. Ring and M. A. Whittam

Institute of Food Research, Colney Lane, Norwich,
Norfolk NR4 7UA, United Kingdom

This article reviews physical and physico - chemical
studies on linear malto - dextrins. The various
crystalline forms of the dextrins are described and the
melting and glass transition behavior discussed. Solution
conformations and the viscous behavior of dilute and
concentrated aqueous solutions are considered. In addition
factors affecting the interaction of small solutes with
the dextrins in aqueous solution are discussed and
compared with the known behavior of cyclodextrins.

Amylodextrins are conveniently prepared by the limited hydrolysis
of starch to produce fragments ranging in degree of polymerization
from 2 to approximately 60. Although branched and cyclic forms may
be produced, in this article we wish to consider the aqueous
solution behavior of linear dextrins. These dextrins may be
produced in a number of ways for example by the limited acid
hydrolysis of solid forms of starch, more particularly starch
granules. Alternatively they may be produced by limited
α-amylolysis of soluble forms of starch. More recently with the
production of the debranching enzymes pullulanase and isoamylase
and with the identification of α-amylases which produce
oligosaccharides of fixed length from their hydrolysis of starch
(1,2) there is increased opportunity for the production of defined
dextrins from starch. In addition there is potential, through the
application of genetic approaches and molecular biology to
manipulate enzyme structure and hence function to obtain a wider
range of products more effectively. These molecules have a variety
of industrial uses both in the food industry and elsewhere. They
may be used as agents for the manipulation of liquid texture and
also the mechanical properties of solid - like materials. In both
of these instances the interaction of the dextrin with water
influences usefulness. Cyclodextrins have very many potential uses
as agents for molecular encapsulation, the encapsulated species
occupying the cyclodextrin cavity. Dextrins can also form molecular
complexes with guest molecules. In addition it is possible to

0097–6156/91/0458–0273$06.25/0
© 1991 American Chemical Society

encapsulate active ingredients in glassy matrices of the dextrin; in these matrices it is possible to protect active species and subsequently obtain release on plasticization. For these applications interaction of the dextrin with water again influences usefulness. Current research on these dextrins is examining the effect of water on material and molecular properties as a function of composition, temperature and degree of polymerization. In this way it is hoped to establish predictive relationships for structure and function which will help target the application of biotechnology to their production.

Crystalline Forms

Solid amylodextrins may be in the glassy/rubbery or crystalline state; considerably more studies have been carried out on the latter, so crystal structures are quite well understood. Whilst amylopectin is thought to be the main crystalline component in starch granules, amylose fractions ranging in degree of polymerization from DP100 to DP2700 have been crystallized from water in the form of fibrils (3). Shorter chain amyloses of DP15 and 35 have been shown to form more perfect single crystals (4). Oligomers of glucose however between heptamer and trimer are very difficult to crystallize. Although oligomers as short as maltohexaose have been co-crystallized in the presence of longer chains, the minimum chain length necessary for double helix formation appears to be DP10 (5,6). Crystal structures of glucose and maltose have long been known, but it is only recently that a crystalline structure for maltotriose has been reported (7). Degree of crystallinity can vary widely, from retrograded amylose which is predominantly amorphous, containing perhaps 10% crystalline material, to highly crystalline single crystals (4) or spherulites (8) prepared from shorter chain amyloses. Similarly, single crystals of V amylose have been reported (4,9) as well as amorphous amylose complexes which do not give rise to X-ray diffraction patterns (10).   Since the first publication of X-ray diffraction patterns of A,B and C starches sixty years ago (11), several models have been proposed for their crystal structures. Until recently, a right handed, parallel stranded arrangement of double helices, with antiparallel packing of the helices themselves was favored for both A and B structures (12,13). In the last two years, however, studies combining X-ray and electron diffraction methods on single crystals of amylose DP15, together with computer energy calculations have been carried out (14,15). From these studies, a left handed, parallel stranded double helix formation seems more likely, with helices packed parallel to one another in the crystal. Left handed or right handed, it is generally accepted that A and B structures consist of similar double helices; the difference lies in the spatial arrangement of these double helices within the crystal. In the B structure, six double helices are hexagonally arranged around a central cavity containing water molecules. A is more compact, the central cavity containing another double helix, leading to a crystal structure including less water than that of B.

Less controversy surrounds the structure of V amylose, now

widely accepted as a six fold single helix, although within the
classification "V" several crystalline variants are found. Vh and
Va, the hydrated and anhydrous forms of complexed amylose, can be
interconverted by manipulating moisture or humidity conditions.
When 1-butanol is used as the complexing agent, a third crystal
structure is observed (16), whilst seven fold and eight fold
helical conformations have been reported for amylose complexes with
branched chain alcohols and α-naphthol respectively (9). One
question which remains to be resolved is that of the exact location
of the complexing agent within the V structure. In solution, there
is a body of evidence to suggest that the amylose helix
encapsulates the complexing "guest" molecule within its core,
giving rise, for example, to the familiar blue color of the
amylose-iodine complex. Evidence from X-ray and electron
diffraction that seven and eight fold helices are formed when
increasingly large complexing molecules are used (9) lends weight
to the inclusion theory. However, an alternative explanation for
the larger unit cells (ascribed to seven and eight fold helices) is
that the complexing molecules are present in the interstices
between the individual helices (16).

X-ray diffraction analysis has given little information on the
location of the complexed molecules. This may be because these
molecules are randomly placed and consequently form no regular
lattice for diffraction, or it may be that despite being confined
within the helices guest molecules possess too much rotational and
translational freedom to be identifiable by X-ray diffraction. If
complexing molecules are removed from the amylose crystals by
drying, the V structure remains intact, indicating that the
complexant is not a necessity for crystal stability. However there
is some evidence from X-ray diffraction together with potential
energy calculations that dimethyl sulphoxide molecules are located
inside the amylose helix with one DMSO molecule for every three
glucose residues (17). Further work is obviously needed to
determine unequivocally where the included molecules are situated
within the V amylose structure.   Computer modelling is
being used increasingly to gain insight into the energies of
different molecular conformations. Helical forms of amylodextrins
have been modelled using the dimer maltose as the starting unit
(18). Minimum energy conformations of the disaccharide are found by
calculating the energy as the two sugar rings are rotated about the
glycosidic linkage. Polymers built up from these low energy maltose
units are found to have helical structures which correspond to
conformations observed for maltodextrins in the crystalline state.
Thus computer studies complement X-ray analysis, and may help in
the future to resolve the question of where complexed molecules are
located within crystalline V amylose.

The polymorphic form of amylose or amylodextrin depends on
several factors. For an amylodextrin of a given length in aqueous
solution, formation of the A crystal type is favored over B at high
temperature and high concentration. Chain length also has an

effect, shorter chains giving rise to A structures and longer chains (including amylose retrograded under ordinary conditions) forming B structures. Intermediate conditions give rise to C-type crystal formation. Below DP10, solutions of amylodextrins do not crystallize even at 4°C and 50% w/w, implying a lower limit of chain length necessary for double helix formation (5). The crystallization, melting and dissolution behavior of amylodextrins may give insight into the behavior of whole starches, Since it is the short chains of amylopectin which give rise to the crystalline regions within starch granules and which are believed to melt and recrystallize during gelatinization and retrogradation respectively. These transitions are important in terms of functional properties of dextrins and starches.

Addition of an alcohol, for example ethanol or iso-propanol, to an amylodextrin solution encourages the formation of A- rather than B- type crystals. Above a certain alcohol concentration, V-type crystals then begin to form preferentially. Whilst it is difficult to prepare highly crystalline samples of A and B structure using long chain amylose, V amylose crystals will form relatively easily under the right conditions. Suitable complexing agents include fatty acids, linear and branched alcohols, monoglycerides and various other organic compounds. Care is required when crystallizing high molecular weight amylose complexes; for example a rapid cooling rate can lead to formation of amorphous as well as crystalline material (10). Successful technique results in platelet-like crystals consisting of stacked lamellae, themselves formed from amylose molecules folded back and forth between the bottom and top surfaces of the lamella (9,10). For synthetic polymer crystallization it is found that lamellar thickness, and hence crystal stability, depends strongly on preparation conditions such as temperature of crystallization and cooling rate. Future studies may indicate similar relationships for these biopolymer crystals.

If it is accepted that the complexed molecules in V amylose are included within the helix, then the complexation process can be thought of as molecular encapsulation. Stability of these complexes has been studied by observation of the melting/dissolution behavior of amorphous and crystalline complexes. Differential scanning calorimetry (DSC) shows an endothermic transition on heating amylose complexes; the value at which it occurs depends on the nature of the complex. This transition corresponds to melting and dissociation of the amylose-guest complex. Amorphous complexes have been shown to melt at lower temperatures than their crystalline counterparts, demonstrating the stabilizing influence of the crystal lattice (Table I).

The nature of the guest molecule also affects complex stability; for example complexes formed using guest molecules of increasing hydrocarbon chain length dissociate at increasingly higher temperatures. Thus the longer hydrocarbon chains confer greater stability on the amylose complex structure.

Table I. Melting temperatures for crystalline and amorphous complexes of amylose with linear alcohols

Alcohol Chain Length	Melting Temperature (K)	
	Crystalline	Amorphous
C4	341	321
C5	354	334
C6	364	344
C7	368	348
C8	372	352
C10	377	358

Attempts have been made to describe the dissolution of starch crystallites on gelatinization using the Flory equation for polymer/diluent systems (19):

$$\frac{1}{T_m^o} - \frac{1}{T_m} = \frac{R}{\Delta H_u} \frac{V_u}{V_1}\left[v_1 - \chi v_1^2 \right] \qquad (1)$$

where T_m is the observed melting temperature, T_m^o is the melting temperature of the undiluted polymer, ΔH_u is the enthalpy of fusion of the repeating unit (glucose), V_u/V_1 is the ratio of the molar volume of the repeating unit (glucose) to that of the diluent (water), v is the volume fraction of the diluent and χ is the Flory-Huggins polymer- solvent interaction parameter. Whilst meeting with limited success, the application of the Flory equation to whole starch must be viewed with caution since it is valid only for equilibrium melting of pure crystalline phases. The presence of glassy or amorphous material within the granule would tend to affect crystallite melting temperature, thus jeopardizing the applicability of the Flory equation. Spherulites of amylodextrins in A or B form may be considered models of amylopectin crystallites, and in this case application of the Flory equation is more justified. Studies are currently in progress on the melting behavior of A- and B-type crystals and the effect of water as a diluent.

Biotechnology will soon permit the breeding of plants with starches of pre-determined molecular composition. In addition to variations in amylose/amylopectin ratio, it will be possible to vary the length of the amylopectin chains. It has already been shown that variation in amylopectin chain length has a significant effect on starch retrogradation behavior (20). In order for technology to exploit these variations fully, it is important to understand the underlying molecular processes. This will permit a rational approach to the use of genetically manipulated starches and will lead to improved understanding of crystallization behavior of amylodextrins in general. An example might be fractionation of amylodextrins using knowledge of their respective crystallization temperatures in a given solution.

Amorphous Forms

While dextrins can often be encountered as crystalline solids in many cases they can be prepared as partially crystalline or wholly amorphous materials. For these materials, the glass transition behavior of the amorphous regions is relevant. A glass has many of the characteristics of a liquid except that of mobility, its importance to technological application stems from the dramatic changes in material properties e.g. mechanical and diffusive behavior, which occur in the region of the glass transition. The technological importance of the glass transition for polymeric materials, inorganic glasses and more recently biomaterials has been reviewed (21-24). An example on how it might affect the usage of dextrins will illustrate some of the principles. Glassy matrices may be used for encapsulation and protection of active ingredients. The mechanism of protection is somewhat different to that obtained from the formation of molecular inclusion complexes and does not rely upon the molecular recognition of active ingredient and host. In the glass, the rate of diffusion of compounds is severely hindered. Hence, the glassy matrix can protect through limiting the diffusional encounter of reacting species. With plasticization of the matrix diffusion will accelerate; through manipulation of plasticization controlled release is possible.

Although a glass is metastable with respect to the crystalline form, it can remain in a glassy form for many years; above the glass transition crystallization is much more rapid. Glasses are therefore prepared by cooling at rates sufficiently fast to avoid crystallization. The required cooling rate depends on the material. For example, to vitrify water, cooling rates in excess of 10^5 Ks^{-1} are necessary (25,26). Polymeric materials crystallize less readily and cooling rates of $<1Ks^{-1}$ are often sufficient for vitrification. As a liquid is supercooled, molecular motion and the local rearrangement of molecules becomes progressively slower and the viscosity increases. For D-glucose the viscosity increases by 13 orders of magnitude on cooling from a melt at 413K to the glass at 290K (27,28). With cooling, the probability of a molecule overcoming an activation energy to escape from its neighbors decreases. At the glass transition there is a sudden "freezing"of motion with a consequent sharp fall in heat capacity. The glass transition temperature, Tg, is affected by cooling rate, the "freezing in" of liquid structure will occur at progressively lower temperatures as the cooling rate decreased, always assuming that crystallization does not intervene. The observed Tg is thus profoundly affected by kinetic factors (23). However, it is argued in various theoretical approaches (29) that underlying the experimentally observed transition is a true phase transition.

The Tg of amorphous materials can be determined by measuring heat capacity as a function of temperature and observing the sharp increase in heat capacity,indicative of a glass transition, which occurs at Tg. At the present time this is conveniently performed using differential scanning calorimetry to determine Tg and ΔCp (figure 1). In addition by varying both heating and cooling rates and observing the change in Tg it is possible to obtain an "activation enthalpy" for the relaxation process at Tg (30,31). This enthalpy can be usefully compared with that from viscosity data (30).

The glass transition behavior of carbohydrates in general is receiving recent attention. The Tg's of amorphous maltooligomers from maltose to maltohexaose was recently determined (32) and ranged from 364K for maltose to 448K for maltohexaose. Data for higher oligomers was not obtainable due to thermal degradation, the high molecular weight limit of Tg obtained by extrapolation was 500±10K. At low diluent concentrations the addition of water strongly depressed Tg, for example the Tg for maltohexaose fell 100K on addition of 10% w/w water. The compositional variation of Tg with water content was investigated at water contents <20% where the Tg of the carbohydrate water mixture was above 273K. From a knowledge of Tg and ΔCp at the glass transition of pure components it is possible to predict the Tg of the mixture using a thermodynamic approach developed by Couchman (29)

$$Tg = \frac{w_1 \Delta Cp_1 Tg_1 + w_2 \Delta Cp_2 Tg_2}{w_1 \Delta Cp_1 + w_2 \Delta Cp_2} \tag{2}$$

where w is the mass fraction and subscripts 1 and 2 refer to the two components, in this case malto-oligomer and water. Agreement between experimental and predicted values was satisfactory for the higher oligomers, theory always over-estimating Tg. For ternary mixtures e.g. maltohexaose / glucose / water agreement between experiment and theory was less good (33).

If a carbohydrate/water mixture is cooled very slowly to well below the freezing point of water both the water and the carbohydrate should eventually crystallize. If cooled very rapidly, for example by spraying onto a liquid helium cooled cryo plate, the solution will vitrify (26). Between these two extremes of cooling rate, most cooling rates which are practically obtainable will result in the formation of a carbohydrate/water glass and ice (34). The composition dependence, usually expressed as the mass fraction of water in this glassy form, and the Tg of this glass have been investigated in detail (35). Some general trends have been noted e.g. Tg increases on going from maltose (253.5K) to maltoheptaose (259.5K). The water contents of these mixtures show no consistent trend with chain length (35,36).

Further research is necessary to put the glass transition behavior of carbohydrates in general, and their mixtures, on a predictive basis. Additionally studies on the relaxation behavior and mechanical properties and factors affecting the sorption of water will give useful information relevant to the technological use of these materials.

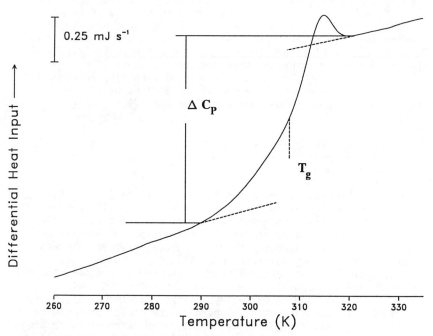

Figure 1. Plot of heat capacity versus temperature obtained at a scanning rate of 10K/min, for amorphous D-glucose, showing position of glass transition temperature (Tg) and heat capacity increment (ΔCp).

Aqueous Solution Behavior

In this section we examine the aqueous solution behavior of linear
α,1-4 linked D-glucans. Important information can be gained by
studying behavior as a function of degree of polymerization and
composition. The aqueous solution behavior of dilute solutions of
amylose has been investigated using a range of physico-chemical
techniques including total intensity light scattering,
quasi-elastic light scattering and viscometry. Research up until
1975 has been reviewed by Banks and Greenwood (37). The amyloses
studied have either been extracted from raw granular starch (38,39)
or more preferably (40) they have been enzymically synthesized. In
the latter case they are monodisperse and linear, thereby avoiding
the interpretive problems caused by polydispersity and limited
chain branching. There is a growing consensus in published work
(41). Firstly, water at ambient temperatures is an indifferent
solvent for amylose, which will precipitate even from dilute
solutions. Secondly the linear dependence of molecular weight, M,
on the square of radius of gyration, Rg and the square root of
diffusion coefficient, Dt, and intrinsic viscosity indicates that
amylose in neutral aqueous solution behaves as a non-free draining
flexible coil. Additionally these measurements provide information
on the local stiffness of the polymer chain. Values of the
characteristic ratio for amylose defined as

$$C = \frac{6\ \langle R_g^2 \rangle}{\langle n \rangle\ l^2} \tag{3}$$

where n is the degree of polymerization and l the length of the
monomer unit, are in the range 4.5 - 5.9. This value is comparable
to that found for synthetic polymers which adopt flexible
conformations in solution. Information on polydispersity can be
obtained (38) by comparing the average quantities Mw, Rg, and D_r.
The recent use of light scattering detectors with gel permeation
chromatography has become the preferred method for obtaining
molecular weight distributions (42). While amylose in neutral
aqueous solution behaves as a flexible coil it is accepted from
molecular modeling studies (43) that successive residues have a
preferred orientation with respect to each other. If the preferred
orientation persists over many residues a helix is generated. At
any instant therefore, in the constantly fluctuating structure of
the amylose chain, parts of this structure may have a helical
conformation.

While there has been sustained research on the solution
properties of amylose, the solution behavior of linear dextrins has
been less investigated. For these molecules light scattering
studies become more difficult as the intensity of the light
scattered by the dextrin is much less than that of the
macromolecule. Viscometric studies have, however, been performed
both on dilute and concentrated solutions. For the dilute solutions
it is possible to obtain by extrapolation values of the intrinsic
viscosity. One approach (44-46) to interpreting this value is to
consider the dextrin as a particle immersed in a continuous fluid.

It can be described by the relationship for the viscosity of dilute suspensions

$$\eta = \eta_0(1 + \nu\phi) \qquad (4)$$

where η is the viscosity of the solution, η_0 the viscosity of the solvent ϕ the volume fraction of the particle and ν a constant depending on particle asymmetry, having a value of 2.5 for spherical particles. It is questionable whether this approach is appropriate when the size of the solute particle is approaching that of the solvent. Nevertheless it forms a useful starting point. Table (II) shows a comparison of the intrinsic viscosities for maltose through to maltohexaose (L.Botham, P.S.Belton, S.G.Ring, unpublished data) and for a dextrin of DP18. These are compared with the values obtained for cellodextrins (<u>44,45</u>).

Table II. Intrinsic viscosities of malto and cello dextrins (<u>39,40</u>)

	$[\eta]$ 25°C (ml/g)		$[\eta]$ 25°C (ml/g)
D-Glucose	2.5		
Maltose	2.6	Cellobiose	2.74
Maltotriose	2.7	Cellotriose	3.03
Maltohexaose	3.6	Cellohexaose	4.70
Dextrin (DP18)	6.0		

Table III. Evaluation of hydrated volumes and asymmetries of malto-dextrins

	Assume Spherical Specific volume/g carbohydrate	Assume no hydration Constant ν, (see Eq. 1)
D-Glucose	1.0	4.0
Maltose	1.04	4.1
Maltotriose	1.08	4.3
Maltohexaose	1.44	5.7

At a fixed DP the intrinsic viscosities of the maltodextrins are less than those of the cellodextrins. If it is assumed that the partial molar volume accurately describes the volume occupied by the "solute particle" then it is possible to obtain information on particle shape through calculation of the value of v. Alternatively a spherical shape may be assumed and a volume calculated which represents the effective hydrodynamic volume of the malto-oligomer and "associated" water molecules (table III). For example, the extrapolated intrinsic viscosity of maltohexaose in water at 25°C is

3.6 ml/g, assuming a spherical shape. The specific volume of the maltohexaose is 1.44 ml/g carbohydrate, indicating that the effective hydrodynamic volume of the hexaose contains 40 - 50 water molecules.

Alternatively, it can be assumed that there is no hydrating water and the increase in intrinsic viscosity over that expected for spherical particles is entirely due to particle asymmetry. For the malto-oligomers the constant v ranges from 4.1 to 5.7, on going from dimer to hexamer and has a value of 4 for the monomer. This would indicate axial ratios in the range 3-4 (<u>47</u>), which while plausible for the higher oligomers would not be consistent with the axial ratio of the glucose molecule (<u>44,45</u>). The increase in specific volume on going from glucose to maltohexaose also indicates that hydration alone cannot account for the observed values of intrinsic viscosity. From the study of diffusive and viscous behavior it has been possible to isolate these effects for the cellodextrins. For the maltodextrins current experiments indicate that hydration and asymmetry influence dilute solution viscosity. The lower values of intrinsic viscosity for maltodextrins when compared with the equivalent cellodextrin indicate that the former have a less extended conformation in solution.

It is also of interest to examine the viscous behavior at higher concentrations. In an investigation of the aqueous solution viscosities of aqueous solutions of fructose, glucose and sucrose (<u>48</u>) it was found that the measured viscosity had an exponential dependence on mole fraction of saccharide. A useful correlation was found of the form

$$\eta(x) = \alpha\exp[\kappa V(x)] \tag{5}$$

where $V(x) = (1-x)V_s + xV_m$ where α and κ are constants and V_s and V_m refer to the molar volumes of solvent and solute respectively. Using the same approach (L.Botham, P.S.Belton, S.G.Ring, unpublished data) the solution behavior of the malto-oligomers was examined and are shown in figure 2 where log viscosity is plotted as a function of $V(x)$. The relationship shown in Eq.(5) describes the viscous behavior of the malto-oligomers and D-glucose. It demonstrates that the main determinant of solution viscosity in this concentration range is related to the volume occupied by the saccharide; particle assymetry has a smaller influence. With increasing concentration the viscosity of the solution will increase; the viscosity of glassy materials is thought to be $\approx 10^{12}$ - 10^{13} Pa s^{-1}. As the glass transition temperature of the malto-oligomers changes by 80K on going from dimer to hexamer it is to be expected that the viscous behavior of the malto-oligomers will diverge, from that presented in figure 2, at higher concentrations.

Complexation

<u>Linear Dextrins</u>. Both linear and cyclic maltodextrins in aqueous solution can form molecular complexes with other small molecules, possibly the most commonly known example being the blue

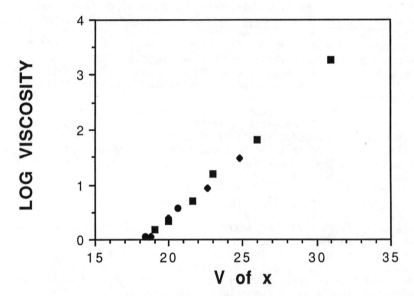

Figure 2.   Plot of log viscosity versus V(x) (for explanation see text) for D-glucose (■), maltose (◆), and maltohexaose (●).

iodine-amylose complex. A wide range of organic molecules exhibit this complexation behavior, including free fatty acids and their salts (49), linear and branched alcohols (50), mono-glycerides and other commercial emulsifiers (51), lysolecithins and even smaller molecules such as dimethylsulphoxide (52), the 4-nitrophenolate ion, the dye methyl orange and α-naphthol (53). The general requirement for a good complexing agent appears to be a hydrophobic, non polar group attached to a more polar moiety, although iodine forms an exception to this rule. Iodine, however is fairly hydrophobic in character and it is thought to be the polyiodide ion rather than iodine itself which complexes with amylose in solution. For a linear oligomer, the requirement for complex formation is a chain length of greater than 5 units (54). Cyclic oligomers (cyclodextrins) do not exist with fewer than 6 glucose units, presumably due to steric effects producing a highly strained structure. The branched molecule amylopectin may complex to a limited extent, but it does not generally exhibit the pronounced changes in molecular properties seen with linear oligomers. In general the complexes are not stable in solution, precipitating out after a period of time which depends on a variety of factors including temperature, concentration and degree of polymerization of amylose oligomer and the nature of the complexing molecule.

The occurence of complexation with certain molecules is immediately apparent. For example, with starch there is an immediate color change on addition of iodine. Similarly the spectrum of methyl orange shifts when a linear maltodextrin DP20 is added (figure 3). For other molecules, complexation in solution is not observable visually and must be inferred using other techniques. The conformation of linear amylose oligomers in aqueous solution is now widely accepted to be that of a flexible coil. Complexation induces a coil-helix transition which can be monitored, for example, by measuring the optical rotation or viscosity of the solution (55). Very early experiments on flow dichroism of amylose-iodine solutions suggested that amylose adopts a helical conformation on complexation, with the iodine molecules occupying the central core of the helix (56). This model has since been confirmed by other techniques such as NMR (53).

Whilst a coil-helix transition is generally believed to occur on complexation due to encapsulation of the guest, other explanations have been put forward to account for the observed changes. One is that amylose exists in solution as loose helices which then contract on entrapment of iodine (57). Another is that the guest molecule affects the solvent structure (in this case water) thus affecting the solute conformation (in this case the linear amylodextrin). The former hypothesis is unlikely in view of the optical rotation change which accompanies the complexation process; the latter,however, should not be disregarded. Interestingly, optical rotation studies appear to favor a right-handed helical structure in solution (58), compared to the left-handed structure predicted by X-ray diffraction of crystalline amylose complexes.

The driving forces behind the complexation process are still not completely understood. The hydrophobic effect would appear to be a major influence. A gain in entropy being achieved by the removal of hydrophobic solutes from the aqueous solvent to the less polar, non-hydrogen bonding helix interior. Surface tension studies support this theory; addition of a linear maltodextrin increases the surface tension of a solution of SDS (sodium dodecyl sulphate) (figure 4), effectively removing the SDS molecules from solution and preventing their surface activity. Another form of interaction between guest and host molecule might be electrostatic in nature, in the form of dipole-dipole interaction (49). The amylose helix must possess a dipole moment as a result of the combined dipole moments of each of its constituent glucose units. Guest molecules might then be oriented within the helix, their own dipoles directed antiparallel to that of the helix. In the case of iodine, which has no permanent dipole, the high molecular polarizability might lead to dipole-induced dipole interactions between helix and guest molecule. Factors affecting stability of complexes may also indicate the type of interactions which lead to complex formation. Several studies have shown that an increase in hydrocarbon chain length of the guest molecule results in a higher melting temperature for the complex. The increased thermal stability may be due either to a greater hydrophobic effect, or perhaps to an increase in van der Waals interaction between the longer hydrocarbon chain and the surrounding helix. The effect of the functional group, or polar part of the guest molecule is less well documented. There is some evidence, however, that it is the electrostatic charge rather than the specific group which determines stability of complexes, neutralized fatty acid complexes showing quite different melting behavior from that of their charged counterparts (50). The oligosaccharide chain length also has an important influence on complex formation. Once again, an immediately observable effect is the color of iodine complexes with amylose oligomers of varying degrees of polymerization. The color varies from red (for DP15) through purple (for DP50) to blue (for DP 100). No color is observed for DP 12 (59). The iodine binding capacity, measured as weight of iodine bound per weight of oligosaccharide increases with oligomer chain length, although above DP100 it changes very slowly (60). Dissociation constants and stoichiometry of the inclusion complexes are difficult to determine for several reasons. Firstly, binding appears to be cooperative for longer oligomers of amylose, so a single binding or dissociation constant is not sufficient to characterize the interaction. Secondly, it is difficult to isolate pure samples of complex to determine stoichiometry since washing the solid precipitate (in order to remove excess complexing agent) is likely also to remove complexed molecules. Recently a surface tension and optical rotation study provided evidence that more than one cooperative process take place on binding of fatty acid to amylose. The first, at low fatty acid concentration, was claimed to induce a complete coil - helix transition; the second mode of binding was less clear (61). Previous potentiometric studies indicated that non-cooperative binding, in this case of sodium dodecyl sulphate, dominated for DP 57 whereas for DP 76 cooperativity was observed (62). Thus it may be possible for future investigations to characterize the binding behavior more fully for a wider variety of

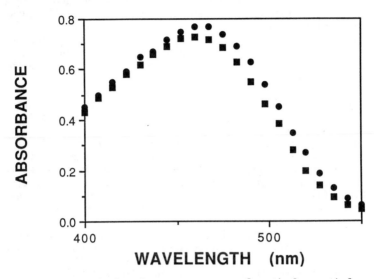

Figure 3.   Plot of absorbance versus wavelength for methyl orange
$(2.9 \times 10^{-5}M)$ with (■) added dextrin (0.8 mM) and without (●).

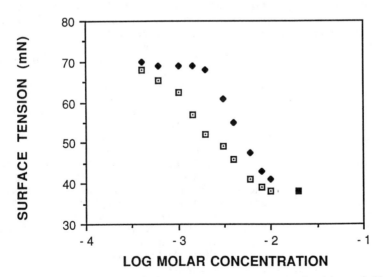

Figure 4.   Plot of surface tension versus concentration of SDS
alone (▫) and with added amylose 5mg/ml (◆).

complexing molecules. Determinations of complex stoichiometry have
been carried out by monitoring viscosity changes with increasing
amounts of fatty acid added to an amylose solution, and using ether
extraction to determine complexed lipid (63). The results showed
that saturation of the amylose helices occured, at a ratio of fatty
acid to amylose which could be predicted from the chain length of
the fatty acid. Such a relationship would tend to support the
theory that the fatty acid molecules are, indeed, present within
the amylose helices.

Cyclodextrins. Cyclodextrins, like linear maltodextrins, are
capable of complexation with a variety of guest molecules in
solution. Unlike their linear counterparts, cyclodextrins have been
the subject of extensive research, driven largely by their
potential as molecular encapsulating agents and as models for the
study of molecular recognition processes. Molecules reported to
form complexes with cyclodextrins include drugs, hormones, dye
molecules, flavorings, pesticides and insecticides. Recent reviews
give an up to date account of the complexation behavior of
cyclodextrins and their technological use in processes such as
decaffeination of coffee and removal of bitter components from
fruit juices (64,65). The most commonly found cyclodextrins are
those containing 6,7 or 8 glucose units joined by the same $\alpha,1-4$
linkages which form linear maltodextrin. The major difference
between cyclic and linear dextrins is that whilst cyclodextrins are
constrained to the well known 'doughnut' conformation with little
intramolecular flexibility, linear oligomers in solution have
considerable conformational mobility.The helical conformation
adopted by linear chains in the presence of complexing molecules
resembles more closely that of the cyclodextrin, where the interior
cavity is less polar than the exterior surface, being more or less
devoid of hydroxyl groups. This similarity suggests that at least
part of the mechanism of interaction between host and guest
molecule may be common to both cyclic and linear glucans. Due to
the wide variety of molecules which are able to form complexes with
cyclodextrins, it is unlikely that a single effect is responsible
for the interaction. A number of interaction mechanisms have been
suggested, the majority involving water either as a solvent or as a
high enthalpy form within the non-polar cyclodextrin cavity. The
possible driving forces for complexation include the hydrophobic
effect (removal of non-polar complexing molecules from the aqueous
solvent producing an increase in entropy of the water molecules)
and the effect of returning the high enthalpy water within the
cavity to the bulk phase and replacing it with the guest molecule.
The hydrophobic effect is likely to be common to both linear and
cyclic maltodextrin complexation but the concept of high enthalpy
water is less valid for the linear dextrin, since in aqueous
solution it forms no permanent cavity.
    The high enthalpy water theory may explain, at least in part,
the reason why cyclodextrins tend to have higher affinities for
guest molecules than do linear dextrins. Another reason may be that
of strain relaxation within cyclodextrin molecules. Saenger has
suggested (66) that in α cyclodextrin the uncomplexed molecule is
in a slightly strained conformation due to rotation of one of the

glucose units. On complexation this strain is released. The strain energy probably has only a small influence on complexation, however, since no other cyclodextrin molecules appear to have a strained conformation, yet their complex forming abilities are comparable to those of a cyclodextrin. In addition to the hydrophobic effect, another interaction mechanism likely to be common to both linear and cyclic dextrins is the dipole-dipole interaction. Quite a large dipole moment has been calculated for a cyclodextrin (67), so guest molecules may orient themselves inside the cyclodextrin cavity with their own dipole moment oriented antiparallel to that of the cyclodextrin molecule, with a resulting attractive dipole - dipole force. In cases where the guest molecule has no permanent dipole moment, its polarizability will determine the size of dipole moment induced by the permanent dipole of the cyclodextrin molecule and thus influence the strength of interaction between the two. Van der Waals' forces, including both dipole - dipole interaction and London dispersion forces, are thus likely to play a part in the inclusion process. Evidence for this comes from the far greater affinity of α cyclodextrin for the 4-nitrophenolate anion than for 4-nitrophenol (68) Linear maltodextrin helices presumably also possess a dipole moment, although the effect might be expected to be smaller than that of the cyclodextrin due to the more extended form of the linear maltodextrin.

A close spatial fit between cyclodextrin and guest molecule is generally recognized as important for good complexation, which supports the evidence that van der Waals' interactions, which are short range, may stabilize the complex. Because of this requirement for a close fit between guest and host, perhaps, cyclodextrins of different sizes complex preferentially with different guest molecules and may be used for the separation of specific molecules from a mixture. Linear maltodextrins, on the other hand, show less specific "recognition" of molecules, and are believed to form helical complexes with 6,7 or 8 glucose units per turn, depending on the size of the guest molecule. No single mode of interaction, therefore, seems to be responsible for complexation, but a variety of interactions are operational, their relative contributions varying with the nature of the guest molecule. This explains why such a wide range of molecules can form inclusion complexes with either cyclic or linear glucans, and may also help to shed light on why, for example the complexing abilities of cyclodextrins are 2-3 orders of magnitude higher than those of the corresponding non-cyclic analogs (69).

Acknowledgments

The authors wish to thank the Ministry of Agriculture, Fisheries and Food for financial support.

Literature Cited

1. Umeki, K.; Yamamoto, T. J. Biochem. 1975, 78, 889-96.

2. Hayashi, T.; Akiba, T.; Horikoshi, K. Appl. Microbiol. Biotechnol. 1988, 52, 443-448.

3. Pfannemuller, B.; Bauer-Carnap, A. Colloid Polymer Sci. 1977, 255, 844-9.

4. Buleon, A.; Duprat, F.; Booy, F. P.; Chanzy, H. Carbohydr. Polym. 1984, 4, 161-73.

5. Gidley, M. J.; Bulpin, P. V. Carbohydr. Res. 1987, 161, 291-300.

6. Pfannemuller, B. Int. J. Biol. Macromol. 1987, 9, 105-8.

7. Pangborn, W.; Langs, D.; Perez, S. Int. J. Biol. Macromol. 1985, 7, 363-9.

8. Ring, S. G.; Miles, M. J.; Morris, V. J.; Turner, R.; Colonna, P. Int. J. Biol. Macromol. 1987, 9, 158-60.

9. Yamashita, Y.; Ryugo, J.; Monabe,K. J. Electron Microsc. 1973, 22(1), 19-26.

10. Whittam, M.; Orford, P. D.; Ring, S. G.; Clark, S. A.; Parker, M. L.; Cairns, P.; Miles, M. J. Int. J. Biol. Macromol. 1989, 11, 400-6.

11. Katz, J. R.; van Italie, T. B. Z. Physik. Chem. A 1930, 150, 90-100.

12. Wu, H. C. H.; Sarko, A. Carbohydr. Res. 1978, 61, 7-25.

13. Wu, H. C. H.; Sarko, A. Carbohydr. Res. 1978, 61, 27-40.

14. Imberty, A.; Chanzy, H.; Perez, S.; Buleon, A., Tran, V. J. Mol. Biol. 1988, 201, 365-78.

15. Imberty, A., Perez, S. Biopolymers 1988, 18, 1205-21.

16. Booy, F. P.; Chanzy, H.; Sarko, A. Biopolymers 1979, 18, 2261-66.

17. Winter, W. T.; Sarko, A. Biopolymers 1974, 13, 1461-82.

18. Perez, S.; Vergelati, C. Polymer Bull. 1987, 17, 141-8.

19. Donovan, J. W. Biopolymers 1979, 18, 263-75.

20. Kalichevsky, M. T.; Orford, P. D.; Ring, S. G. Carbohydr. Res. 1989, 193, 196-201.

21. Jackle, J. Rep. Prog. Phys. 1986, 49, 171.

22. Angell, C.A. J. Phys. Chem. Solids 1988, 49(8), 863.

23. Fredrickson, G.H. <u>Ann. Rev. Phys. Chem</u>. 1988, 39, 149.

24. Slade, L.; Levine, H. In Frontiers in Carbohydrate Research - 1: Food Applications; Millane, R.P.; BeMiller, J.N.; Chandrasekaran, R., Eds; Elsevier Applied Science, 1989; p215.

25. Johari, G. P.; Hallbrucker, A.; Mayer, E. <u>Nature</u> 1987, 330, 552.

26. Hofer, K.; Hallbruker, A.; Mayer, E.; Johari, G. P. <u>J. Phys. Chem</u>. 1989, 93, 4674.

27. Parks, G. S.; Barton, L. E.; Spaght, M. E.; Richardson, J. W. <u>Physics</u> 1934, 8, 193.

28. Parks, G. S.; Huffman, H. M.; Cattoir, F. R. <u>J. Phys. Chem</u>. 1928, 32, 1366.

29. Couchman, P. R. <u>Macromolecules</u> 1978, 11, 1156 & Couchman, P. R. <u>Macromolecules</u> 1987, 20, 1712.

30. Angell, C. A.; Stell, R. C.; Sichina, W. <u>J. Phys. Chem</u>. 1982, 86, 1540.

31. Moynihan, C. T.; Easteal, A. J.; Wilder, J.; Tucker, J. <u>J. Phys. Chem</u>. 1974, 78, 2673.

32. Orford, P. D.; Parker, R.; Ring, S. G.; Smith, A. C. <u>Int. J. Biol. Macromol</u>. 1989, 11, 91.

33. Orford, P. D.; Parker, R.; Ring, S. G. <u>Carbohydr. Res</u>. 1989, 85, 23.

34. Mackenzie, A. P. <u>Phil, Trans. R. Soc. Lond. B</u>. 1977, 278, 167.

35. Levine, H.; Slade, L. In Water and Food Quality; Hardman, T. M., Ed.; Elsevier Applied Science: New York, 1989.

36. Levine, H.; Slade, L. <u>Carbohydr. Polym</u>. 1986, 6, 213.

37. Banks, W.; Greenwood, C. T. Starch and its Components, Edinburgh University Press, 1975.

38. Ring, S. G.; I'Anson, K.; Morris, V. J. <u>Macromolecules</u> 1985, 18, 182

39. Banks, W.; Greenwood, C. T.; Sloss, J. <u>Carbohydr. Res</u>. 1969, 11, 399.

40. Burchard, W. <u>Makromol. Chem</u>. 1963, 59, 16.

41. Kitamura, S.; Kuge, T. <u>Food Hydrocolloids</u> 1989, 3, 313.

42. Takeda, Y.; Hizukuri, S.; Juliano, B. O. Carbohydr. Res.
    1986, 148, 299.

43. Ha, S. N.; Madsen, L. J.; Brady, J. W. Biopolymers 1988,
    27, 1927.

44. Ihnat, M.; Goring, D. A. I. Can. J. Chem. 1967, 45, 2353.

45. Ihnat, M.; Goring, D. A. I. Can. J. Chem. 1967, 45, 2363.

46. Miyajima, K.; Sauada, M.; Nakagaki, M.
    Bull. Chem. Soc. Jpn. 1983, 56, 1620.

47. Tanford, C. Physical Chemistry of Macromolecules; John
    Wiley, 1961, p335.

48. Soesanto, T.; Williams, M. C. J. Phys. Chem. 1981, 85,
    3338.

49. Mikus, F. F.; Hixon, R. M.; Rundle, R. E.
    J. Am. Chem. Soc. 1946, 68, 1115-23.

50. Kowblansky, M. Macromolecules 1987, 18, 1776-9.

51. Biliaderis, C.G.; Galloway, G. Carbohydr. Res. 1989, 189, 31-
    48.

52. Simpson, T.D.; Dintzis, F.R.; Taylor, N.W. Biopolymers 1972,
    11, 2591-2600.

53. Jane, J. L.; Robyt, J. F.; Huang, D. H. Carbohydr. Res.
    1985, 140, 21-35.

54. Komiyama, M.; Hidefuni, H. Macromol. Chem. Rapid Commun.
    1986, 7, 739-42.

55. Whittam, M. A.; Ring, S. G.; Orford, P. D. In Gums and
    Stabilisers for the Food Industry Vol.3;
    Phillips, G. O.; Wedlock, D. J.; Williams, P. A. Eds;
    Elsevier: 1986, p555.

56. Rundle, R. E.; Baldwin, R. R. J. Am. Chem. Soc. 1943, 65,
    554-8.

57. Senior, M. B.; Hamori, E. Biopolymers 1973, 12, 65-78.

58. Bulpin, P. V.; Welsh, E. J.; Morris, E. R. Staerke
    1982, 34, 335-9.

59. Bailey, J. M.; Whelan, W. J. J. Biol. Chem. 1961, 236,
    969-73.

60. Banks, W.; Greenwood, C. T.; Khan, K. M. Carbohydr. Res.
    1971, 17, 25-33.

61. Bulpin, P. V.; Cutler, A. N.; Lips, A. <u>Macromolecules</u>
    1987, 20, 44-9.

62. Yamamoto, M.; Sano, T.; Harada, S.; Yasunaga, T.
    <u>Bull. Chem. Soc. Jpn</u> 1983, 56, 2643-46.

63. Karkalas, J.; Raphaelides, S. <u>Carbohydr. Res</u>. 1986, 157,
    215-34.

64. Szejtli, J. Cyclodextrin Technology; Kluwer Academic
    Publishers, 1988.

65. Clarke, R. J.; Coates, J. H.; Lincoln, S. F. In Advances
    in Carbohydrate Chemistry and Biochmistry Vol.46;
    Tipson, R. S.; Horton, D. Eds; Academic Press Inc.: 1988

66. Saenger, W.; Noltemeyer, M.; Manor, P. C.; Hingerty,
    B.E.; Klar, B. <u>Bioorg. Chem</u>. 1976, 5, 187-95.

67. Kitagawa, M.; Hoshi, H.; Sakurai, M.; Inoue, Y.;
    Chujo, R. <u>Bull. Chem. Soc. Jpn</u> 1988, 61, 4225-9.

68. Bergeron, R. J.; Channing, M. A.; Gibeily, G. J.;
    Pillor, D. M. <u>J. Am. Chem. Soc</u>. 1977, 99, 5146-51.

69. Komiyama, M.; Hirai, H.; Kobayashi, K.
    <u>Makromol. Chem. Rapid Commun</u>. 1986, 7, 739-42.

RECEIVED November 9, 1990

APPLICATIONS

Chapter 19

Molecular Specificity of Cyclodextrin Complexation

Ching-jer Chang[1], Hee-Sook Choi[1], Yu-Chien Wei[1], Vivien Mak[1],
Adelbert M. Knevel[1], Kathryn M. Madden[2], Gary P. Carlson[2],
David M. Grant[3], Luis Diaz[3], and Frederick G. Morin[3]

[1]Department of Medicinal Chemistry and Pharmacognosy and [2]Department
of Pharmacology and Toxicology, School of Pharmacy and Pharmacal
Sciences, Purdue University, West Lafayette, IN 47907
[3]Department of Chemistry, University of Utah, Salt Lake City, UT 84112

The most remarkable molecular feature of cyclodextrins is their ability to
form inclusion complexes with numerous guest molecules without a
covalent bond being formed. Molecular specificity of cyclodextrin inclusion
complexation was elucidated by determining the dynamics of molecular
motion and relaxation, and local electric field gradient in the solid state as
well as by analyzing the chemical shifts and couplings of high resolution
NMR, and fast atom bombardment mass spectral data. The molecular
specificities with benzaldehyde and tolbutamide provided fundamental
information for understanding the molecular mechanisms involved in the
enhancement of *in vitro* antitumor activity of benzaldehyde in human tumor
cell lines and *in vivo* hypoglycemic effects of tolbutamide in rabbits.

It is widely recognized that regiospecificity and stereospecificity of most chemical and
biochemical reaction specificities are governed by a prior "complexation" process (host-guest
or intermolecular recognition). In recent years, our understanding of the molecular specificity
in the intermolecular recognition has been greatly enhanced by many newly developed
physical and chemical methods. In particular, NMR spectroscopy has developed into a
perpetually expanding and exciting research method by the remarkable improvements in both
hardware and software designs which take advantage of the versatile computer capabilities
and high field superconducting technology.

Studies on the intermolecular complexations in solution have, however, often
encountered some intrinsic difficulties. Firstly, the complexation processes in solution can be
highly complicated due to dynamic chemical exchanges, solvent interference and
conformational fluctuation. Secondly, the solubility of complexes may be limited which
greatly reduces the sensitivity of detection. Thirdly, chemical decomposition and
rearrangement may occur in solution. Most of these problems can be circumvented/simplified
by studying the complexation process in the solid state. This restricts stringently the
intramolecular motion and intermolecular exchange, and simultaneously overcome solubility
and stability problems. Therefore, one of the primary methods employed by us in
investigating the molecular specificity of cyclodextrin complexation was solid state carbon-13
nuclear magnetic resonance (^{13}CNMR) spectroscopy using the combined techniques of high
power proton decoupling, magic angle spinning (MAS) and carbon-hydrogen cross-
polarization (CP) (1, 2).

SOLID STATE C-13 NMR of α-CYCLODEXTRIN

The merits of this new methodology can be clearly illustrated in the ^{13}CNMR analyses of cyclodextrin-water complexes. The solution C-13 spectrum of α-cyclodextrin shows only six signals for the six carbon resonances of the glucose units. This indicates the rapid motion of α-cyclodextrin and fast exchange of water molecules. Thus it provides a non-specific and time-averaged spectrum. However, this motion and exchange are largely restricted in the solid state. The solid state CP/MAS C-13 spectrum of α-cyclodextrin (recrystallized from water) distinctly displays resolved C-1 and C-4 resonances from different α-(1 → 4)-linked glucose units in the macrocycle, indicative of the molecular asymmetry and specificity of α-cyclodextrin-water complex (Figure 1A). One of the C-1 (100.1 ppm) and one of the C-4 (80.3 ppm) signals occurring at relatively higher field are likely attributable to the resonances of a conformationally strained glucosidic bond (3). The earlier work of Saito *et al.* (4) and Inoue *et al.* (5) described spectra of limited resolution. The high resolution spectra of α-cyclodextrin were recently reported by Furo *et al.* (6) and Gidley and Bociek (7). Our spectrum is significantly different from those reported spectra in the C-1, C-4 and C-6 resonance regions. These differences probably resulted from the polymorphic variation and degree of hydration. Thus far, at least three different crystalline forms of α-cyclodextrin have been determined by x-ray or neutron diffractions (3,8,9). Our studies suggest that the C-13 CP/MAS spectral pattern is intimately dependent on the degree of hydration (Figure 1B). The spectrum of this dehydrated α-cyclodextrin is also surprisingly different from that of the anhydrous α-cyclodextrin (6). It appears evident that the solid state C-13 NMR spectra of cyclodextrins are much more intricate than we expected. A precaution might have to be taken to prevent partial dehydration occurring during high speed spinning. A direct comparison of samples used in the solid state NMR measurements and those utilized in single-crystallographic analyses would be helpful in the elucidation of the molecular structure specificity of cyclodextrin in the solid state.

Our studies of the molecular host-guest interactions are primarily directed toward understanding the cyclodextrin inclusion effect on the molecular motion, local electron density and molecular relaxation.

MOLECULAR MOTION

A stereospecific acceleration of the hydrolysis of nitrophenol acetates by cyclodextrins has generated a great deal of interest in studying cyclodextrins as enzyme models (10-12). The crystal structure of α-cyclodextrin-*p*-nitrophenol 1:1 complex was determined by the x-ray method (13). This served as a molecular basis for analyzing solid state C-13 NMR spectral data of cyclodextrin inclusion complexes by Inoue *et al.* (14) and by us. The most extraordinary change for the spectrum of *p*-nitrophenol (Figure 2) is the disappearance of the protonated aromatic carbons after formation of the cyclodextrin inclusion complex. One of the plausible interpretations for this spectral change is that it is due to the increase of carbon-proton dipole-dipole coupling from the free form to the complex form. A higher decoupling power (40 KHz → 56 KHz) resulted in the same spectral feature, suggesting that dipole-dipole coupling itself may not be the direct cause. An alternative explanation was suggested by Inoue *et al.* (14) based on the earlier studies of Rothwell and Waugh (15). The C-13 resonances could be greatly broadened by inefficient decoupling because the motion frequency of the aromatic ring is close to the nutation frequency of the proton decoupling field. X-ray crystallographic data clearly indicate that *p*-nitrophenol cannot undergo 180° rotation within the cavity (3). Thus, the flip motion must take place within a small range between two C-3 protons or C-5 protons of the neighboring glucoses.

This observation has profound implications in the studies of host-guest (enzyme-substrate/inhibitor) recognition by solid state C-13 NMR. Further substantiation of this implication was therefore conducted. The most direct approach to perturb the dynamic effect

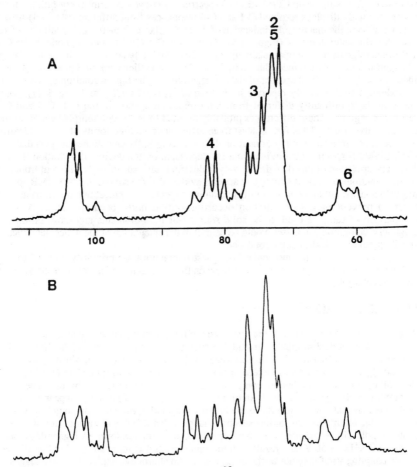

Figure 1.   Solid state CP/MAS 25 MHz ^{13}CNMR spectra of (A) α-cyclodextrin, recrystallized from water and, (B) dehydrated α-cyclodextrin, 50°C under vacuum for 10 hr.

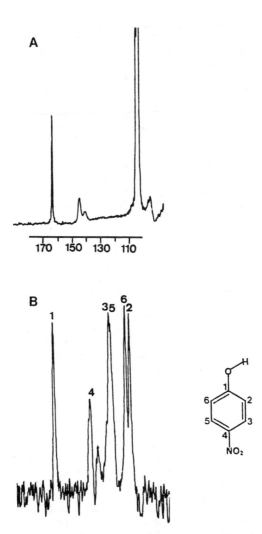

Figure 2. Solid state CP/MAS 25 MHz ^{13}CNMR spectra of (A) α-cyclodextrin-*p*-nitrophenol inclusion complex I and, (B) *p*-nitrophenol, recrystallized from water.

of molecular motion is to freeze the freedom of motion. Indeed, the protonated carbon signals were recovered by conducting the NMR experiment at -120° (Figure 3).

LOCAL ELECTRONIC DENSITY

A second α-cyclodextrin-*p*-nitrophenol 1:1 powdery complex was also obtained during the preparation of the crystalline inclusion complex. Its solution C-13 NMR spectra were identical. However, the solid state C-13 NMR spectrum (Figure 4) appears remarkedly different. Most of the protonated carbon resonance signals are still retained whereas the nitro-attached carbon is changed from a doublet to a singlet. The initial splitting of this carbon resonance is due to the C-13 and N-14 quadrupole coupling which cannot be fully reduced in the magic angle spinning mode at the magnetic field strength of 2.35T (25MHz). The magnitude of this splitting is inversely proportional to the ratio (Z/A) of the Zeeman frequency (Z) to the quadrupole coupling constant ($A=e^2Qq/h$) ([16]). Since the spectra of *p*-nitrophenol and its complex are measured at constant field, the Zeeman frequency remains unchanged. Then the reduction of this quadrupole splitting is most likely ascribed to the reduction of the quadrupole coupling constant. This surprising finding indicates the decrease of the local electric field gradient (q) after the inclusion complex with α-cyclodextrin is formed because all other factors (e: electronic charge, Q: N-14 electric quadrupole moment, and h: Planck's constant) involved in the quadrupole coupling constant are invariable. This directly manifests that the perturbation of the local electric field gradient in the non-covalent host-guest complex can be detected by a direct NMR method.

One of the possible complexation probabilities is that the second complex may be a non-specific adsorbed complex formed primarily by intermolecular hydrogen-bonding. An independent procedure was designed for preparing the adsorbed complex by mixing a methanol solution of *p*-nitrophenol directly with α–cyclodextrin and then quickly removing methanol under reduced pressure. The solid state C-13 NMR spectrum of the resulting adsorbed complex (Figure 5A) looks somewhat similar to that of the second complex. However, *p*-nitrophenol in this adsorbed complex is completely removed by washing with cold ether or methylene chloride (Figure 5B) since *p*-nitrophenol is not included in the cavity. A similar washing process failed to remove *p*-nitrophenol from the second complex, strongly suggesting that the second complex is very likely an inclusion complex. The most plausible structure for the second inclusion complex is that, in contrast to the first inclusion complex, the nitro group of *p*-nitrophenol is oriented toward the outside of the α-cyclodextrin cavity.

MOLECULAR RELAXATION

The degree of free motion of the substrate in the active site is critical in determining the stereospecificity of a biochemical reaction. It is thus important to study the freedom of motion of the guest molecule in the inclusion complex. Adamantane has been shown to possess plastic crystal structures with high freedom of motion in the solid state. Thus, its carbon signals in the solid state are extremely sharp ($W_{1/2} < 1Hz$) ([17]). The solid state spectrum of the β-cyclodextrin-1-adamantanecarboxylic acid complex revealed the asymmetric structure of the dimeric carboxylic acid ([18]) in the inclusion complex (Figure 6A). A unique way for quantitatively assessing the freedom of motion is to measure the relative rate of losing carbon magnetization after the C-13/H-1 spin-locked process is terminated. This dipolar dephasing experiment is performed by turning off the proton decoupler for a short period prior to the acquisition of carbon signals ([19]). The dipolar dephasing spectrum of the complex shows that all the cyclodextrin peaks are almost diminished whereas all the 1-adamantenecarboxylic acid signals are still retained (Figure 6B). This result indicates that the guest molecule in the cavity of β-cyclodextrin has much greater freedom of motion than that of the host molecule. This technique provides a new approach to monitor selectively the variations of guest or substrate/inhibitor in the biochemical intermolecular complexes.

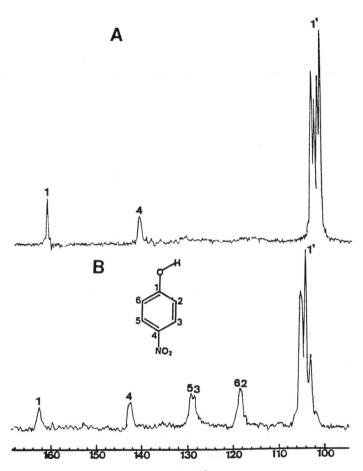

Figure 3. Solid state CP/MAS 50 MHz ^{13}CNMR spectra of α-cyclodextrin-*p*-nitrophenol inclusion complex I at (A) room temperature and, (B) -120°C.

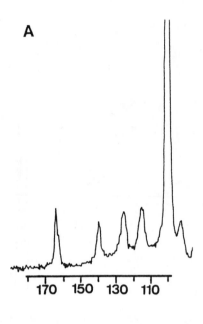

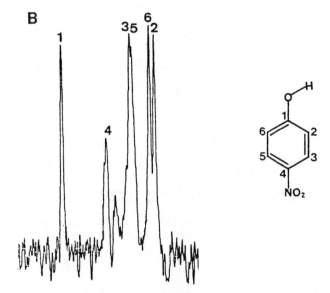

Figure 4.   Solid state CP/MAS 25 MHz ^{13}CNMR spectra of (A) α-cyclodextrin-*p*-nitrophenol inclusion complex II and, (B) *p*-nitrophenol, recrystallized from water.

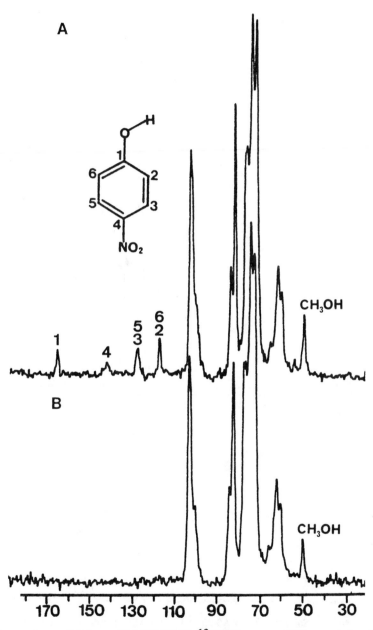

Figure 5.   Solid state CP/MAS 25MHz ^{13}CNMR spectra of (A) α-cyclodextrin-*p*-nitrophenol adsorbed complex and, (B) after washed with cold ether.

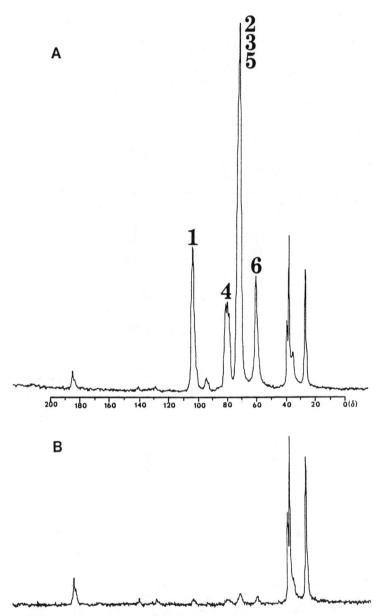

Figure 6. Solid state CP/MAS 25 MHz ^{13}CNMR spectra of β-cyclodextrin-1-adamantane carboxylic acid inclusion complex; (A) Regular acquisition, (B) Dipolar dephasing acquisition (dephasing time: 40 μsec).

MOLECULAR ENCAPSULATION AND BIOAVAILABILITY

Each guest molecule in a cyclodextrin inclusion complex is encapsulated by cyclodextrin. This molecular encapsulation can profoundly modify the chemical and physical properties of the guest molecules. Its potential applications have been shown in (1) stabilization of air- or light sensitive substances, (2) enhancement of water solubility, (3) suppression of unpleasant taste or odor and (4) improvement of bioavailability (20, 21). Our primary interest focuses on elucidating the structure specificity in cyclodextrin complexation and thereby providing a definite basis for understanding their molecular mechanisms.

INCLUSION COMPLEXATION OF BENZALDEHYDE

Previous interest in the antitumor activity of fig (*Ficus carica* L.) fruit led to the isolation of benzaldehyde as a major active component (22). Its carcinostatic effect was attributed to selective inhibition of the uptake of nucleosides and carbohydrates, and the reduction of intracellular adenosine 5'-triphosphate level (23). However, benzaldehyde is an oilly liquid and only sparingly water soluble. Its instability in air and light presented considerable problems in the delivery and formulation. Takeuchi *et al.* (24) first prepared the complex with α-, β- and γ-cyclodextrin. More importantly, the x-ray structure of α-cyclodextrin-benzaldehyde 1:1 complex was subsequently determined (25) , providing a working basis for uncovering its structure in solution.

NMR ANALYSIS: The 470 MHz ^{1}HNMR spectra of benzaldehyde before and after the formation of α-cyclodextrin displayed a first-order pattern. The results are summarized in Table I. Spectra of α-cyclodextrin before and after inclusion of benzaldehyde clearly showed a second-order pattern (Figure 7). A rigorous computer spin-simulation allowed us to determine their chemical shifts and coupling constants (Tables I and II). The calculated values appear in reasonable agreement with those reported by Wood *et al.* (26) except for the H-6' protons. The discrepancy which appears is presumably due to the limited resolution of the previously measured 220 MHz ^{1}HNMR spectrum. A 500 MHz spectrum of α-cyclodextrin recently reported by Yamamoto and Inoue (27) is almost identical to our spectrum. However, no attempt was reported to calculate the accurate chemical shifts and coupling constants.

INCLUSION STRUCTURES: The single-crystal x-ray analysis of the α-cyclodextrin-benzaldehyde (1:1) complex (25) indicates that the phenyl ring of benzaldehyde is the leading group inserted into the center of the α-cyclodextrin cavity from the broader end and the aldehyde group protrudes from the cavity (structure A). However, the *ortho* proton chemical shift change ($\Delta\delta$=32.4 Hz) is larger than the *para* ($\Delta\delta$=21.6 Hz) or *meta* ($\Delta\delta$=21.2 Hz) proton chemical shift change upon formation of α-cyclodextrin inclusion complex (Table I). This suggests that in solution the aldehyde group is the leading group included into the cavity

(structure B). This tentative structure is corroborated by the solution ^{13}CNMR results. The upfield shift for the aldehyde carbon (-23.8 Hz) and C-1 (-16.1 Hz) resonances is a good indication for the changes of dielectric environment from the high dielectric water medium to the low dielectric cavity of cyclodextrin. Furthermore, the nuclear Overhauser effect (Figure 8) between the aldehyde proton and the H-5' protron of α-cyclodextrin strongly favors structure B for the α-cyclodextrin-benzaldehyde inclusion complex in solution.

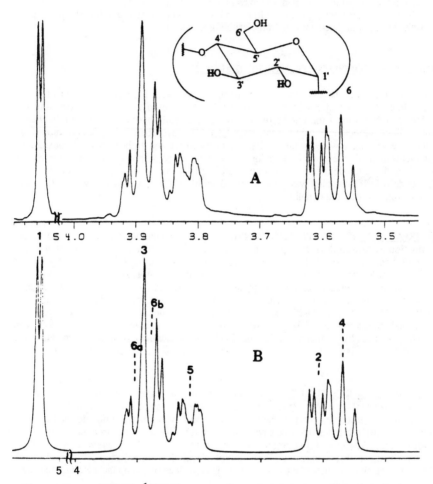

Figure 7.   470 MHz ^{1}HNMR spectra of α-cyclodextrin-benzaldehyde inclusion complex in pD 7.4 phosphate buffer solution . (A) measured, (B) calculated based on Raccoon spin simulation program.

Table I. 470 MHz ^{1}HNMR chemical shift of benzaldehyde before and after complexation with α-cyclodextrin in pD 7.4 phosphate buffer solution (DMSO-d_6 was used as an external reference)

	CHO	
	1	
6		2
5		3
	4	

Protons	α-Cyclodextrin benzaldehyde 1:1 Complex (ppm)	α-Cyclodextrin (ppm)	Benzaldehyde (ppm)	Difference (Hz)
CHO	9.9680		9.9350	15.5
2,6	8.0280		7.9590	32.4
3,5	7.7670		7.6250	21.2
4	7.8060		7.7600	21.6
1'	5.0330	5.0510		- 8.5
2'	3.6080	3.6300		-10.3
3'	3.8880	3.9820		-44.2
4'	3.5703	3.5830		- 6.0
5'	3.8110	3.8370		-12.2
6'a	3.9000	3.9070		- 3.3
6'b	3.8535	3.8640		- 4.9

Table II. 470 MHz ^{1}HNMR coupling constants (Hz) of α-cyclodextrin and α-cyclodextrin-benzaldehyde complex in pD 7.4 phosphate buffer solution

Coupling	α-Cyclodextrin	α-Cyclodextrin-Benzaldehyde
J_{12}	3.4	3.5
J_{15}	-0.7	-0.5
J_{23}	10.1	9.8
J_{34}	9.2	9.2
J_{45}	9.5	9.4
J_{46a}	-0.6	-0.7
J_{46b}	-0.5	-0.7
J_{56a}	2.0	1.8
J_{56b}	4.4	4.3
J_{6a6b}	-11.5	-12.5

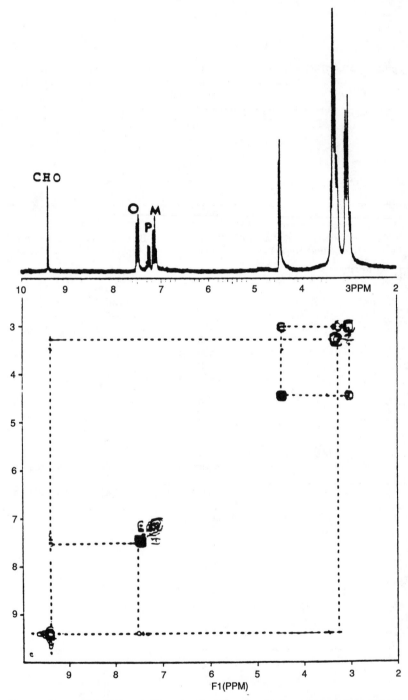

Figure 8.   200 MHz two-dimensional nuclear Overhauser enhancement spectrum of α-cyclodextrin-benzaldehyde complex in pD 7.4 phosphate buffer solution.

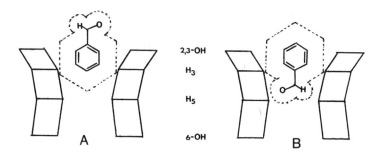

AUTOXIDATION: The autoxidation rates of free benzaldehyde and its inclusion complex with α-cyclodextrin in neutral solution at 70°C were measured by HPLC. The half-life for the α-cyclodextrin-benzaldehyde was 6.2 hr., 3.6 times longer than that for benzaldehyde alone (1.7 hr.), indicating that the aldehyde group is included into the cyclodextrin cavity which prevents oxidation by air.

ANTITUMOR CYTOTOXICITY: The stabilization effect of benzaldehyde by α-cyclodextrin is also reflected in the *in vitro* antitumor cytotoxicity against human tumor cell lines: A-549 (non-small-cell lung carcinoma), MCF-7 (breast adenocarcinoma) and HT-29 (colon adenocarcinoma) (Table III). α-Cyclodextrin-benzaldehyde inclusion complex is about 5-8 fold more cytotoxic than free benzaldehyde.

Table III. *In vitro* antitumor cytotoxicity against human tumor cell lines

Compound	ED_{50} (μ mole/ml)		
	A-549 (lung)	MCF-7 (breast)	HT-29 (colon)
Benzaldehyde	0.19	0.68	0.50
α-Cyclodextrin	0.98	0.88	0.59
α-Cyclodextrin-Benzaldehyde	0.04	0.08	0.06

INCLUSION COMPLEXATION OF TOLBUTAMIDE

Tolbutamide [1-butyl-3(*p*-tolylsulfonyl)urea] is a first-generation hypoglycemic drug used clinically in the treatment for insulin-dependent diabetic patients in whom the pancreas retains the capacity to secrete insulin (28) . Its poor water solubility and dissolution rate are considered to be the rate-limiting steps in the gastrointestinal absorption. The low water solubility and dissolution rate are likely due to the strong intermolecular hydrogen-bonding between the sulfonylurea groups, which is then encircled by two hydrophobic end groups. Modification of tolbutamide dissolution properties by water-soluble polymers has been used to improve oral bioavailability (29-32). Cyclodextrin complexation of tolbutamide was first reported in 1978 to enhance its water solubility (33) . The improvement of the bioavailability of tolbutamide by β-cyclodextrin was recently demonstrated by Vila-Jato *et al.* (34).

Results of most previous experiments suggested that tolbutamide forms inclusion complexes with β-cyclodextrin in a 1:1 ratio (34-36). However, tolbutamide consists of two hydrophobic end groups (tolyl and butyl) and can potentially form a 2:1 complex which should provide maximum enhancement of water solubility. The complex was prepared by dissolving both β-cyclodextrin (2 mmoles) and tolbutamide (1 mmole) in pH 11.0, 0.05M phosphate buffer solution (160 ml) at room temperature. Slow precipitation occurred after neutralization to pH 7.0 and then concentration by vacuum evaporation. The precipitate was further recrystallized from water. The ^{1}HNMR spectra of the initial precipitate and the recrystallized solid (mp: 268°C) showed a molar ratio of 2:1.

FAST ATOM BOMBARDMENT MASS SPECTROMETRIC ANALYSIS: Using dithiothreitol/dithioerythritol (3:1) as a matrix, the protonated molecular ion (m/z: 2540) for the recrystallized β-cyclodextrin-tolbutamide complex, together with strong potassium adduct (m/z: 2578) and weak sodium adduct (m/z: 2562) peaks are clearly detected (Figure 9). This is a strong indication of a 2:1 complex which is consistent with the integration data of ^{1}HNMR.

FOURIER TRANSFORM INFRARED SPECTRAL ANALYSIS: The ureido functional group of tolbutamide shows two characteristic IR absorptions (Figure 10), at 1662.4 (carbonyl stretching band) and 1559.6 (NH bending band) cm^{-1}. This indicates a strong intermolecular hydrogen-bonding as is shown by the x-ray crystallographic data (37). This hydrogen-bonding can be disrupted by the formation of cyclodextrin complex as is revealed by a shift in the carbonyl stretching band and NH bending band to 1701.2 and 1542.1 cm^{-1}, respectively. The absorption intensity of the complex is also significantly reduced. It probably results from some part of tolbutamide being bound to cyclodextrin, thereby restricting its freedom of vibration. However, these changes do not exclusively indicate the formation of an "inclusion" complex. A more informative variation should be furnished by the changes of the phenyl or butyl group. However, the IR absorptions of these hydrophobic groups cannot be unambiguously determined due either to the weak absorption or the overlap with the cyclodextrin absorption.

NMR ANALYSIS: Inoue *et al.* recently reported the relative chemical shift and coupling constant assignments of β-cyclodextrin at pD 3 and pD 10 with the aid of two-dimensional 500 MHz ^{1}H COSY experiments (38) . These data provide initial values for computer spin calculation (Figure 11) although only approximate data based on the first-order analysis were shown. All spectral data were measured in 0.2N NaOD solution because of the limited solubility in neutral solution. Upon complexation with β-cyclodextrin marked spectral changes were observed. The precise chemical shifts and coupling constants calculated from computer spin simulation are summarized in Tables IV and V.

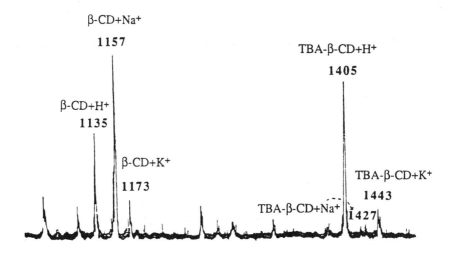

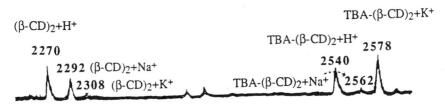

Figure 9. Fast atom bombardment mass spectrum of β-cyclodextrin-tolbutamide 2:1 inclusion complex using dithiothreitol/dithioerythritol (3:1) matrix.

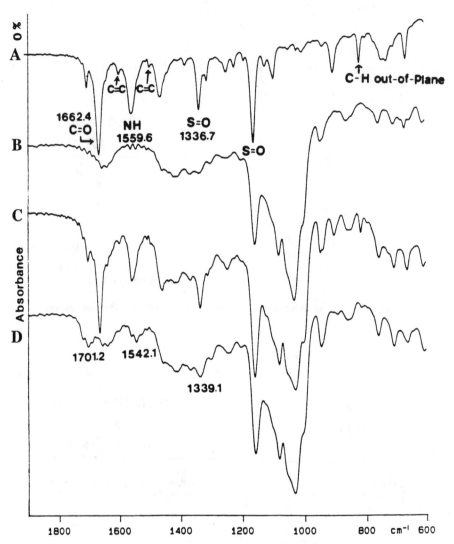

Figure 10. Fourier transform IR spectra of (A) tolbutamide, (B) β-cyclodextrin, (C) β-cyclodextrin and tolbutamide mixture (2:1) and, (D) β-cyclodextrin-tolbutamide 2:1 inclusion complex.

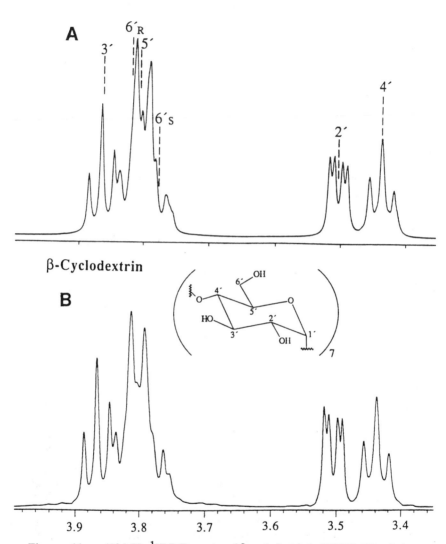

Figure 11. 470 MHz ^{1}HNMR spectra of β-cyclodextrin in 0.2N NaOD solution. (A) calculated based on Raccoon simulation porogram, (B) measured.

Table IV.   470 MHz ^{1}HNMR chemical shifts (ppm) of tolbutamide, β-cyclodextrin
and β-cyclodextrin-tolbutamide in 0.2N NaOD Solution

Protons	β-Cyclodextrin-Tolbutamide (2:1) (ppm)	β-Cyclodextrin (ppm)	Tolbutamide (ppm)	Chemical shift Differences (Hz)
1'	4.9480	4.9460		0.9
2'	3.5095	3.5040		2.6
3'	3.8470	3.8640		-8.0
4'	3.4440	3.4370		3.3
5'	3.7420	3.8060		-30.0
6'a	3.8172	3.8200		-1.3
6'b	3.7600	3.7775		-8.2
1	0.8170		0.8030	6.6
2	1.2180		1.2000	8.4
3	1.3370		1.3250	5.6
4	2.9650		2.9560	4.2
7	7.6595		7.6720	-5.9
8	7.2760		7.3255	-23.2
9	2.3690		2.3590	4.7

Table V.   470 MHz ^{1}HNMR coupling constants (Hz) of β-cyclodextrin and β-
cyclodextrin-tolbutamide complex in 0.2N NaOD Solution

Coupling	β-Cyclodextrin	β-Cyclodextrin-Tolbutamide (2:1)
J_{12}	3.5	3.8
J_{15}	-0.6	-0.7
J_{23}	9.6	9.4
J_{34}	9.3	9.4
J_{45}	9.2	8.7
J_{46a}	-1.0	-0.7
J_{46b}	-1.3	-1.4
J_{56a}	2.0	2.0
J_{56b}	3.8	4.4
J_{6a6b}	-12.0	-11.0

INCLUSION STRUCTURE: The upfield shifts for the H-3' and H-5' of β-cyclodextrin
upon the formation of tolbutamide inclusion complex are attributed to the anisotropic shielding
of the phenyl ring. This anisotropic shielding offers a direct approach to elucidate the spatial
disposition of the aromatic group in the complex by comparing the relative magnitude of
chemical shift changes between H-3' and H-5' (27,39). The larger shift of H-5' (30 Hz),
relative to the shift of H-3' (8.0 Hz), suggests that the aromatic ring extends from the primary
hydroxyl rim or from the secondary hydroxyl rim but penetrates deeply into the cavity. In
0.2N NaOD solution, both the ureido group of tolbutamide (pKa: 5.4) (40) and the secondary
hydroxyl groups of β-cyclodextrin (pKa: 12.2) (26,41,42) should all be ionized. The electric
repulsion between the ureido group and the secondary hydroxyl groups should hinder the
tolbutamide entry in β-cyclodextrin past its secondary hydroxyl rim. Therefore, it is more
likely that tolbutamide enters β-cyclodextrin from the primary hydroxyl rim. The hydrogen-
bonding between the sulfonyl group and the primary hydroxyl group may further strengthen

complex formation. Previous NMR studies also suggested this favorable structure (33,35). The chemical shift changes for the aromatic protons induced by cyclodextrin complexation are very different from those observed for other aromatic substrates, such as benzaldehyde. In most cases, the aromatic protons undergo a downfield shift (27). The upfield shift for the *ortho*-protons (H-7) may be ascribed to steric interference by cyclodextrin of the coplanarity between the sulfonyl group and the aromatic ring. However, this steric effect could not account for the large upfield shift (23.2 Hz) of the *meta*-protons (H-8). One of the most plausible explanations may be the anisotropic shielding of the ether oxygen (O-4') of β-cyclodextrin.

HYPOGLYCEMIC EFFECT: The modulation of the bioavailability of tolbutamide by β-cyclodextrin was monitored by measuring its hypoglycemic effect with and without β-cyclodextrin in male New Zealand white rabbits. Seven rabbits were used in order to eliminate individual differences. Before administering the drug, rabbits were denied food for 24 hours. Blood samples were taken before administering the drug, 30 minutes after oral administration into the stomach directly, and at 1 hour intervals for eight hours. A Beckman glucose analyzer was used to assay the plasma glucose level. The plasma glucose levels versus time after administration of the drugs are shown in Figure 12. When tolbutamide was given alone, the lowest plasma glucose level was obtained after six to seven hours, while tolbutamide administered in the β-cyclodextrin inclusion complex resulted in the lowest plasma glucose level after four to five hours. The decrease in plasma glucose was significantly greater when the rabbits were treated with β-cyclodextrin complex than when given the tolbutamide alone. Clearly, β-cyclodextrin can significantly improve the bioavailability of tolbutamide. These results were similar to the previously reported results of Vila-Jato *et al.* (34) except our dosage was only half the size of theirs.

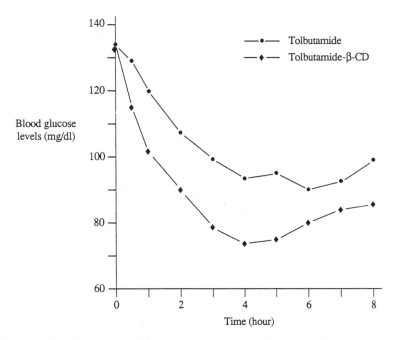

Figure 12. Modulation of plasma glucose levels in rabbits after oral administration of tolbutamide and its β-cyclodextrin inclusion complex.

Acknowledgment

This research was supported partially by the Purdue Research Foundation, Purdue University, West Lafayette, Indiana, and partially by Grant GM08521-29 from the Institute of General Medical Sciences of the National Institutes of Health, PHS.

LITERATURE CITED

1. Fyfe, C.A. *Solid State NMR for Chemist;* CFC Press, Guelph, Ontario, 1983.
2. Bartuska, V.J.; Freeland, S.J.; Sindorf, D.Q.; Dalbow, D.G. *Amer. Lab.* 1987, 139.
3. Manor, P.C.; Saenger, W. *J. Am. Chem. Soc.* 1974, *96*, 3630.
4. Saito, H.; Izumi, G.; Mamizuka, T.; Suzuki, S.; Tabeta, R. *J. Chem. Soc. Chem. Commun.* 1982, 1386.
5. Inoue, Y.; Okuda, T.; Kuan, F.; Chujo, *R. Carbohyd. Res.* 1983, *116*, c5.
6. Furo, I.; Pocsik, I.; Tompa, K.; Teeaar, R.; Lippmaa, E. *Carbonhydr. Res.* 1987, *166*, 27.
7. Gidley, M.J.; Bociek, S.M. *J. Am. Chem. Soc.* 1988, *110*, 3820.
8. Klar, B.; Hingerty, B.; Saenger, W. *Acta Crystallogr.* 1980, *B36*, 1154.
9. Lindner, K.; Saenger, W. *Acta Crystallogr.* 1982, *B38*, 203.
10. Van Etten, R.L.; Sebastian, J.F.; Clowes, G.A.; Bender, M.L. *J. Am. Chem. Soc.* 1965, *89*, 3243.
11. Van Etten, R.L.; Clowes, G.A.; Sebastian, J.F.; Bender, M.L. *J. Am. Chem. Soc.* 1965, *89*, 3253.
12. Breslow, R. *Adv. Enzymol.* 1986, *58*, 1.
13. Harata, K. *Bull. Chem. Soc. Japan* 1977, *50*, 1416.
14. Inoue, Y.; Okuda, T.; Chujo, R. *Carbohyd. Res.* 1985, *141*, 179.
15. Rothwell, W.P.; Waugh, J.S. *J. Chem. Phys.* 1981, *74*, 2721.
16. Zumbulyadis, N.; Henrich, P.M.; Young, R.H. *J. Chem. Phys.* 1981, *75*, 1603.
17. Amoureux, J.P.; Bee, M. *Acta Crystallogr.* 1980, *B36*, 2636.
18. Hamilton, J.A.; Sabesan, M.N. *Acta Crystallogr.* 1982, *B38*, 3063.
19. Opella, S.J.; Frey, M.H. *J. Am. Chem. Soc.* 1979, *101*, 5854.
20. Uekama, K.; Otagiri, M.; *CRC Critical Review in Therapeutic Drug Carrier System*, 1987, *3*, 1.
21. Szejtli, J. *Controlled Drug Bioavailability*, Smolen, V.F.; Ball, L.A. Eds., Wiley: New York, 1985, Vol. 3, 365.
22. Sakaguchi, R.; Hayase, E. *Agric. Biol. Chem.* 1979, *43*, 1775.
23. Watamuki, M.; Sakaguchi, K. *Cancer Res.* 1980, *40*, 2574.
24. Takeguchi, M.; Kochi, M. *Japan Tokkyo Koho*, Japanese Patent 1157737, 1979.
25. Harata, K.; Uekama, K.; Ogino, H. *Bull. Chem. Soc. Jpn.* 1981, *54*, 1954.
26. Wood, D.J.; Hruska, F.E.; Saenger, W. *J. Am. Chem. Soc.* 1977, *99*, 1735.
27. Yamamoto, Y.; Inoue, Y. *J. Carbohyd. Chem.* 1989, *8*, 29.
28. Campbell, G.D. *Oral Hypoglycemic Agents*, Academic, New York, 1969.
29. Sekikawa, H.; Naganuma, T.; Fujiwara, J.; Nakano, M.; Arita, T. *Chem. Pharm. Bull.* 1979, *27*, 31.
30. Kaur, R.; Grant, D.; Eaves, D. *J. Pharm. Sci.* 1980, *69*, 1321.
31. Uekama, K.; Figinaga, T.; Otagiri, M. *J. Pharm. Dyn.* 1981, *4*, 735.
32. Millares, M.; Ginity, J.; Martin, A. *J. Pharm Sci.* 1982, *71*, 301.
33. Uekama, K.; Hirayama, F. *Chem. Lett.* 1978, 70?.
34. Vila-Jato, J.I.; Blanco, J.; Torres, J.J. *Il Farmaco* 1988 *43*, 37.
35. Ueda, H.; Nagai, T. *Chem. Pharm. Bull.* 1980, *28*, 1415.
36. Ueda, H.; Nagai, T. *Chem. Pharm. Bull.* 1981, *29*, 2710.
37. Nirmala, K.A.; Sake Gowda, D.S. *Acta Crystallogr.* 1981, *B37*, 1597.
38. Inoue, Y.; Takahashi, Y.; Chujo, R. *Carbohyd. Res.* 1986, *148*, 109.
39. Komiyama, M.; Hirai, H. *Polymer J.* 1981, *13*, 171.
40. Beyer, W.F.; Jensen, E.H. *Anal. Profiles Drug Sub.* 1974, *3*, 513.
41. Thakkar, A.L.; Demarco, P.V. *J. Pharm. Sci.* 1971, *60*, 653.
42. Inoue, Y.; Okuda, T.; Miyata, Y.; Chujo, R. *Carbohyd. Res.* 1984, *125*, 65.

RECEIVED November 20, 1990

Chapter 20

Preparation and Characterization of Cyclodextrin Complexes of Selected Herbicides

Oliver D. Dailey, Jr.

Agricultural Research Service, U.S. Department of Agriculture, Southern Regional Research Center, P.O. Box 19687, New Orleans, LA 70179

Recently, concern over the contamination of groundwater by pesticides has increased. The β-cyclodextrin (BCD) complexes of a number of herbicides most frequently implicated in groundwater contamination have been prepared in an attempt to develop formulations that prevent entry of the chemical into the groundwater while maintaining effective pest control. In this paper, the methods of preparation of the BCD complexes will be described. In addition, evidence confirming the formation of true inclusion complexes (solubility properties, elemental analyses, UV spectra) will be presented. Two of the herbicides studied failed to form BCD complexes. Computer model studies indicated that these potential guest molecules are too bulky to fit even partially into the BCD central cavity. Formation of complexes of these molecules with the larger γ-cyclodextrin (GCD) was investigated, and the results will be presented.

Recently, concern over the contamination of groundwater by pesticides has mounted. In 1986, the U. S. Environmental Protection Agency disclosed that at least 17 pesticides used in agriculture had been found in groundwater in 23 states (1). According to a 1988 interim report, 74 different pesticides have been detected in the groundwater of 38 states from all sources. Contamination attributable to normal agricultural use has been confirmed for 46 different pesticides detected in 26 states (2). The chief objectives of our research are to develop pesticide formulations that will maintain or increase efficacy on target organisms when applied and that will not adversely impact on the environment or groundwater while maintaining effective pest control.

Cyclodextrins are macrocyclic torus-shaped polymers consisting of six or more D-glucose residues. They are formed by enzymatic starch degradation. β-Cyclodextrin (BCD) is composed of seven D-glucose units connected by acetal bonds between the 1 and 4 carbon atoms of adjacent glucose units. The high electron density internal cavity (inside diameter approximately 7.8 A) consists of glycosidic oxygen atoms and axial protons. Seven primary hydroxyl groups project from one outer edge of the BCD molecule, and fourteen secondary hydroxyl groups from the other. Consequently, the BCD molecule has a hydrophobic cavity and relatively hydrophilic outer surface. The γ-cyclodextrin (GCD) molecule consists of eight D-glucose monomers and has an inside diameter of about 9.5 A. In aqueous solution, the cyclodextrin molecule can readily accept a guest molecule in its hydrophobic central cavity, forming a stable complex. It is necessary for only a portion of the molecule to fit in the cavity for an inclusion complex to form (3-7).

Many synthetic pesticides can form inclusion complexes with cyclodextrins, often resulting in improvements in the properties of the complexed substances. Cyclodextrins have found particular application for the formulation of poorly water soluble, volatile, or unstable herbicides. Among the advantages of cyclodextrin complexes of pesticides are enhanced stabilization, reduced volatility, masked bad odor, enhanced wettability, solubility and bioavailability, and controlled release properties. Of the cyclodextrins, BCD is the only one available at a reasonable price, and its use may be economically feasible within a few years (4). Several herbicides that have been frequently implicated in groundwater contamination (1-2) were selected as candidates for complexation with BCD in an attempt to develop formulations that could prevent entry of the chemical into the groundwater while maintaining effective weed control.

Materials and Methods

Preparation of BCD Complexes. The following herbicides were selected for complexation with BCD: atrazine, simazine, metribuzin, alachlor, and metolachlor (Fig. 1). Typical reaction conditions for the formation of BCD complexes of these herbicides are shown in Table I.

TABLE I. Cyclodextrin complexation of selected herbicides

Herbicide	Complex Formation?	Reaction Conditions (Aqueous solution)	Analysis[a]
Metribuzin	Yes	65-75 °C, 20 min.-6 h	1:1:5
Atrazine	No	60 °C, 45 min.	...
Atrazine	Yes	100 °C, 7-18 days	1:1:5
Alachlor	No	100 °C, 7 days	...
Simazine	Yes	100 °C, 11 days; 100 °C, 2 days after adding 10% (v/v) dioxane	2:1:5

[a]Herbicide:BCD:water molar ratio as determined by elemental analysis. (Analysis of the BCD used in the preparations indicated the presence of 5 moles of water per mole of BCD).

The BCD complex of atrazine was prepared by adding technical grade atrazine to a solution of an equimolar amount of BCD in water at 60-80 °C and then refluxing under an argon atmosphere until all solid had dissolved. The reaction mixtures were allowed to cool to room temperature, and any precipitate was filtered. The solid BCD complex was isolated from the filtrate by removal of water at 1-2 torr and 30-40 °C) for characterization purposes. Reaction times varied, with longer periods of time required for more concentrated solutions.

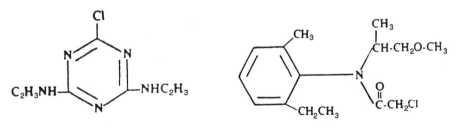

Figure 1.   Herbicides frequently found in groundwater.

In the preparation of the BCD complex of metribuzin, a solution of 1.28 g (6.00 mmol) of metribuzin in 2.5 ml of methanol (60 °C) was added to a solution of 6.81 g (5.56 mmol) of BCD in 350 ml of water at 70 °C. The mixture was heated at 65-75 °C under argon for 20 min., at which time all solids had dissolved. After cooling overnight, the solvent was removed from the clear solution in vacuo. Some material was lost during solvent removal. The isolated white solid was dried further over phosphorus pentoxide in a vacuum desiccator at 1 torr, yielding 6.97 g of crude complex, mp 280 °C (dec). Recrystallation from water gave the analytical sample, mp 281.4-284.0 °C (dec). A mechanical mixture of metribuzin and BCD showed partial localized melting at 125 °C [mp of metribuzin: 125.5-126.5 °C (8)] and complete melting with decomposition at ca. 300 °C.

In the preparation of the BCD complex of simazine, a mixture of 5.00 g (24.8 mmol) of technical grade simazine and 30.4 g (24.8 mmol) of BCD in 500 ml of water was heated at 100 °C for 11 days. There was no significant dissolution of the simazine. Fifty ml of 1,4-dioxane was added as a co-solvent, and the mixture heated at 100 °C for 2 days, yielding a clear solution. After cooling to room temperature 392 mg of white precipitate (unreacted simazine) was isolated; after refrigeration of the filtrate at 4 °C for 5 days, 12.99 g of unreacted BCD came out of solution. Removal of solvent from the remaining solution followed by vacuum drying provided 22.12 g of BCD complex, mp 150-162 °C. Elementary analysis (Table I) of the material indicated a 2:1:5 simazine/BCD/H$_2$O molar ratio (24.8% simazine).

Preparation of GCD Complexes. Gamma-cyclodextrin (GCD) complexes of alachlor and metolachlor were prepared as outlined in Table II.

Table II. Preparation of gamma-cyclodextrin complexes

Herbicide	Reaction Conditions (Aqueous solution)	Analysis[a]
Alachlor	70-85 °C, 16 hours	1:1:2
Metolachlor	60-90 °C, 16 hours	1:1:3

[a]Herbicide:GCD:water molar ratio as determined by elemental analysis. (Analysis of the GCD used in the preparations indicated the presence of 7 moles of water per mole of GCD).

Alachlor 62.4 mg (0.231 mmol) was added to a solution of 300 mg (0.211 mmol) of GCD·7H$_2$O in 30 ml of water. The stirred mixture was heated under argon to 85 °C over a 1.5 h period, then at 70-85 °C for 16 h. The resulting clear solution was allowed to cool to room temperature, and 212.1 mg (62.8%) of complex, mp 226-228 °C (dec), was obtained upon filtration and drying. The filtrate yielded an additional 31.8 mg (9.4%) of complex upon cooling to 4 °C.

A mixture of 65.2 mg (0.230 mmol) of metolachlor and 298 mg (0.209 mmol) of GCD·7H$_2$O in 25 ml of water was stirred under argon at room temperature for 1 h, during which time the mixture remained milky cloudy. The mixture was then heated at 60-90 °C for 16 h. The resulting clear solution was allowed to cool to room temperature, and 249.2 mg (72.8%) of complex, mp 225-231 °C (dec), was obtained upon filtration and drying.

<u>Characterization of BCD and GCD Complexes.</u> All of the BCD and GCD complexes were characterized by physical properties, NMR, IR, and UV spectra, and elemental analysis (Tables I and II). All of the BCD complexes contained 5 molecules of water per molecule of BCD. Elemental analysis of the BCD used in the complex preparations also showed a 5:1 water/BCD ratio. Elemental analysis of the GCD indicated a 7:1 water/GCD ratio.

   Proton nuclear magnetic resonace (^{1}H NMR) spectra were recorded in DMSO-d$_6$ or trifluoroacetic acid (TFA-d$_1$) solution on a Varian EM-360L 60-MHz NMR spectrometer; chemical shifts are reported in parts per million from internal tetramethylsilane (TMS). Ultraviolet (UV) spectra were recorded on a Gilford Response UV-visible spectrometer using deionized water as solvent. Infrared (IR) spectra were recorded on a Beckman AccuLab 8 spectrometer and were calibrated with the 3027.9, 1601.8, and 1028.3 cm^{-1} bands of polystyrene. Potassium bromide (KBr) disks of herbicide-cyclodextrin complexes and of proportional mechanical mixtures of herbicide and cyclodextrin were prepared for the analysis.

Results and Discussion

In the preparation of the BCD complexes, there was excellent correlation of the experimental results with computer molecular model studies (Dailey, O. D.; French, A. D., unpublished data). These studies indicated that the entire metribuzin molecule can be accommodated inside the BCD central cavity with relative ease. Indeed, a metribuzin-BCD complex was formed under mild conditions. Model studies of the atrazine-BCD complex indicated that at least part of the atrazine molecule and probably the entire molecule (with some crowding) can fit inside the BCD central cavity. One would predict formation of an atrazine-BCD complex, but with greater difficulty than the metribuzin complex. Experimentally, atrazine was recovered unchanged under mild reaction conditions (60 °C, 45 min.) but a complex was formed under more forcing conditions (100 °C, 7 days). Model studies of the alachlor-BCD complex indicated that at best only a small portion of the alachlor molecule can fit inside the BCD central cavity. Experimentally, no BCD complex was formed under the most forcing conditions (100 °C, 7 days).

<u>Physical Properties of BCD and GCD Complexes.</u> A number of physical properties were taken into account in determining whether or not the aforementioned herbicides did indeed form true cyclodextrin inclusion compounds.

   In general, BCD complexes are less soluble in water than either BCD or the guest molecule. The BCD complexes of atrazine and metribuzin, however, were considerably more soluble in water. In a typical preparation of the atrazine-BCD complex, a mixture of 5.00 g (23.2 mmol) of atrazine and 28.4 g (23.2 mmol) of BCD in 500 ml of water was heated at 100 °C for nine days. Upon cooling to 21 °C, only 1.05 g of insoluble material was isolated, and the presumed complex remained in solution. The solubility of BCD in water at 25 °C is 1.85 g/100 ml (5,6) and that of atrazine is 28 mg/l (8). Based on these solubilities, one would expect only 14 mg of atrazine and 9.25 g of BCD to remain in solution. In a large scale preparation of the metribuzin-BCD complex, a mixture of 5.00 g (23.3 mmol) of metribuzin and 28.58 g (23.3 mmol) of BCD in 500 ml of water was heated at 70-75 °C for 5.75 h resulting in a clear solution. Upon cooling to 21 °C, only 54 mg of insoluble material was isolated, and the presumed complex remained in solution. The solubility of metribuzin is 1.2 g/l at 20 °C (8). Based on solubilities, one would expect only 0.60 g of metribuzin and 9.25 of BCD to remain in solution. The concentration of BCD in the above preparations is approximately 5.6%. Turbidity studies show that crystallization of BCD from a less concentrated 4% aqueous solution begins at 28 °C and is essentially complete at 26 °C (9). The pronounced changes in the solubilities of atrazine, metribuzin, and BCD are consistent with complex formation.

There is the possibility of herbicide decomposition under the conditions of BCD complex formation, particularly hydrolysis of atrazine to its hydroxy derivative (8). Accordingly, in separate control experiments, a solution of 5.68 g of BCD in 100 ml of water and a suspension of 1.00 g of atrazine in 100 ml of water were heated at reflux under argon for seven days (conditions identical to those under which the presumed atrazine-BCD complex was formed). The BCD was recovered unchanged, and 92% of the atrazine was recovered unchanged upon cooling to room temperature and filtration of the undissolved solid.

Further evidence of the formation of a metribuzin-BCD complex stems from the observation that metribuzin has a pronounced odor and the presumed complex was odorless.

Infrared (IR) spectra. In general, IR spectroscopy is not useful in the characterization of cyclodextrin complexes owing to little or no observable change due to complex formation (6,7). However, in the cases of the BCD complexes of metribuzin, atrazine, and simazine, there are definite observable changes. Figure 2 gives a comparison of the IR spectra of a mechanical mixture of metribuzin and BCD and the metribuzin-BCD complex. The differences are generally small with the most significant ones occurring in the 1300-1460 cm^{-1} range. As can be seen in Figure 3, there are considerable differences between the IR spectra of the atrazine-BCD mixture and the atrazine-BCD complex. Most significantly, a band present at 795 cm^{-1} in the spectrum of the mixture is absent in the spectrum of the complex, and a prominent band at 1740 cm^{-1} appears in the spectrum of the complex. The most dramatic differences occur between the IR spectra of the simazine-BCD mixture and the simazine-BCD complex (Figure 4). A prominent band at 800 cm^{-1} in the spectrum of the mixture is absent in the spectrum of the complex, and a band appears at 1735 cm^{-1} in the spectrum of the complex. IR spectra of an alachlor-GCD mixture and alachlor-GCD complex were also recorded, but there were no observable differences between the two.

Ultraviolet (UV) spectra. The results of the UV analyses of the BCD complexes of metribuzin, atrazine, and simazine and the GCD complexes of alachlor and metolachlor are summarized in Table III. All solutions were prepared in deionized water and analyzed 2-3 days later unless otherwise noted. Metribuzin samples were scanned over the 200-400 nanometer (nm) range at 2 nm increments. The UV spectra of metribuzin, a metribuzin-BCD mixture, and metribuzin-BCD complex were essentially identical. A solution of metribuzin-BCD complex analyzed immediately after preparation (15 min.) afforded the same spectrum as the 3-day sample. These results indicate that either the metribuzin-BCD complex dissociates rapidly and completely in dilute aqueous media or BCD complexation has no effect on the UV spectrum of metribuzin. Atrazine samples were scanned over the 200-325 nm wavelength range at 0.5 nm increments. A composite of the UV spectra of the atrazine samples is shown in Figure 5. The UV spectra of atrazine and an atrazine-BCD mixture were essentially identical with a large broad peak at 220-222.5 nm and a smaller peak at 264.5 nm. Two different samples of the atrazine-BCD complex exhibited the same spectra with peaks at 204.5 and 240.5 nm, indicating reproducibility in the preparation of the complex. A solution of atrazine-BCD complex analyzed immediately after preparation (15 min.) afforded the same spectrum as the 3-day sample. Simazine, alachlor, and metolachlor samples were scanned over the 190-290 nm wavelength range. As was the case with the atrazine samples, the UV spectrum of simazine was markedly different than that of the simazine-BCD complex. These large differences suggest that the BCD complexes of atrazine and simazine are unusually strong. There were only small differences

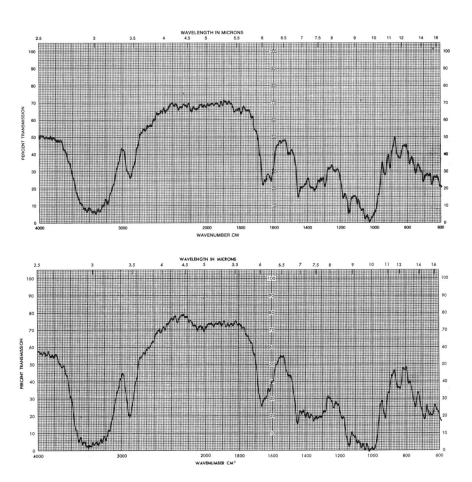

Figure 2. Comparison of the IR spectra of metribuzin–BCD mixture (upper) and metribuzin–BCD complex (lower).

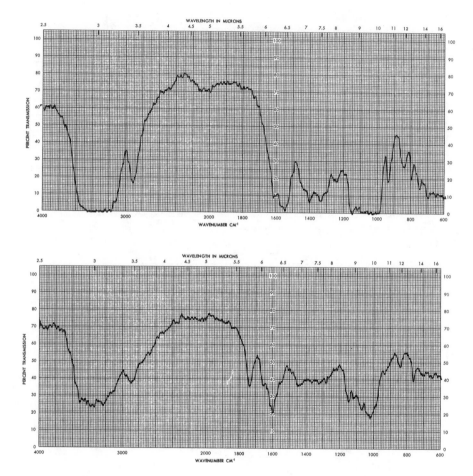

Figure 3. Comparison of the IR spectra of atrazine–BCD mixture (upper) and atrazine–BCD complex (lower).

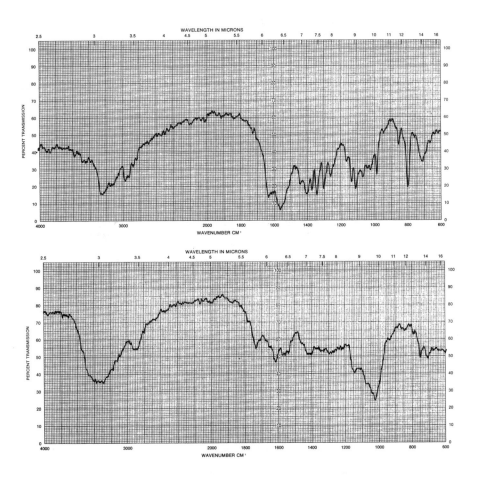

Figure 4. Comparison of the IR spectra of simazine–BCD mixture (upper) and simazine–BCD complex (lower).

Table III. UV Spectra of Selected Herbicides and Their
            BCD or GCD Complexes

Sample (Concentration)	Wavelength, nm (Lambda max)	Absorbance
Metribuzin (0.14 mM)	214	1.5019
	230	1.4420
	294	1.2212
Metribuzin-BCD Complex (0.14 mM) (3 days)	212	1.3973
	230	1.3036
	294	1.0963
Metribuzin-BCD Complex (0.14 mM) (15 min.)	212	1.3696
	230	1.2739
	294	1.0655
Metribuzin-BCD Mixture (0.14 mM)	214	1.8655
	230	1.7803
	296	1.5042
Atrazine (0.14 mM)	220.0	3.5243
	264.5	0.5032
Atrazine-BCD Complex 1 (0.14 mM) (3 days)	205.0	1.6870
	240.0	2.1408
Atrazine-BCD Complex 1 (0.14 mM) (15 min.)	204.5	1.7042
	240.5	2.1621
Atrazine-BCD Complex 2 (0.14 mM)	204.5	2.5042
	240.5	3.0272
Atrazine-BCD Mixture (0.14 mM)	222.5	3.5322
	264.5	0.5048
Simazine (0.042 mM)	222.2	1.0837
Simazine-BCD Complex (0.025 mM)	205.2	0.8144
	239.6	1.0157
Alachlor (0.11 mM)	201.4	3.0630
Alachlor-GCD Complex (0.11 mM)	199.0	2.9203
Metolachlor (0.11 mM)	196.4	2.5875
Metolachlor-GCD Complex (0.10 mM)	199.6	3.0204

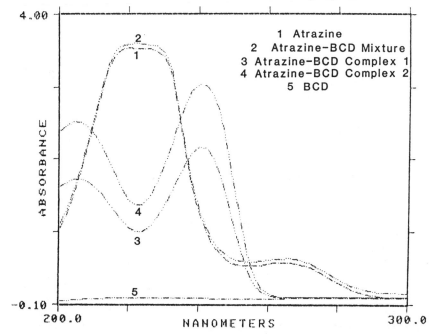

Figure 5. Composite of UV spectra of atrazine and its BCD complex.

between the UV spectra of alachlor and its GCD complex and between the UV spectra of metolachlor and its GCD complex, but these differences are thought to be real since they are reproducible.

Proton NMR Spectra. The proton NMR spectra of BCD and GCD in DMSO-d_6 at 60 °C have been reported by Vincendon (10) and were recorded at 38 °C in this work. Pertinent data from the two studies are given in Table IV.

Table IV. Proton NMR Spectra of BCD and GCD in DMSO-d_6[a]

	H-1	OH-2	OH-3	OH-6	Remaining H
BCD[b]	4.82	5.51	5.50	4.26	3.29-3.64
BCD[c]	4.84	5.77	5.66	4.45	3.32-3.65
GCD[b]	4.89	5.55	5.61	4.36	3.32-3.65
GCD[c]	4.92	5.72	5.79	4.57	3.56 (broad)

[a]Chemical shifts are recorded in ppm from internal TMS.
[b]Data of Vincendon (10). [c]This work.

   In order to further substantiate complex formation, the proton NMR spectra of metribuzin, atrazine, simazine, alachlor, and metolachlor and their BCD or GCD complexes were recorded. The solvent for all except the simazine samples was DMSO-d_6. The spectra of BCD, GCD, and all complexes were complicated by the presence of water peaks which appeared at 3.5-4.2 ppm in the spectra of BCD-containing samples and at 3.5-3.6 ppm in the spectra of GCD-containing samples. In general, changes in the chemical shifts of protons of the guest molecules were small (<0.05 ppm) but reproducible. In the spectra of complexes, the signals of the BCD and GCD protons were usually shifted upfield, the most dramatic shifts being those of the OH-2 and OH-3 protons. An upfield shift is indicative of complexation with BCD (11).
   In the NMR spectrum of atrazine, the methyl group protons were observed at 1.10 and 1.14 ppm and the NH signal appeared at 7.60 ppm. In the spectrum of the BCD complex, the corresponding signals appeared at 1.14, 1.20, and 8.33 ppm and the BCD signals were shifted upfield to the 3.07-5.03 ppm range. The spectrum of a mechanical mixture of BCD and atrazine exhibited no change in the signals of either the BCD or atrazine protons. In the NMR spectrum of metribuzin, the CH_3 signals appeared at 1.37 and 2.48 ppm. In the spectrum of the BCD complex, the CH[3] signals appeared at 1.34 and 2.44 ppm (a slight upfield shift) and the OH-2, OH-3, H-1, and OH-6 BCD protons appeared at 5.80, 5.69, 4.81, and 4.58 ppm, respectively (a slight downfield shift).
   As can be seen in Table V, GCD complexation of alachlor and metolachlor has only a slight influence on the chemical shifts of the protons of the guest molecule, with the most significant change being a 0.05 ppm upfield shift of the OCH_3 protons of alachlor. However, the GCD proton chemical shifts were considerably different, with broad peaks appearing at 3.53, 4.57, and 4.92 ppm in the spectrum of the alachlor-GCD complex and at 3.55, 4.29, and 4.94 ppm in the spectrum of the metolachlor-GCD complex.
   Finally, the NMR spectra of simazine and its BCD complex were recorded, using TFA-d_1 as solvent. In the spectrum of the complex, the simazine CH_3 signal w:.s

shifted from 1.35 to 1.30 ppm and the CH$_2$ signal was shifted from 3.68 to 3.62 ppm. The spectrum of BCD exhibited broad signals at 4.07, 4.22, and 5.28 ppm. The spectrum of the complex showed broad signals at 3.87, 4.00, and 5.12 ppm, a general upfield shift.

Table V. NMR Spectra of Alachlor and Metolachlor and GCD Complexes[a]

Alachlor	1.20 (t)	2.53 (q)	3.43 (s)	3.81 (s)	4.87 (s)	7.30 (m)
GCD Complex	1.18	2.53	3.38	3.82	4.87	7.33
Metolachlor	1.06 (d)	1.20 (t)	2.22 (s)	2.54 (q)	3.17 (s)	3.37 (d)
GCD Complex	1.06	1.20	2.20	2.52	3.17	b
Metolachlor	3.69 (s)	4.15 (m)	7.26 (s)			
GCD Complex	3.71	b	7.29			

[a]Chemical shifts are recorded in ppm from internal TMS. s = singlet; d = doublet; t = triplet; q = quartet; m = multiplet. [b]Obscured by GCD proton signals.

Conclusions

The BCD complexes of atrazine, metribuzin, and simazine and GCD complexes of alachlor and metolachlor have been prepared. BCD complexes of atrazine and simazine were prepared only under forcing reaction conditions over lengthy periods of time. Both atrazine and simazine have very low solubilities in water. Apparently, these compounds must first dissolve in water before complexing with BCD, accounting for the long reaction times. Once formed, the BCD complexes of atrazine and simazine are highly stable and impervious to disassociation. Metribuzin forms a complex with BCD with relative ease under mild reaction conditions. The BCD and GCD complexes were characterized by their physical properties and their IR, UV, and NMR spectra in order to establish them as true inclusion complexes. Although there were some atypical results (such the UV and NMR spectra of the metribuzin-BCD complex), the bulk of the evidence supports the formation of true inclusion complexes of all five herbicides.

As demonstrated by the present article, not all potential guest molecules readily form complexes. Furthermore, when a stable complex is formed, not only is it not necessarily crystalline, but it also may be highly water soluble.

Acknowledgments

The author thanks James V. Kelly, James A. Miller, and James R. Smith for technical assistance. We also acknowledge with appreciation American-Maize Products Company, Hammond, Indiana, for supplying samples of BCD and GCD, Ciba-Geigy, Greensboro, North Carolina for samples of atrazine, simazine, and metolachlor, Mobay Chemical Corp., Kansas City, Missouri, for samples of metribuzin, and Monsanto, St. Louis, Missouri, for a sample of alachlor.

Mention of a trademark, proprietary product or vendor does not constitute a guarantee or warranty of the product by the U. S. Department of Agriculture and does not imply its approval to the exclusion of other products or vendors that may also be suitable.

Literature Cited

1. Cohen, S. Z.; Eiden, C.; Lorber, M. N. In Evaluation of Pesticides in Ground Water; Garner, W. Y.; Honeycutt, R. C.; Nigg, H. N., Eds., ACS Symposium Series No. 315; American Chemical Society: Washington, D. C., 1986; pp 170-196.
2. Williams, W. M.; Holden, P. W.; Parsons, D. W.; Lorber, M. N. "Pesticides in Ground Water Data Base: 1988 Interim Report," U.S.E.P.A., Office of Pesticide Programs, Environmental Fate and Effects Division: Washington, D. C., 1988.
3. Pagington, J. S. Chem. Br. 1987, 23, 455-458.
4. Szejtli, J. Starch 1985, 37, 382-386.
5. Pszczola, D. E. Food Technol. 1988, 42(1), 96-100.
6. Saenger, W. Angew. Chem. Int. Ed. Engl. 1980, 19, 344-362.
7. Szejtli, J. Cyclodextrins and Their Inclusion Complexes; Akademiai Kiado: Budapest, 1982.
8. The Agrochemicals Handbook, Second Edition; Hartley, D.; Kidd, H., Eds.; The Royal Society of Chemistry: The University, Nottingham, England, 1987.
9. Szejtli, J.; Budai, Zs. Acta Chim. Acad. Sci. Hung. 1979, 99, 433-446.
10. Vincendon, M. Bull. Soc. Chim. Fr. 1981, (3-4, Pt. 2), 129-134.
11. Demarco, P. V.; Thakkar, A. L. J. Chem. Soc. Chem. Commun. 1970, 2-4.

RECEIVED September 9, 1990

Author Index

Affiliation Index

Subject Index